VOLUME ONE HUNDRED AND SIX

ADVANCES IN PROTEIN CHEMISTRY AND STRUCTURAL BIOLOGY

Chromatin Remodelling and Immunity

VOLUME ONE HUNDRED AND SIX

ADVANCES IN PROTEIN CHEMISTRY AND STRUCTURAL BIOLOGY

Chromatin Remodelling and Immunity

Edited by

ROSSEN DONEV

Biomed Consult Ltd
Swansea, United Kingdom

ELSEVIER

ACADEMIC PRESS

An imprint of Elsevier
elsevier.com

Academic Press is an imprint of Elsevier
50 Hampshire Street, 5th Floor, Cambridge, MA 02139, United States
525 B Street, Suite 1800, San Diego, CA 92101-4495, United States
The Boulevard, Langford Lane, Kidlington, Oxford OX5 1GB, United Kingdom
125 London Wall, London, EC2Y 5AS, United Kingdom

First edition 2017

Notices
Knowledge and best practice in this field are constantly changing. As new research and experience broaden our understanding, changes in research methods, professional practices, or medical treatment may become necessary.

Practitioners and researchers must always rely on their own experience and knowledge in evaluating and using any information, methods, compounds, or experiments described herein. In using such information or methods they should be mindful of their own safety and the safety of others, including parties for whom they have a professional responsibility.

To the fullest extent of the law, neither the Publisher nor the authors, contributors, or editors, assume any liability for any injury and/or damage to persons or property as a matter of products liability, negligence or otherwise, or from any use or operation of any methods, products, instructions, or ideas contained in the material herein.

ISBN: 978-0-12-812392-8
ISSN: 1876-1623

Publisher: Zoe Kruze
Acquisition Editor: Alex White
Senior Editorial Project Manager: Helene Kabes
Production Project Manager: Surya Narayanan Jayachandran
Senior Cover Designer: Maria Ines Cruz

Typeset by SPi Global, India

CONTENTS

CONTRIBUTORS

W.F. Carson IV
University of Michigan Medical School, Ann Arbor, MI, United States

W. Chen
State Key Laboratory for Conservation and Utilization of Subtropical Agro-Bioresources, College of Life Sciences, South China Agricultural University, Guangzhou, China

O. Fogel
Université Paris Descartes, Hôpital Cochin—Assistance Publique—Hôpitaux de Paris, EULAR Centre of Excellence, Paris; Laboratory for Epigenetics and Environment, Centre National de Génotypage, CEA-Institut de Génomique, Evry, France

A.F. Holloway
School of Medicine, University of Tasmania, Hobart, TAS, Australia

S.L. Kunkel
University of Michigan Medical School, Ann Arbor, MI, United States

Y. Liu
State Key Laboratory for Conservation and Utilization of Subtropical Agro-Bioresources, College of Life Sciences, South China Agricultural University, Guangzhou, China

Y. Luo
Faculty of Medical Technology, Teikyo University, Itabashi, Japan

T. Nakaoka
Division of Pharmacotherapeutics, Keio University Faculty of Pharmacy, Minato-ku, Japan

D.S. Pisetsky
Duke University Medical Center, VA Medical Center, Durham, NC, United States

J.S. Rawlings
Furman University, Greenville, SC, United States

C. Richard-Miceli
Université Paris Descartes, Hôpital Cochin—Assistance Publique—Hôpitaux de Paris, EULAR Centre of Excellence; Immunoregulation Unit, Unité Mixte Pasteur/APHP, Institut Pasteur, Paris, France

H. Saito
Division of Pharmacotherapeutics, Keio University Faculty of Pharmacy, Minato-ku, Japan

Y. Saito
Division of Pharmacotherapeutics, Keio University Faculty of Pharmacy, Minato-ku, Japan

J.L. Schultze
Genomics & Immunoregulation, LIMES-Institute, University of Bonn; Single Cell Genomics and Epigenomics Unit, German Center for Neurodegenerative Diseases, University of Bonn, Bonn, Germany

K. Suzuki
Faculty of Medical Technology, Teikyo University, Itabashi, Japan

J. Tost
Laboratory for Epigenetics and Environment, Centre National de Génotypage, CEA-Institut de Génomique, Evry, France

A.M. Woodworth
School of Medicine, University of Tasmania, Hobart, TAS, Australia

Q. Zhang
State Key Laboratory for Conservation and Utilization of Subtropical Agro-Bioresources, College of Life Sciences; Key Laboratory of Plant Functional Genomics and Biotechnology of Guangdong Provincial Higher Education Institutions, College of Life Sciences, South China Agricultural University, Guangzhou, China

Q. Zhu
State Key Laboratory for Conservation and Utilization of Subtropical Agro-Bioresources, College of Life Sciences, South China Agricultural University, Guangzhou, China

PREFACE

The immune system is exquisitely poised to identify, respond to, and eradicate pathogens from the body, as well as to produce a more rapid and augmented response to a subsequent encounter with the pathogen. These cellular responses rely on the highly coordinated and rapid activation of gene expression programs as well as the ability of the cell to retain a memory of the initial gene response. It is now clear that chromatin structure and epigenetic mechanisms play a critical role in determining these gene responses, and in fact the immune system has proved an instructive model for investigating the multifaceted mechanisms through which the chromatin landscape contributes to gene expression programs. These mechanisms include modifications to the DNA and histone proteins, the positioning, composition and remodeling of nucleosomes, as well as the formation of higher order chromatin structures (Woodworth & Holloway, 2017).

This thematic volume of the *Advances in Protein Chemistry and Structural Biology* is dedicated to the role of different chromatin-remodeling mechanisms in controlling immune responses in mammals and plants. Chapter 1 explores in detail the interplay between activation of monocytes and macrophages, the transcriptional consequences, and the epigenetic regulation, so far mainly studied on the level of chromatin landscapes and histone modifications. Chromatin remodeling is an important mechanism during cellular differentiation of monocytes and macrophages from adult stem cells and during embryogenesis. However, changes in the chromatin landscape are also associated with differential gene expression of monocytes and tissue macrophages under homeostatic conditions, mainly due to the integration of dominant local signals from the microenvironment. Furthermore, even stress signal-induced changes in gene expression are guided by significant remodeling of the chromatin. While the current multidimensional model of macrophage activation is mainly based on global transcriptional analysis, Chapter 1 proposes that specific epigenetic differences are also part of this model. Furthermore, epigenetic regulation in monocytes and macrophages has been reviewed as a new field for therapeutic intervention in acute and chronic inflammatory diseases associated with dysfunctional monocytes and macrophages. Understanding this sophisticated epigenetic regulatory network in monocytes and macrophages will open up new avenues toward tissue- and disease-specific therapeutic strategies in many of the chronic inflammatory diseases our societies are currently facing.

Chapter 2 focuses on the role of epigenetic alterations specifically in T lymphocytes (T cells) development and function. T cells comprise a critical component of the immune system charged with diverse functions during an immune response. As a function of maturation in the thymus, T cells become quiescent and remain so until they participate in an immune response in the periphery. Recent work indicates that the control of T cell proliferation is mediated, at least in part, by chromatin architecture. Quiescent T cells possess a condensed chromatin, whereas proliferating T cells have a more open chromatin configuration. The structural maintenance of chromosome (SMC) complexes, which include Cohesin and Condensin, have long been known to play roles in modulating chromatin architecture during cell division; however, they are now known to have additional roles during interphase biology. These roles include the large-scale reorganization of chromatin as well as the regulation of specific gene loci. This review focuses particularly on the roles that SMC complexes play in T cell development and function.

It is now established that epigenetic mechanisms also provide an instrument by which cells can retain memory of the initial transcriptional response, "priming" the genome so that it can respond more quickly to subsequent exposure to the same signal. Chapter 3 uses the immune system as a model to demonstrate the complex interplay between transcription factors and the chromatin landscape required to orchestrate precise gene responses to external stimuli and further to demonstrate how these interactions can establish memory of past transcriptional events. Authors focus on the multilayered mechanisms that contribute to epigenetic and transcriptional memory in the immune system, keeping in mind that the lessons that we have learnt from the immune system are informing our understanding of other biological systems and also disease processes.

Chapter 4 in this volume focuses on the role of chromatin remodeling, particularly remodeling that involves histone acetylation, in the constitutive and inducible regulation of major histocompatibility complex (MHC) class II gene expression. MHC class II molecules are essential for processing and presenting exogenous pathogen antigens to activate $CD4^{+}$ T cells. Given their central role in adaptive immune responses, MHC class II genes are tightly regulated in a tissue- and activation-specific manner. The regulation of MHC class II gene expression involves various transcription factors that interact with conserved proximal *cis*-acting regulatory promoter elements, as well as MHC class II transactivator that interacts with a variety of chromatin-remodeling machineries. Recent studies also identified distal regulatory elements within MHC class II gene locus that provide enormous insight into the long-range coordination of MHC class II gene expression. Novel therapeutic modalities that can modify MHC class II genes at the epigenetic level are also discussed in this chapter.

Chapter 5 considers the remodeling processes that affect chromatin during cell death and their immunological consequences. Chromatin remodeling is usually conceptualized as an intranuclear (and certainly intracellular) event. However, remodeling can have a broader context and involve extensive and far-reaching molecular movement. Indeed, depending on the cell process, chromatin can translocate entirely from the inside to the outside of the cell. Outside the cell, the functional activities of chromatin can undergo a dramatic transformation since components of chromatin can, when present in the extracellular space, display powerful immunological activities. With the breakdown of cells, many intracellular components—both large and small—can enter the extracellular space to serve as signaling elements to drive inflammation and to counteract events challenging the host. Chromatin and its components show such mobility and immune activity. As these translocation events of chromatin can be regulated and mediated by enzymes, they can be considered a facet of chromatin remodeling and repurposing albeit in the setting of cell death.

Next three chapters review the role of epigenetic changes in different diseases. The number of people diagnosed with inflammatory diseases has increased noteworthy in the last 40 years and this fast-growing number has been impossible to explain by Mendelian inheritance only. Among them, spondyloarthritis, inflammatory bowel diseases, and psoriasis are the most frequent chronic inflammatory diseases, resulting from a combination of genetic predisposition and environmental factors. For example, in ankylosing spondyloarthritis, only 25% of heritability can be explained by currently known risk loci, suggesting that 75% of the disease risk remains unknown. This lack of knowledge is called the missing heritability. Among other hypotheses, the epigenome might hold at least part to the answer of this missing heritability. Chapter 6 reviews in detail the reported changes in patterns of DNA methylation and posttranslational histone modifications in chronic inflammatory diseases, highlighting potential connections between disease-associated pathophysiological changes such as the dysbiosis of the microbiome or genetic variations associated with disease susceptibility and the epigenome. It also discusses important parameters of meaningful epigenetic studies such as the use of well-defined, disease-relevant cell populations and eludes on the potential future of engineering of the epigenome in inflammatory diseases. Chapter 7 focuses on the role of histone modifications in governing immune responses in severe sepsis, with an emphasis on specific leukocyte subsets and the histone modifications observed in these cells during chronic stages of sepsis. Additionally, the expression and function of chromatin modifying enzymes are discussed in the context of severe sepsis, as potential mediators of epigenetic regulation of gene expression in sepsis responses. In summary, this review argues for the use of chromatin modifications and chromatin-modifying enzyme expression in leukocytes as potential biomarkers of immunosuppression in patients with severe sepsis. Chapter 8 reviews a new molecular mechanism underlying the antitumor effect of DNA methylation inhibitors that involves induction of an antiviral immune response. This new mechanism was recently proposed: induction of interferon-responsive genes via double-stranded RNAs derived from ERVs. The available data indicate that DNA demethylation suppresses the proliferation of cancer-initiating cells by inducing an antiviral response including activation of interferon-responsive genes.

In the absence of a circulatory system and mobile immune cells that vertebrates have, plants rely on a multilayered innate immune system to achieve highly specific, adaptive immunity and establish immune memory. Increasing evidence indicates a critical role of chromatin remodeling in the establishment of long-lasting, even transgenerational immune memory in plants, as DNA methylation, histone deacetylation, and methylation prime the promoters of immune-related genes required for disease defense. The role of chromatin remodeling in plant immunity is discussed in Chapter 9 in this thematic volume of the *Advances in Protein Chemistry and Structural Biology* with particular emphasis on the links between Snf2-mediated ATP-dependent chromosome remodeling and plant immunity, and the mechanisms that might support its involvement in disease resistance.

Nonetheless, chromatin-remodeling factors remain a relatively underexploited group of proteins for therapeutic interventions on immune system responses. A number of recent advances in technology and biomedical knowledge suggest that these proteins are promising targets for future therapeutic development. Therefore, the aim of this volume is to promote further research in this very exciting and central field for the molecular control of immune responses to environmental factors. Advancing our knowledge on this fundamental topic would allow further design of therapeutic strategies to control immune system misbehavior, for example, in chronic inflammatory disorders, sepsis, cancer, etc.

Dr. Rossen Donev
Biomed Consult
United Kingdom

REFERENCE

Woodworth, A. M., & Holloway, A. F. (2017). The role of epigenetic regulation in transcriptional memory in the immune system. *Advances in Protein Chemistry & Structural Biology*, *106*, 43–69.

CHAPTER ONE

Chromatin Remodeling in Monocyte and Macrophage Activation

J.L. Schultze[1]

Genomics & Immunoregulation, LIMES-Institute, University of Bonn, Bonn, Germany
Single Cell Genomics and Epigenomics Unit, German Center for Neurodegenerative Diseases, University of Bonn, Bonn, Germany
[1]Corresponding author: e-mail address: j.schultze@uni-bonn.de

Contents

Abstract

Increasing evidence collected during the last years supports the idea that monocyte and macrophage activation is not only associated with transcriptional changes but also changes in the chromatin landscape. Moreover, the introduction of a multidimensional model of macrophage activation allows a more precise description of monocytes and macrophages under homeostatic and pathophysiological conditions. Monocytes and macrophages are masters of integrating microenvironmental signals, thereby reshaping their chromatin landscape and as a consequence their transcriptional and functional programs. Albeit these cells share a large number of epigenetic landmarks, their chromatin is significantly shaped by environmental signals. The chromatin landscape of any given tissue macrophage is a rather specific fingerprint of these cells, which is directly linked to tissue-specific functions of these cells. Moreover, chromatin remodeling in response to stress signals also seems to be an important mechanism of these cells to increase their readiness for future stressors. Understanding this sophisticated epigenetic regulatory network in monocytes and macrophages will open up new avenues toward tissue- and disease-specific therapeutic strategies in many of the chronic inflammatory diseases our societies are currently facing.

Advances in Protein Chemistry and Structural Biology, Volume 106
ISSN 1876-1623
http://dx.doi.org/10.1016/bs.apcsb.2016.09.001

1. INTRODUCTION

Monocytes and macrophages are central to tissue homeostasis, but they are also important cells during host defense against viruses, bacteria, and other pathogens (Wynn, Chawla, & Pollard, 2013). Moreover, they are involved in chronic inflammatory diseases including atherosclerosis, diabetes, cancer, neurodegenerative diseases, obesity, and skin diseases (De Felice & Ferreira, 2014; Heneka, Kummer, & Latz, 2014; Tall & Yvan-Charvet, 2015; Viola & Soehnlein, 2015). Changes in function of monocytes and macrophages have been associated with cellular activation as a consequence of transcriptional regulation (Mosser, 2003; Mosser & Edwards, 2008; Murray & Wynn, 2011; Sica & Mantovani, 2012). During acute infections, such transcriptional activation is seen as being beneficial and required to overcome infections (Wynn et al., 2013), while activation in context of chronic inflammatory responses is often judged as a dysfunctional state (Pollard, 2009; Qian & Pollard, 2010; Wynn et al., 2013). About 15 years ago, a simplistic model describing both pro- and antiinflammatory activation as two opposite activation conditions was ascribed to macrophages (Mantovani, Sozzani, Locati, Allavena, & Sica, 2002; Mantovani, Vecchi, & Allavena, 2014; Sica & Mantovani, 2012). Enormous progress has been made since, particularly by applying high-content transcriptome, but also epigenome approaches to monocyte and macrophage biology, thereby extending the previous bipolar model of macrophage activation to a multidimensional model (Ginhoux, Schultze, Murray, Ochandi, & Biswas, 2016; Schultze, 2015; Schultze, Freeman, Hume, & Latz, 2015; Schultze & Schmidt, 2015; Xue et al., 2014). In fact, macrophages are also specifically activated under homeostatic conditions (Gosselin et al., 2014; Lavin et al., 2014) and their transcriptional programs are further changed in response to physiological and nonphysiological stress signals (Ghisletti et al., 2010; Heinz et al., 2010; Ostuni et al., 2013; Xue et al., 2014). Here, I will describe the interplay between activation of monocytes and macrophages, the transcriptional consequences, and the epigenetic regulation, so far mainly studied on the level of chromatin landscapes and histone modifications.

2. THE MULTIDIMENSIONAL MODEL OF MONOCYTE AND MACROPHAGE ACTIVATION

The application of transcriptome and epigenome analysis to monocytes and tissue macrophages has challenged our view on how we define activation in these myeloid cells. While former models suggested that monocytes and macrophages are activated by signals such as LPS and other pathogen-associated molecular patterns leading to a proinflammatory state, while other signals, for example, the cytokine IL-4, lead to an anti-inflammatory condition, assessment of activation on a genome-wide level argued against such models. In fact, when comparing tissue macrophages derived from different tissues, it became rather clear that the transcriptional programs differ significantly. Moreover, as further outlined later, the chromatin landscapes were also rather different. Similarly, when stimulating monocyte-derived cells with a larger number of potential stress signals, it also became clear that these cells integrate external signals in a rather signal-specific fashion. There was no evidence that monocytes or macrophages follow an activation model simply described by being either pro- or antiinflammatory (Xue et al., 2014).

This led us to define the new multidimensional model of macrophage activation (Ginhoux et al., 2016; Schultze, 2015; Schultze et al., 2015; Schultze & Schmidt, 2015). In principle, monocytes and macrophages are never in a resting state, since they always integrate signals from their micro-environment. Changes in the spectrum of these signals lead to epigenetic and transcriptional changes. According to different qualities of input signals, the state of activation and its epigenetic and transcriptional consequences can be categorized.

Recent elegant work concerning the ontogeny of monocytes and macrophages has revealed different routes of development and it also seems that every adult tissue has another spectrum of cells either being derived from the yolk sac or from adult hematopoiesis (for excellent reviews, see Chorro & Geissmann, 2010; Geissmann et al., 2010; Ginhoux & Jung, 2014; Gomez Perdiguero, Schulz, & Geissmann, 2013). Some tissues lose their yolk sac-derived cell populations over time during adulthood and these cells are then replaced by monocyte-derived cells. Since epigenetic regulation plays an enormous role during cellular differentiation, it is obvious to speculate that different developmental paths of myeloid cells might also lead to differential epigenetic landscapes. However, formal proof of differential epigenetic landscapes of yolk sac- vs monocyte-derived macrophages of any given tissue has not been provided yet, although this might be of great importance in light of increasing inflammatory responses in aging individuals. If monocyte-derived cells would retain a more proinflammatory epigenetic landscape, their higher inflammatory tonus might be a critical part of the increased tissue inflammation observed during aging. Nevertheless, differences in the epigenetic promoter and enhancer landscapes of tissue macrophages derived from different tissues have been clearly observed (Gosselin et al., 2014; Lavin et al., 2014). So far, these differences have been mainly explained by signals derived from the local environment. However, whether the origin of the cells also contributes to differences in the epigenetic landscape is not yet completely excluded.

In addition to tissue-derived signals, monocytes and macrophages can also be exposed to acute stress signals. Using a reductionist in vitro model, we recently demonstrated that monocytes and macrophages sense stress signals, integrate them, and reprogram their epigenetic and transcriptional landscapes to fulfill situation-adapted functions (Schmidt et al., 2016; Xue et al., 2014). Similarly, chronic exposure to stress-related signals, as they are present under chronic inflammatory conditions (Dey, Allen, & Hankey-Giblin, 2014), leads to changes in the chromatin landscape of these cells.

But also physiological processes such as tissue repair require macrophages to fulfill specific functions such as restoration of tissue integrity, structure, and function, which can dramatically differ at different locations (Chazaud, 2014; Mantovani, Biswas, Galdiero, Sica, & Locati, 2013). Furthermore, depending on specific reprogramming of the cells the outcome of tissue repair ranges from complete restoration of tissue function to fibrosis with complete loss of function. It is not entirely surprising that macrophage activity and function differ in respect to the overall outcome. Another important activation program is associated with the phase of resolution, for example, at the end of an acute inflammatory response. For example, macrophages can express higher levels of CD163, CD206, and TGFβ during the resolution phase, arguing that gene regulation must have changed at least for these hallmark genes (Mounier et al., 2013). While phenotypic, functional, and regulatory mechanisms have been linked to this particular state of activation, the role of epigenetic changes during resolution has not been fully understood.

A special form of activation, according to the multidimensional model, is the suppression of particular functions, whether they are homeostatic or stress signal induced. While the suppression of both programs might be linked, it will be important to learn, whether it is possible, for example, to suppress only stress signal-induced epigenetic and transcriptional programs, while keeping homeostatic functions of monocytes and macrophages intact. Suppressive signals from the local microenvironment need to be seen in context of tissue integrity, thereby reducing the risk of overt induction of proinflammatory effector molecules in tissue macrophages. The expression of TGFβ in the brain reducing the inflammatory tonus of microglia cells is probably a good example (Butovsky et al., 2014). However, such mechanisms can be hijacked, which is particularly eminent in tumor-associated macrophages (Colegio et al., 2014). Collectively, monocytes and macrophages exist at very different activation states and the regulation of activation involves epigenetic and transcriptional mechanisms.

3. EPIGENETICS OF TISSUE MACROPHAGES IN HOMEOSTASIS

Here, I focus on epigenetic regulation under steady-state conditions and only shortly mention that dramatic epigenetic changes occur during myeloid cell differentiation, which has been previously reviewed (Alvarez-Errico, Vento-Tormo, Sieweke, & Ballestar, 2015). In brief, the principle concept of adult myelopoiesis is a gradual change of the epigenetic landscape of hematopoietic stem cells toward differentiated cell types, which is associated with open chromatin space at cell lineage-specific genes, while genes necessary for self-renewal and stemness show a closed chromatin landscape in differentiated cells. While the traditional model of myelopoiesis suggested several distinct cell types serving as precursors of several lineages, for example, granulocyte-monocyte progenitors capable of differentiating into granulocytes and monocytes (Alvarez-Errico et al., 2015), more recent evidence from single-cell RNA-seq experiments suggest a differentiation model with early lineage commitment at the hematopoietic multipotent progenitor state without late precursors capable of differentiating into several independent lineages (Paul et al., 2015; Schultze & Beyer, 2016). Irrespective of this new concept, pioneering and lineage-associated transcription factors (TFs) play an essential role during such differentiation programs (Heinz et al., 2010; Smale, Tarakhovsky, & Natoli, 2014). For monocytes and macrophages, PU.1 (Spi1) is both an essential pioneering and lineage-associated TF (Heinz et al., 2010; Smale et al., 2014). These TFs are capable of inducing changes in the chromatin landscape by recruiting respective histone-modifying enzymes, which open up chromatin space at lineage-specific enhancer sites. During the establishment of the chromatin landscape of the cell lineage, TFs such as PU.1 can then bind to these newly derived enhancers, thereby securing the cell phenotype on epigenetic level. Moreover, during this chromatin remodeling process, other TFs such as the CEBPs are required for guiding the differentiation process toward a particular cell lineage (Dahl et al., 2003; Laiosa, Stadtfeld, Xie, de Andres-Aguayo, & Graf, 2006).

Under homeostatic conditions, fully differentiated macrophages exist in basically every organ. Very recently, it was demonstrated that macrophages from different tissues vary significantly in their transcriptional makeup. Moreover, these transcriptional differences are accompanied by quite significant differences in their chromatin landscape, particularly their enhancer landscape (Amit, Winter, & Jung, 2016; Gosselin et al., 2014; Lavin et al., 2014). In a recent study, it was demonstrated that macrophages, monocytes, and neutrophils share more than 82% of the ~10,800 myeloid cell-associated active promoters defined by trimethylation at lysine 4 of histone 3 (H3K4me3). For active enhancers, defined by monomethylation at lysine 4 of H3 (H3K4me1) and acetylation at lysine 27 of H3 (H3K27ac), only a small fraction was shared among all tissue macrophages analyzed. Furthermore, this strong difference in *cis*-enhancer usage was associated with differential gene expression, strongly arguing that the epigenetic differences at enhancer sites guide transcription and therefore function of tissue macrophages. In addition to the lineage-determining TF PU.1, other TFs were predicted to bind to these tissue macrophage-specific enhancers, suggesting that PU.1 together with other more tissue-specific TFs acts in combination, thereby guiding tissue-specific transcriptional programs. A good example is the expression of the TF Spi-C in red pulp macrophages (Haldar et al., 2014; Kohyama et al., 2009). Spi-C regulates important genes regulating iron homeostasis, which is a major function of red pulp macrophages. Furthermore, it has been shown recently that an environmental factor inducing Spi-C in monocytes is heme itself and as a consequence such monocytes are then differentiating toward iron-recycling macrophages (Haldar et al., 2014; Kohyama et al., 2009). Further support for epigenetic remodeling of monocytes and macrophages by tissue-related factors was derived from the identification of so-called super enhancers (SEs) that were also present in a tissue macrophage-specific manner. In general, SEs can be defined as epigenetic regions with long stretches of disproportionately high histone acetylation, particularly H3K27Ac, which allows strong binding of TFs at these sites (Hnisz et al., 2013; Whyte et al., 2013). The cooperative accessibility of SEs and promoters of genes in close proximity often leads to elevated expression of these genes. Interestingly, for lineage-associated genes such as Spi1, Cebpa, or Csf1r, SEs were identified in macrophages from different tissues, while an SE for Gata6 was only identified in peritoneal macrophages, further supporting that SEs are an epigenetic element guiding lineage- but also tissue-specific gene expression in the myeloid cell compartment (Gosselin et al., 2014). This was further corroborated by findings in large peritoneal macrophages that lost SEs when cultured in vitro in the absence of retinoic acid, a major microenvironmental signal shaping chromatin landscapes and gene transcription in peritoneal macrophages, but not other tissue macrophages.

4. EPIGENETIC REGULATION DURING STRESS SIGNAL-INDUCED ACTIVATION

The concept of the multidimensional model of monocyte and macrophage activation includes the necessity to shape the chromatin landscape prior to changes in gene transcription. In fact, there has been clear evidence that epigenetic regulation occurs during stress signal-mediated activation (Lawrence & Natoli, 2011; Natoli, 2010; Schmidt et al., 2016; Smale, 2010). Two hallmark papers by the Natoli and the Glass group in 2010 clearly revealed that macrophages remodel their chromatin landscape in response to LPS (Ghisletti et al., 2010; Heinz et al., 2010). The lineage-determining TF PU.1 and the signal-associated TFs NF-kB, IRF, and AP-1 are acting in concert at a defined set of enhancers following macrophage activation. These findings clearly indicated that microenvironmental stress signals lead to chromatin remodeling, thereby contributing to inducible gene transcription. Furthermore, gene loci of LPS-inducible genes were defined by properties related to particular requirements for SWI/SNF-dependent nucleosome remodeling (Li, Carey, & Workman, 2007; Ramirez-Carrozzi et al., 2009; Smale et al., 2014). While SWI/SNF-independent gene loci predominantly contain CpG island promoters, most of the SWI/SNF-dependent gene loci lack CpG islands at their promoter sites (Hargreaves, Horng, & Medzhitov, 2009; Li et al., 2007; Ramirez-Carrozzi et al., 2009; Smale et al., 2014). Stress signal-induced chromatin remodeling is not only characterized by changes in the activation state of the chromatin. More recently, the Natoli group could show that new enhancers are formed during LPS exposure of murine macrophages (Ostuni et al., 2013). The induction of these so-called latent enhancers was also seen in response to other stimuli including a TLR2-ligand (MALP2), a TLR9-ligand (CpGs), or cytokine exposure including IL1β, IL-4, TNF, IFNγ, or TGFβ. An interesting feature of these latent enhancers was their persistence even after the stress signal exposure had been seized. Moreover, second exposures to stress stimuli were also associated with a faster and stronger response, suggesting that the changed requirements for chromatin remodeling also impacted on the kinetics and magnitude of the cellular response toward such signals, a concept related to "trained immunity" or "innate memory," which gained quite some attention recently (Netea et al., 2016; see later). Importantly, concept identified in the murine system is also applying to human monocyte subsets (Schmidl et al., 2014) and monocyte-derived macrophages stimulated by M-CSF (Pham et al., 2012), or GM-CSF and stress signals (Schmidt et al., 2016). Monocytes seem to possess an epigenetic peculiarity, which might explain their enormous functional plasticity and fast reactivity toward stress signals. We characterized the chromatin landscape of genes involved in epigenetic and transcriptional regulation and observed that the large majority of genes that can be induced by one or the other stress signal are already characterized by an open chromatin landscape even before they encounter such stress signals. Similar to latent enhancers, which allow a faster and more robust response toward secondary signals, the open chromatin landscape at key regulatory genes might be responsible for the very flexible and fast responsiveness of monocytes to a myriad of potential stress signals. Epigenetic regulation might therefore be a hallmark for the plasticity of monocytes and macrophages both during homeostasis and even more important in acute and chronic inflammatory reactions.

5. EVIDENCE FOR EPIGENETIC REGULATION DURING TRAINED IMMUNITY AND TOLERANCE

It has been observed that innate immune cells including macrophages show an elevated reactivity after an initial encounter of a stress signal and this phenomenon was recently linked to epigenetic regulatory mechanisms (Bekkering, Joosten, van der Meer, Netea, & Riksen, 2013; Morris, Gilliam, & Li, 2014; Netea et al., 2016; Netea & van Crevel, 2014; Smale et al., 2014). Different terms for this phenomenon have been suggested such as "innate memory" or "trained immunity" (Netea et al., 2016). Albeit there is still debate about the overall concept, two recent studies in humans provided evidence that epigenetic regulation might indeed play a role in the development of trained immunity in innate immune cells (Kleinnijenhuis et al., 2012; Quintin et al., 2012). In these studies, it was demonstrated that BCG vaccination against *Candida albicans* infection can lead to modified monocyte activation, which was associated with changes in histone trimethylation landscapes. These findings are in line with earlier work demonstrating that murine bone marrow-derived macrophages are epigenetically reprogrammed following stress signals including IFNg. In these cells, additional enhancers, so-called latent enhancers, were identified, which lost histone acetylation marks and TFs binding to these enhancers upon removal of the stress signals, but kept H3K4me1 enhancer marks (Ostuni et al., 2013). In human monocyte-derived macrophages, β-glucan (BG) exposure leads to genome-wide changes in H3K4 mono- and trimethylation as well as H3K27 acetylation, open chromatin structure, and global gene transcription (Saeed et al., 2014). Of particular interest, gene loci coding for genes involved in glucose metabolism were characterized by an open chromatin landscape in cells exposed to BG, while this was not the case prior to BG exposure (Cheng et al., 2014). The epigenetic changes were associated with altered cellular functions including a high ratio of nicotinamide adenine dinucleotide (NAD^+) to its reduced form NADH, high lactate production, and high glucose consumption. Mechanistically, this shift toward elevated glycolysis was related to activation of several signaling cascades involving CLEC7a (Dectin-1), AKT, mTOR (mechanistic target of rapamycin), and HIF-1α (Cheng et al., 2014). It was suggested that trained immunity in human monocytes is linked to a metabolic switch toward aerobic glycolysis, which then results in epigenetic regulation. Certainly, more work is required to better understand such link. Furthermore, the concept of epigenetic regulation as a hallmark for trained immunity has been mainly discussed in context of BCG vaccination (Netea et al., 2016; Netea & van Crevel, 2014) and atherosclerosis (Bekkering et al., 2013; Bekkering, Joosten, van der Meer, Netea, & Riksen, 2015). Whether this will also be true in other situations is currently unknown.

While trained immunity of monocytes and macrophages is associated with enhanced functionality, the loss of functionality has also been observed, particularly in response to LPS exposure. This phenomenon has been termed LPS tolerance (for review, see Biswas & Lopez-Collazo, 2009; Morris et al., 2014). Most work has been focused on differential usage of signal transduction cascades following either MyD88-dependent or TRIF-dependent gene regulation following TLR4 signaling; however, some evidence exists that epigenetic regulation at a subset of LPS-induced genes leads to a repressive chromatin structure after an initial activation (Foster, Hargreaves, & Medzhitov, 2007; Yan et al., 2012), indicating that repressive epigenetic mechanisms can shape the transcriptional response of some genes during a second stimulation. Further studies are necessary to determine which role epigenetic mechanisms are playing during LPS-induced tolerance.

6. CONCLUSION

Chromatin remodeling is an important mechanism during cellular differentiation of monocytes and macrophages from adult stem cells and during embryogenesis. However, changes in the chromatin landscape are also associated with differential gene expression of monocytes and tissue macrophages under homeostatic conditions, mainly due to the integration of dominant local signals from the microenvironment. Moreover, even stress signal-induced changes in gene expression are guided by significant remodeling of the chromatin. While our current multi-dimensional model of macrophage activation is mainly based on global transcriptional analysis, we propose that specific epigenetic differences are also part of this model. Different combinations of TFs are a major part of this myriad of activation states, both under homeostatic conditions and stress situations. A major task ahead will be to identify dominant environmental signals that guide this TF combinatorics and to perform gain- and loss-of-function experiments for particularly important TFs in a tissue-specific manner.

Furthermore, epigenetic regulation in monocytes and macrophages might also be a new field for therapeutic intervention in many acute and chronic inflammatory diseases associated with dysfunctional monocytes and macrophages. In a recent publication, it was demonstrated that the antiviral immune response of macrophages relies on the function of the DNA methyltransferase Dnmt3a via maintaining elevated expression of the histone deacetylase HDAC9, which directly maintained the deacetylation status of the pattern-recognition receptor signaling molecule TBK1 (Li et al., 2016). TBK1 activity was required for the production of type I interferons. Another example of a clinically relevant epigenetic regulation was reported by Fan et al. who clearly indicated that the co-repressor complex containing G protein pathway suppressor 2 (GPS2) crucially controls the macrophage epigenome during activation by metabolic stress and that this reversible disease mechanism might be a target for suppressing adipose tissue inflammation and the development of type 2 diabetes (Fan et al., 2016). Collectively, we are at the verge of a new era of macrophage biology with epigenetics not only being a central regulatory hub of macrophage activation but also a new field to discover novel therapeutic concepts in many of the chronic inflammatory diseases of our societies.

ACKNOWLEDGMENTS

This work was supported by Sonderforschungsbereiche SFB645 and SFB704 to J.L.S. J.L.S. is a member of the Excellence Cluster ImmunoSensation.

REFERENCES

Alvarez-Errico, D., Vento-Tormo, R., Sieweke, M., & Ballestar, E. (2015). Epigenetic control of myeloid cell differentiation, identity and function. *Nature Reviews. Immunology, 15*(1), 7–17. http://dx.doi.org/10.1038/nri3777.

Amit, I., Winter, D. R., & Jung, S. (2016). The role of the local environment and epigenetics in shaping macrophage identity and their effect on tissue homeostasis. *Nature Immunology, 17*(1), 18–25. http://dx.doi.org/10.1038/ni.3325.

Bekkering, S., Joosten, L. A., van der Meer, J. W., Netea, M. G., & Riksen, N. P. (2013). Trained innate immunity and atherosclerosis. *Current Opinion in Lipidology, 24*(6), 487–492. http://dx.doi.org/10.1097/MOL.0000000000000023.

Bekkering, S., Joosten, L. A., van der Meer, J. W., Netea, M. G., & Riksen, N. P. (2015). The epigenetic memory of monocytes and macrophages as a novel drug target in atherosclerosis. *Clinical Therapeutics, 37*(4), 914–923. http://dx.doi.org/10.1016/j.clinthera.2015.01.008.

Biswas, S. K., & Lopez-Collazo, E. (2009). Endotoxin tolerance: New mechanisms, molecules and clinical significance. *Trends in Immunology, 30*(10), 475–487. http://dx.doi.org/10.1016/j.it.2009.07.009.

Butovsky, O., Jedrychowski, M. P., Moore, C. S., Cialic, R., Lanser, A. J., Gabriely, G., … Weiner, H. L. (2014). Identification of a unique TGF-beta-dependent molecular and functional signature in microglia. *Nature Neuroscience, 17*(1), 131–143. http://dx.doi.org/10.1038/nn.3599.

Chazaud, B. (2014). Macrophages: Supportive cells for tissue repair and regeneration. *Immunobiology, 219*(3), 172–178. http://dx.doi.org/10.1016/j.imbio.2013.09.001.

Cheng, S. C., Quintin, J., Cramer, R. A., Shepardson, K. M., Saeed, S., Kumar, V., … Netea, M. G. (2014). mTOR- and HIF-1alpha-mediated aerobic glycolysis as metabolic basis for trained immunity. *Science, 345*(6204), 1250684. http://dx.doi.org/10.1126/science.1250684.

Chorro, L., & Geissmann, F. (2010). Development and homeostasis of 'resident' myeloid cells: The case of the Langerhans cell. *Trends in Immunology, 31*(12), 438–445. http://dx.doi.org/10.1016/j.it.2010.09.003.

Colegio, O. R., Chu, N. Q., Szabo, A. L., Chu, T., Rhebergen, A. M., Jairam, V., … Medzhitov, R. (2014). Functional polarization of tumour-associated macrophages by tumour-derived lactic acid. *Nature, 513*, 559–563. http://dx.doi.org/10.1038/nature13490.

Dahl, R., Walsh, J. C., Lancki, D., Laslo, P., Iyer, S. R., Singh, H., & Simon, M. C. (2003). Regulation of macrophage and neutrophil cell fates by the PU.1:C/EBPalpha ratio and granulocyte colony-stimulating factor. *Nature Immunology, 4*(10), 1029–1036. http://dx.doi.org/10.1038/ni973.

De Felice, F. G., & Ferreira, S. T. (2014). Inflammation, defective insulin signaling, and mitochondrial dysfunction as common molecular denominators connecting type 2 diabetes to Alzheimer disease. *Diabetes, 63*(7), 2262–2272. http://dx.doi.org/10.2337/db13-1954.

Dey, A., Allen, J., & Hankey-Giblin, P. A. (2014). Ontogeny and polarization of macrophages in inflammation: Blood monocytes versus tissue macrophages. *Frontiers in Immunology, 5*, 683. http://dx.doi.org/10.3389/fimmu.2014.00683.

Fan, R., Toubal, A., Goni, S., Drareni, K., Huang, Z., Alzaid, F., ... Treuter, E. (2016). Loss of the co-repressor GPS2 sensitizes macrophage activation upon metabolic stress induced by obesity and type 2 diabetes. *Nature Medicine*, *22*, 780–791. http://dx.doi.org/10.1038/nm.4114.

Foster, S. L., Hargreaves, D. C., & Medzhitov, R. (2007). Gene-specific control of inflammation by TLR-induced chromatin modifications. *Nature*, *447*(7147), 972–978. http://dx.doi.org/10.1038/nature05836.

Geissmann, F., Manz, M. G., Jung, S., Sieweke, M. H., Merad, M., & Ley, K. (2010). Development of monocytes, macrophages, and dendritic cells. *Science*, *327*(5966), 656–661. http://dx.doi.org/10.1126/science.1178331.

Ghisletti, S., Barozzi, I., Mietton, F., Polletti, S., De Santa, F., Venturini, E., ... Natoli, G. (2010). Identification and characterization of enhancers controlling the inflammatory gene expression program in macrophages. *Immunity*, *32*(3), 317–328. http://dx.doi.org/10.1016/j.immuni.2010.02.008.

Ginhoux, F., & Jung, S. (2014). Monocytes and macrophages: Developmental pathways and tissue homeostasis. *Nature Reviews. Immunology*, *14*(6), 392–404. http://dx.doi.org/10.1038/nri3671.

Ginhoux, F., Schultze, J. L., Murray, P. J., Ochandi, J., & Biswas, S. K. (2016). New insights into the multidimensional concept of macrophage ontogeny, activation and function. *Nature Immunology*, *17*(1), 34–40. http://dx.doi.org/10.1038/ni.3324.

Gomez Perdiguero, E., Schulz, C., & Geissmann, F. (2013). Development and homeostasis of "resident" myeloid cells: The case of the microglia. *Glia*, *61*(1), 112–120. http://dx.doi.org/10.1002/glia.22393.

Gosselin, D., Link, V. M., Romanoski, C. E., Fonseca, G. J., Eichenfield, D. Z., Spann, N. J., ... Glass, C. K. (2014). Environment drives selection and function of enhancers controlling tissue-specific macrophage identities. *Cell*, *159*(6), 1327–1340. http://dx.doi.org/10.1016/j.cell.2014.11.023.

Haldar, M., Kohyama, M., So, A. Y., Kc, W., Wu, X., Briseno, C. G., ... Murphy, K. M. (2014). Heme-mediated SPI-C induction promotes monocyte differentiation into iron-recycling macrophages. *Cell*, *156*(6), 1223–1234. http://dx.doi.org/10.1016/j.cell.2014.01.069.

Hargreaves, D. C., Horng, T., & Medzhitov, R. (2009). Control of inducible gene expression by signal-dependent transcriptional elongation. *Cell*, *138*(1), 129–145. http://dx.doi.org/10.1016/j.cell.2009.05.047.

Heinz, S., Benner, C., Spann, N., Bertolino, E., Lin, Y. C., Laslo, P., ... Glass, C. K. (2010). Simple combinations of lineage-determining transcription factors prime cis-regulatory elements required for macrophage and B cell identities. *Molecular Cell*, *38*(4), 576–589. http://dx.doi.org/10.1016/j.molcel.2010.05.004.

Heneka, M. T., Kummer, M. P., & Latz, E. (2014). Innate immune activation in neurodegenerative disease. *Nature Reviews. Immunology*, *14*(7), 463–477. http://dx.doi.org/10.1038/nri3705.

Hnisz, D., Abraham, B. J., Lee, T. I., Lau, A., Saint-Andre, V., Sigova, A. A., ... Young, R. A. (2013). Super-enhancers in the control of cell identity and disease. *Cell*, *155*(4), 934–947. http://dx.doi.org/10.1016/j.cell.2013.09.053.

Kleinnijenhuis, J., Quintin, J., Preijers, F., Joosten, L. A., Ifrim, D. C., Saeed, S., ... Netea, M. G. (2012). Bacille Calmette-Guerin induces NOD2-dependent nonspecific protection from reinfection via epigenetic reprogramming of monocytes. *Proceedings of the National Academy of Sciences of the United States of America*, *109*(43), 17537–17542. http://dx.doi.org/10.1073/pnas.1202870109.

Kohyama, M., Ise, W., Edelson, B. T., Wilker, P. R., Hildner, K., Mejia, C., ... Murphy, K. M. (2009). Role for Spi-C in the development of red pulp macrophages and splenic iron homeostasis. *Nature*, *457*(7227), 318–321. http://dx.doi.org/10.1038/nature07472.

Laiosa, C. V., Stadtfeld, M., Xie, H., de Andres-Aguayo, L., & Graf, T. (2006). Reprogramming of committed T cell progenitors to macrophages and dendritic cells by C/EBP alpha and PU.1 transcription factors. *Immunity*, *25*(5), 731–744. http://dx.doi.org/10.1016/j.immuni.2006.09.011.

Lavin, Y., Winter, D., Blecher-Gonen, R., David, E., Keren-Shaul, H., Merad, M., ... Amit, I. (2014). Tissue-resident macrophage enhancer landscapes are shaped by the local microenvironment. *Cell*, *159*(6), 1312–1326. http://dx.doi.org/10.1016/j.cell.2014.11.018.

Lawrence, T., & Natoli, G. (2011). Transcriptional regulation of macrophage polarization: Enabling diversity with identity. *Nature Reviews. Immunology*, *11*(11), 750–761. http://dx.doi.org/10.1038/nri3088.

Li, B., Carey, M., & Workman, J. L. (2007). The role of chromatin during transcription. *Cell*, *128*(4), 707–719. http://dx.doi.org/10.1016/j.cell.2007.01.015.

Li, X., Zhang, Q., Ding, Y., Liu, Y., Zhao, D., Zhao, K., ... Cao, X. (2016). Methyltransferase Dnmt3a upregulates HDAC9 to deacetylate the kinase TBK1 for activation of antiviral innate immunity. *Nature Immunology*, *17*(7), 806–815. http://dx.doi.org/10.1038/ni.3464.

Mantovani, A., Biswas, S. K., Galdiero, M. R., Sica, A., & Locati, M. (2013). Macrophage plasticity and polarization in tissue repair and remodelling. *The Journal of Pathology*, *229*(2), 176–185. http://dx.doi.org/10.1002/path.4133.

Mantovani, A., Sozzani, S., Locati, M., Allavena, P., & Sica, A. (2002). Macrophage polarization: Tumor-associated macrophages as a paradigm for polarized M2 mononuclear phagocytes. *Trends in Immunology*, *23*(11), 549–555.

Mantovani, A., Vecchi, A., & Allavena, P. (2014). Pharmacological modulation of monocytes and macrophages. *Current Opinion in Pharmacology*, *17*, 38–44. http://dx.doi.org/10.1016/j.coph.2014.07.004.

Morris, M. C., Gilliam, E. A., & Li, L. (2014). Innate immune programing by endotoxin and its pathological consequences. *Frontiers in Immunology*, *5*, 680. http://dx.doi.org/10.3389/fimmu.2014.00680.

Mosser, D. M. (2003). The many faces of macrophage activation. *Journal of Leukocyte Biology*, *73*(2), 209–212.

Mosser, D. M., & Edwards, J. P. (2008). Exploring the full spectrum of macrophage activation. *Nature Reviews. Immunology*, *8*(12), 958–969. http://dx.doi.org/10.1038/nri2448.

Mounier, R., Theret, M., Arnold, L., Cuvellier, S., Bultot, L., Goransson, O., ... Chazaud, B. (2013). AMPKalpha1 regulates macrophage skewing at the time of resolution of inflammation during skeletal muscle regeneration. *Cell Metabolism*, *18*(2), 251–264. http://dx.doi.org/10.1016/j.cmet.2013.06.017.

Murray, P. J., & Wynn, T. A. (2011). Protective and pathogenic functions of macrophage subsets. *Nature Reviews. Immunology*, *11*(11), 723–737. http://dx.doi.org/10.1038/nri3073.

Natoli, G. (2010). Maintaining cell identity through global control of genomic organization. *Immunity*, *33*(1), 12–24. http://dx.doi.org/10.1016/j.immuni.2010.07.006.

Netea, M. G., Joosten, L. A., Latz, E., Mills, K. H., Natoli, G., Stunnenberg, H. G., ... Xavier, R. J. (2016). Trained immunity: A program of innate immune memory in health and disease. *Science*, *352*(6284), aaf1098. http://dx.doi.org/10.1126/science.aaf1098.

Netea, M. G., & van Crevel, R. (2014). BCG-induced protection: Effects on innate immune memory. *Seminars in Immunology*, *26*(6), 512–517. http://dx.doi.org/10.1016/j.smim.2014.09.006.

Ostuni, R., Piccolo, V., Barozzi, I., Polletti, S., Termanini, A., Bonifacio, S., ... Natoli, G. (2013). Latent enhancers activated by stimulation in differentiated cells. *Cell*, *152*(1–2), 157–171. http://dx.doi.org/10.1016/j.cell.2012.12.018.

Paul, F., Arkin, Y., Giladi, A., Jaitin, D. A., Kenigsberg, E., Keren-Shaul, H., ... Amit, I. (2015). Transcriptional heterogeneity and lineage commitment in myeloid progenitors. *Cell*, *163*, 1663–1677. http://dx.doi.org/10.1016/j.cell.2015.11.013.

Pham, T. H., Benner, C., Lichtinger, M., Schwarzfischer, L., Hu, Y., Andreesen, R., … Rehli, M. (2012). Dynamic epigenetic enhancer signatures reveal key transcription factors associated with monocytic differentiation states. *Blood, 119*(24), e161–e171. http://dx.doi.org/10.1182/blood-2012-01-402453.
Pollard, J. W. (2009). Trophic macrophages in development and disease. *Nature Reviews. Immunology, 9*(4), 259–270. http://dx.doi.org/10.1038/nri2528.
Qian, B. Z., & Pollard, J. W. (2010). Macrophage diversity enhances tumor progression and metastasis. *Cell, 141*(1), 39–51. http://dx.doi.org/10.1016/j.cell.2010.03.014.
Quintin, J., Saeed, S., Martens, J. H., Giamarellos-Bourboulis, E. J., Ifrim, D. C., Logie, C., … Netea, M. G. (2012). Candida albicans infection affords protection against reinfection via functional reprogramming of monocytes. *Cell Host & Microbe, 12*(2), 223–232. http://dx.doi.org/10.1016/j.chom.2012.06.006.
Ramirez-Carrozzi, V. R., Braas, D., Bhatt, D. M., Cheng, C. S., Hong, C., Doty, K. R., … Smale, S. T. (2009). A unifying model for the selective regulation of inducible transcription by CpG islands and nucleosome remodeling. *Cell, 138*(1), 114–128. http://dx.doi.org/10.1016/j.cell.2009.04.020.
Saeed, S., Quintin, J., Kerstens, H. H., Rao, N. A., Aghajanirefah, A., Matarese, F., … Stunnenberg, H. G. (2014). Epigenetic programming of monocyte-to-macrophage differentiation and trained innate immunity. *Science, 345*(6204), 1251086. http://dx.doi.org/10.1126/science.1251086.
Schmidl, C., Renner, K., Peter, K., Eder, R., Lassmann, T., Balwierz, P. J., … Rehli, M. (2014). Transcription and enhancer profiling in human monocyte subsets. *Blood, 123*, e90–e99. http://dx.doi.org/10.1182/blood-2013-02-484188.
Schmidt, S. V., Krebs, W., Ulas, T., Xue, J., Baßler, K., Günther, P., … Schultze, J. L. (2016). The transcriptional regulator network of human inflammatory macrophages is defined by open chromatin. *Cell Research, 26*(1), 151–170. http://dx.doi.org/10.1038/cr.2016.1.
Schultze, J. L. (2015). Transcriptional programming of human macrophages: On the way to systems immunology. *Journal of Molecular Medicine (Berlin, Germany), 93*, 589–597. http://dx.doi.org/10.1007/s00109-015-1286-y.
Schultze, J. L., & Beyer, M. (2016). Myelopoiesis reloaded: Single-cell transcriptomics leads the way. *Immunity, 44*(1), 18–20. http://dx.doi.org/10.1016/j.immuni.2015.12.019.
Schultze, J. L., Freeman, T., Hume, D. A., & Latz, E. (2015). A transcriptional perspective on human macrophage biology. *Seminars in Immunology, 27*, 44–50. http://dx.doi.org/10.1016/j.smim.2015.02.001.
Schultze, J. L., & Schmidt, S. V. (2015). Molecular features of macrophage activation. *Seminars in Immunology, 27*(6), 416–423. http://dx.doi.org/10.1016/j.smim.2016.03.009.
Sica, A., & Mantovani, A. (2012). Macrophage plasticity and polarization: In vivo veritas. *The Journal of Clinical Investigation, 122*(3), 787–795. http://dx.doi.org/10.1172/jci59643.
Smale, S. T. (2010). Selective transcription in response to an inflammatory stimulus. *Cell, 140*(6), 833–844. http://dx.doi.org/10.1016/j.cell.2010.01.037.
Smale, S. T., Tarakhovsky, A., & Natoli, G. (2014). Chromatin contributions to the regulation of innate immunity. *Annual Review of Immunology, 32*, 489–511. http://dx.doi.org/10.1146/annurev-immunol-031210-101303.
Tall, A. R., & Yvan-Charvet, L. (2015). Cholesterol, inflammation and innate immunity. *Nature Reviews. Immunology, 15*(2), 104–116. http://dx.doi.org/10.1038/nri3793.
Viola, J., & Soehnlein, O. (2015). Atherosclerosis—A matter of unresolved inflammation. *Seminars in Immunology, 27*, 184–193. http://dx.doi.org/10.1016/j.smim.2015.03.013.
Whyte, W. A., Orlando, D. A., Hnisz, D., Abraham, B. J., Lin, C. Y., Kagey, M. H., … Young, R. A. (2013). Master transcription factors and mediator establish super-enhancers at key cell identity genes. *Cell, 153*(2), 307–319. http://dx.doi.org/10.1016/j.cell.2013.03.035.

Wynn, T. A., Chawla, A., & Pollard, J. W. (2013). Macrophage biology in development, homeostasis and disease. *Nature*, *496*(7446), 445–455. http://dx.doi.org/10.1038/nature12034.

Xue, J., Schmidt, S. V., Sander, J., Draffehn, A. M., Krebs, W., Quester, I., ... Schultze, J. L. (2014). Transcriptome-based network analysis reveals a spectrum model of human macrophage activation. *Immunity*, *40*(2), 274–288. http://dx.doi.org/10.1016/j.immuni.2014.01.006.

Yan, Q., Carmody, R. J., Qu, Z., Ruan, Q., Jager, J., Mullican, S. E., ... Chen, Y. H. (2012). Nuclear factor-kappaB binding motifs specify Toll-like receptor-induced gene repression through an inducible repressosome. *Proceedings of the National Academy of Sciences of the United States of America*, *109*(35), 14140–14145. http://dx.doi.org/10.1073/pnas.1119842109.

CHAPTER TWO

Roles of SMC Complexes During T Lymphocyte Development and Function

J.S. Rawlings[1]
Furman University, Greenville, SC, United States
[1]Corresponding author: e-mail address: jason.rawlings@furman.edu

Contents

Abstract

T lymphocytes (T cells) comprise a critical component of the immune system charged with diverse functions during an immune response. As a function of maturation in the thymus, T cells become quiescent and remain so until they participate in an immune response in the periphery. Recent work indicates that the control of T cell proliferation is mediated, at least in part, by chromatin architecture. Quiescent T cells possess a condensed chromatin, whereas proliferating T cells have a more open chromatin configuration. The structural maintenance of chromosome (SMC) complexes, which include Cohesin and Condensin, have long been known to play roles in modulating chromatin architecture during cell division; however, they are now known to have additional roles during interphase biology. These roles include the large-scale reorganization of chromatin as well as the regulation of specific gene loci. This review focuses on the roles that SMC complexes play in T cell development and function.

Advances in Protein Chemistry and Structural Biology, Volume 106
ISSN 1876-1623
http://dx.doi.org/10.1016/bs.apcsb.2016.08.001

1. OVERVIEW OF CHROMATIN DYNAMICS DURING T CELL DEVELOPMENT AND ACTIVATION

T cells play multiple critical roles in immunity, ranging from the ability to kill infected cells to maintaining peripheral tolerance. Control of T cell proliferation is absolutely essential for their proper development and function. Arising from precursors in the bone marrow, immature T cells enter the thymus with relatively high proliferative capacity and a decondensed chromatin configuration. Following maturation, T cells exit the thymus in a naïve, quiescent state, where they may remain for decades. In this state, they have an extremely low metabolic profile, rudimentary organelles, and an extremely condensed chromatin. Upon activation during an immune response, their chromatin decondenses, metabolism increases dramatically, and they regain the ability to proliferate in response to cytokines (Frauwirth & Thompson, 2004; Morley, Rau, Kay, & Pain, 1993; Rawlings, Gatzka, Thomas, & Ihle, 2011). How T cell quiescence is established during development in the thymus, maintained in the periphery, and ultimately overcome during an immune response is not clear.

The chromatin architecture of developing thymocytes undergoes dramatic changes that correlate with key events in thymocyte development that include the capacity to proliferate (Fig. 1 and Table 1). First, chromatin decondenses as the cells progress from the double-negative 1 (DN1) to DN2 stage. This decondensation is significant, as at the DN2 stage, rearrangement of the β chain of the TCR begins. At the DN3 stage, chromatin begins to condense. At this stage, the developing T cells are subjected to a process known as β selection, whereby only those cells for which a successful β chain rearrangement occurred can survive. Failure to generate a functional β chain results in death by neglect. As the cells progress to the DN4 stage, condensation of chromatin stalls, correlating with the rearrangement of the α chain of the TCR. Chromatin then condenses to its most condensed state as the cells enter the double-positive (DP) stage. At this stage, the cells will undergo both positive and negative selection, such that only those cells bearing a TCR (consisting of both rearranged α and β chains) that not only recognizes self-MHC but is also not autoreactive will survive. Chromatin remains in this condensed configuration through the completion of thymocyte development and exit into the periphery as naïve CD4+ or CD8+ single-positive (SP) T cells (Fig. 1). The testing of the signaling capacity of the TCR during β selection and +/− selection underpins

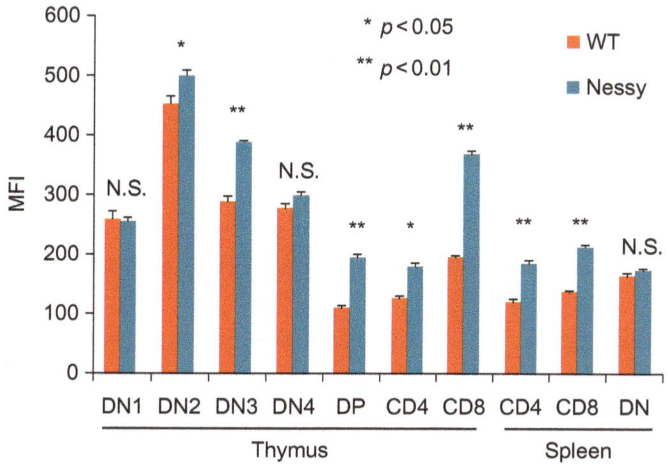

Fig. 1 Proper chromatin condensation during T cell development is dependent on the Condensin II complex. Thymocytes (Thy1.2+ cells) from 6-week-old *nessy* (Kleisin-β mutant) and wild-type littermates were analyzed for surface expression of CD4, CD8, CD25, and CD44 to determine thymocyte subpopulations (Table 1). Histone accessibility, as measured by mean fluorescence intensity (MFI) of intracellular Histone H3K4me1 staining, was used to determine chromatin condensation status. *NS*, not significant difference between WT and *nessy*. *Figure taken with permission from Rawlings, J. S., Gatzka, M., Thomas, P. G., & Ihle, J. N. (2011). Chromatin condensation via the condensin II complex is required for peripheral T-cell quiescence.* EMBO Journal, 30*(2), 263–276. http://dx.doi.org/10.1038/emboj.2010.314.*

Table 1 Stages of Thymocyte Development

Developmental Stage	Surface Markers Used for Determining Developmental Stage: CD44	CD25	CD4	CD8	Cell Proliferation	TCR Loci Rearrangements/ Selection Events
DN1	+	–	–	–	Yes	
DN2	+	+	–	–	Yes	TCRβ rearrangement begins
DN3	–	+	–	–	No	β Selection
DN4	–	–	–	–	Yes	TCRα rearrangement begins
DP			+	+	No	+/– Selection begins
CD4+ SP			+	–	No	
CD8+ SP			–	+	No	

The stages of thymocyte development are determined by expression of select surface markers. *DN*, double-negative; *DP*, double-positive, referring to CD4 and CD8 expression.

both of these critical developmental processes. In β selection, the newly formed β chain is paired with a surrogate pre-Tα chain and the ability of this pre-TCR to transduce signals is tested. Those cells that can successfully transduce the signal will live, while those that cannot will succumb to apoptosis. In +/− selection, the completed TCR comprising of rearranged α and β chains is tested and the strength of the signal will determine if the cell lives or dies. These observations suggest that the chromatin condensation state of the cell could affect the outcome of these critical developmental events.

Concomitant with +/− selection, the DP cell loses its ability to proliferate, primarily through the downregulation of cytokine signaling. Proliferation during thymocyte development is dependent on interleukin-7 (IL-7), which signals via STAT5 to regulate cell cycle progression and promote survival (Rawlings, Rosler, & Harrison, 2004; von Freeden-Jeffry, Solvason, Howard, & Murray, 1997). As thymocytes transition to the DP stage, the expression of SOCS1 (suppressor of cytokine signaling) is upregulated (Chong et al., 2003). The SOCS family of genes have been shown to downregulate cytokine signaling, including IL-7, in a number of contexts and are important for proper thymocyte development and function (Fletcher & Starr, 2005; Marine et al., 1999). Thus, it seems that the restriction of IL-7/STAT5 signaling is imperative for the transition to quiescence at the DP stage. Consistent with this, chromatin is at its most condensed state at the DP stage (Fig. 1), suggesting that in addition to SOCS1 expression, chromatin architecture could potentially restrict IL-7/STAT5 signaling, promoting quiescence. This hypothesis is consistent with the chromatin-mediated regulation of STAT5 signaling seen in the periphery (described later).

In the periphery, proliferation is also controlled by cytokine signaling, specifically IL-2 which signals via STAT5 to regulate expression of genes required for cell cycle progression (Moriggl, Sexl, Piekorz, Topham, & Ihle, 1999; Moriggl, Topham, et al., 1999; Rawlings et al., 2004). Interestingly, IL-2 is unable to induce the proliferation of naïve T cells (those cells that have not encountered their cognate antigen); however, as a consequence of activation, T cells gain competence to respond to IL-2. The result is only those cells that have engaged their TCR will proliferate in response to IL-2, resulting in the clonal expansion of antigen-specific T cells during an immune response. The differential response to IL-2 is not limited to cell cycle genes, as it appears that IL-2 cannot induce the expression of any known STAT5 target genes in naïve T cells. This is true despite the fact that naïve T cells possess all of the needed components for IL-2-dependent gene expression (Gatzka, Piekorz, Moriggl, Rawlings, & Ihle, 2006). The discovery that naïve and activated T cells possess different IL-2 receptors, suggested a possible mechanism for the differential responses to IL-2. Naïve T cells possess an intermediate affinity receptor, composed of IL-2Rβ and IL-2Rγ chains, while activated T cells express an additional receptor chain (IL-2Rα) that increases ligand–receptor affinity (Lin & Leonard, 1997). However, the addition of exogenous IL-2 at concentrations that would engage the intermediate affinity receptor found on naïve T cells fails to induce expression of STAT5 target genes and subsequent proliferation. This observation suggests that there must be a mechanism, downstream of the IL-2 receptor, which prevents STAT5 target gene expression in naïve peripheral T cells. The repression of gene expression in naïve peripheral T cells may not be limited to IL-2 signaling, as transcription appears to be globally reduced in naïve T cells compared to activated T cells (Jaehning, Stewart, & Roeder, 1975). Our recent work suggests that this repression may be mediated by global chromatin organization. We discovered that the condensed nature of chromatin in naïve T cells prevents activated, nuclear STAT5 from engaging DNA. Upon activation, chromatin decondenses, permitting STAT5-DNA engagement (Fig. 2; Rawlings et al., 2011), resulting in the clonal proliferation of only those T cells that have been activated. Given that T cells retain the ability to respond to IL-2 after the TCR stimulus is removed for many cell cycles suggests that the regulation of IL-2 competence could be epigenetic in nature.

2. EPIGENETIC PHENOMENA IN T CELL DEVELOPMENT AND FUNCTION

A number of well-defined epigenetic phenomena relating to chromatin organization could explain how quiescence is regulated in T cells. These include the methylation of DNA and modification of histones. Methylation of CpG dinucleotides has been shown to be involved in the silencing of gene expression, including genes known to be involved in T cell development (Fitzpatrick et al., 1998; Lee, Agarwal, & Rao, 2002). While most studies have focused on individual loci under specific conditions (e.g., T_H lineage commitment), some studies investigated whether global changes in DNA methylation patterns occur in T cells. Investigators have discovered differences in these patterns between differentiated T cells, memory cells as well as

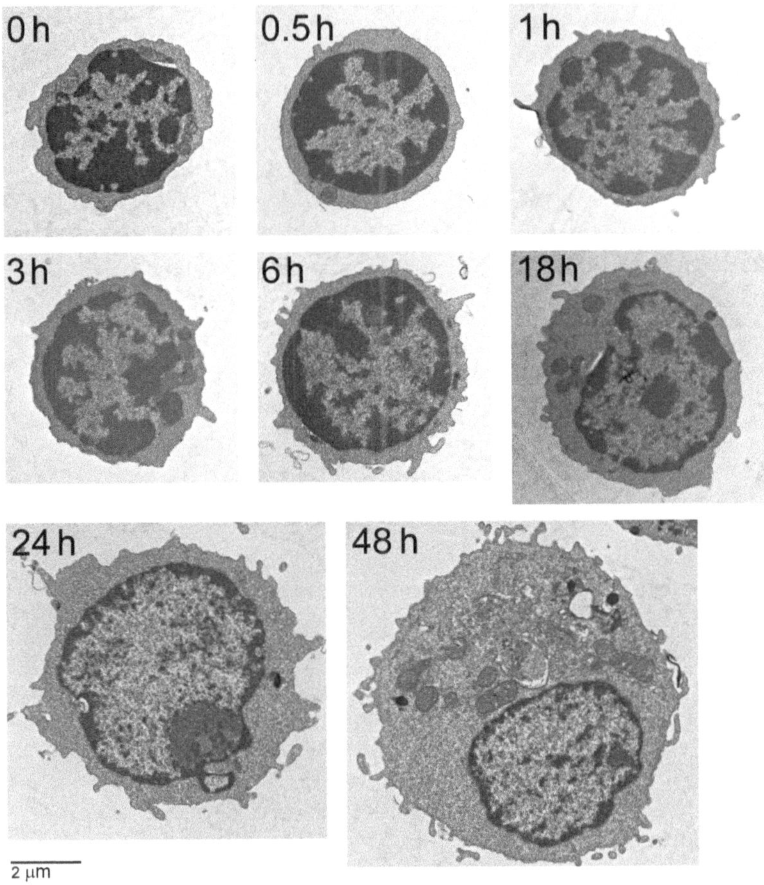

Fig. 2 Chromatin decondenses as a consequence of T cell activation. Transmission electron microscopy (TEM) was used to visualize chromatin in purified naïve (CD25–) peripheral T cells (0 h) or cells activated with 1 μg/mL anti-CD3 antibodies for the times indicated. Cells (1×10^7) were fixed with cacodylate buffer containing 2.5% glutaraldehyde (Tousimis Research Corporation) for 2 h at room temperature. Cells were then washed with 0.2 M cacodylate buffer supplemented with 5% sucrose, post-fixed in 1% OsO_4, embedded, sectioned, and viewed by TEM. Representative images are shown. *Images taken with permission from Rawlings, J. S., Gatzka, M., Thomas, P. G., & Ihle, J. N. (2011). Chromatin condensation via the condensin II complex is required for peripheral T-cell quiescence.* EMBO Journal, *30(2), 263–276. http://dx.doi.org/10.1038/emboj.2010.314.*

during T cell development (Komori, Hart, LaMere, Chew, & Salomon, 2015; Rodriguez et al., 2015; Scharer, Barwick, Youngblood, Ahmed, & Boss, 2013). Additionally, one recent study described a global reduction of DNA methylation during T cell activation (Li et al., 2012).

The amino termini of histone proteins have been shown to be post-translationally modified in a number of ways and many of the enzymes responsible for catalyzing the reactions that produce these modifications have been identified (Kouzarides, 2007). Many of these modifications have been associated with gene activity and/or repression and the combination of various histone modifications at a particular locus constitute a "histone code" that could potentially determine the activity state of a given gene (Jenuwein & Allis, 2001). With the advent of techniques to assess histone marks on a genome-wide scale (e.g., ChIP-seq), it became possible to determine if there is a global modification of histones that takes place as T cells transition into or out of the quiescent state (Schones & Zhao, 2008). Strikingly, many genes known to be induced during T cell activation have histone marks associated with active transcription *prior* to stimulation (i.e., in quiescent T cells), although the genes are not actively transcribed at this time. Furthermore, RNA polymerase II has been shown to already be resident on the promoters of these genes in naïve T cells (Barski et al., 2009; Lim et al., 2009). These observations have led to the conclusion that these loci are in a "poised" state in naïve, quiescent T cells (Lim, Shannon, & Hardy, 2010). The mechanism that holds these loci in this "poised" state is not known, nor is it known how these loci are released from this state, permitting gene expression, as a consequence of T cell activation.

While the above epigenetic phenomena are certainly important in T cell biology, there are also a number of protein complexes (e.g., SWI/SNF) that can alter the architecture of chromatin at or above the nucleosome level. This review will focus on the structural maintenance of chromosome (SMC) complexes. These complexes, in addition to their classical roles in condensing chromatin during cell division, have been implicated in regulating chromatin architecture at the global level as well as regulating specific gene loci during interphase. Recent work, described here, has indicated that the SMC complexes also have specific roles in T cell biology.

3. STRUCTURE OF SMC COMPLEXES

The SMC family of protein complexes has widespread roles in the compaction of chromatin in various aspects of cell biology. In addition to three distinct Condensin complexes (Condensin I, Condensin II, and Condensin I^{DC}), this family also includes the Cohesin and Smc5/6 Complexes. While Condensins are involved in chromatin compaction in a number of contexts, Cohesin is best known for its role in sister chromatid cohesion

Table 2 Components of SMC Complexes in Humans and Yeast

	H. sapiens		*S. cerevisiae*	
Complex	**SMC Subunits**	**Non-SMC Subunits**	**SMC Subunits**	**Non-SMC Subunits**
Cohesin	SMC1a SMC3	RAD21 (α-kleisin) STAG1, STAG2 PDS5a/b	Smc1 Smc3	Scc1 (α-kleisin) Irr1/Scc3 Pds5
Condensin I	SMC2 SMC4	NCAPD2 NCAPG NCAPH	Smc2 Smc4	Ycs4 Ycg1/Ycs5 brn1
Condensin II	SMC2 SMC4	NCAPD3 NCAPG2 NCAPH2 (kleisin-β)	n/a	n/a
Smc5/6	SMC5 SMC6	NSMCE1 NSMCE2 NSMCE3 NSMCE4A	Smc5 Smc6	Nse1 Mms21/Nse2 Nse3 Nse4/Qri2 Nse5 Kre29/Nse6

Gene names of components of SMC complexes according to *Saccharomyces* Genome Database or HGNC. Yeasts do not possess a Condensin II complex.

during meiosis and the Smc5/6 complex is known for its roles in DNA repair (for review, see Aragon, Martinez-Perez, & Merkenschlager, 2013; Michaelis, Ciosk, & Nasmyth, 1997; Strom & Sjogren, 2007). All of the SMC complexes contain a heterodimer of SMC proteins that forms a V-shaped structure consisting of a hinge at one end and an ATP-binding motif located at the distal end of each arm (Anderson, Losada, Erickson, & Hirano, 2002; Melby, Ciampaglio, Briscoe, & Erickson, 1998). Additional non-SMC subunits complete each complex, forming a ring-like structure large enough to hold a double-stranded DNA (for review, see Aragon et al., 2013; Hirano, 2005; Jeppsson, Kanno, Shirahige, & Sjogren, 2014) (Table 2).

4. CLASSICAL FUNCTIONS OF SMC COMPLEXES

The first SMC gene, *SMC1*, was identified in *Saccharomyces cerevisiae* and was shown to be required for proper chromosome segregation (Strunnikov, Larionov, & Koshland, 1993). Later, it was discovered that this protein was part of the Cohesin complex (Losada, Hirano, & Hirano, 1998; Sumara, Vorlaufer, Gieffers, Peters, & Peters, 2000; Toth et al., 1999). This complex consists of a Smc1a/Smc3 heterodimer and two non-SMC components: a Kleisin family protein known as Scc1 (Rad21 in humans) and Scc3 (either STAG1 or STAG2 in humans) (Table 2). Cohesin acts as a molecular glue that holds sister chromatids together during the cell cycle beginning with DNA synthesis until anaphase of mitosis when one of the cohesion subunits, Scc1, is cleaved by Separase, a cysteine protease. This cleavage results in the release of Cohesin from sister chromatids, permitting chromosome segregation to occur (for review, see Peters, Tedeschi, & Schmitz, 2008). Because of its role in sister chromatid cohesion, it is not surprising that the Cohesin complex also has roles in DNA damage repair, specifically postreplicative double-strand break repair (Sjogren & Nasmyth, 2001).

In addition to a Smc2/Smc4 heterodimer, the Condensin I and Condensin II complexes contain two HEAT repeat containing proteins and one Kleisin subunit (Table 2). Although first discovered in 1982 as a chromosome scaffolding protein (then termed ScII), Smc2 was not characterized as a Smc protein until 1994 (Lewis & Laemmli, 1982; Saitoh, Goldberg, Wood, & Earnshaw, 1994). By 1997, the other subunits of the Condensin I complex were discovered and the complex itself was described in Xenopus egg extracts (Hirano, Kobayashi, & Hirano, 1997). The Condensin I complex functions to condense chromosomes during mitosis and is also required for proper chromosome segregation. The function of the Condensin I complex is regulated, in part, by the fact that the complex is sequestered in the cytosol during interphase. Once the nuclear envelope breaks down during prophase, the Condensin I complex can access DNA and condense it in preparation for completion of mitosis (Hirota, Gerlich, Koch, Ellenberg, & Peters, 2004; Ono, Fang, Spector, & Hirano, 2004). A second Condensin complex, found only in higher eukaryotes, was discovered more recently (Ono et al., 2003; Yeong et al., 2003). This complex, termed Condensin II, possesses the same Smc2/Smc4 heterodimer as the Condensin I complex; however, utilizes different non-SMC components (Table 2). It was immediately discovered that Condensin I and Condensin II bind different regions of chromosomes, suggesting that they contribute to chromatin architecture in distinct ways (Ono et al., 2003). Like Condensin I, Condensin II also functions in chromatin compaction and in segregation; however, unlike Condensin I, Condensin II can be found in the nucleus during interphase (Hirota et al., 2004; Ono et al., 2004). This observation signaled the possibility that higher-order chromosome condensation mediated by Condensin II is not limited to cell division and that Condensin II could play roles in interphase biology in higher eukaryotes.

The third Condensin complex, termed Condensin I^{DC}, is a part of the dosage compensation complex (DCC) found only in *Caenorhabditits elegans* and unlike the other Condensins, it contains a heterodimer of DPY-27 and Smc2. As its name implies, the primary role of the DCC is to mediate dosage compensation of X chromosome genes in hermaphrodites (Chuang, Albertson, & Meyer, 1994; Csankovszki et al., 2009; Lieb, Albrecht, Chuang, & Meyer, 1998; Lieb, Capowski, Meneely, & Meyer, 1996; Tsai et al., 2008). Unlike dosage compensation in mammals which is achieved by the random silencing of a single X chromosome, the DCC interacts with both X chromosomes to downregulate gene expression by half (Ercan & Lieb, 2009; Heard & Disteche, 2006).

The Smc5/6 complex is perhaps the least understood of the SMC complexes. The first component of this complex (Smc6) was originally identified because it was able to rescue a radiation-sensitive mutant strain of *Schizosaccharomyces pombe* (Nasim & Smith, 1975; Phipps, Nasim, & Miller, 1985). In addition to the Smc5/6 heterodimer, this complex contains four non-SMC components. Sequence analysis of these subunits reveals that the complex may have E3 ubiquitin ligase activity as well as SUMO ligase activity (Lehmann, 2005). Other than its known roles in DNA repair, the functions of the Smc5/6 complex remain elusive (Lehmann et al., 1995; Verkade, Bugg, Lindsay, Carr, & O'Connell, 1999). However, recent work suggests that there is at least crosstalk between the Smc5/6 complex and the Cohesin complex during chromosome segregation (for review, see Tapia-Alveal, Lin, & O'Connell, 2014).

5. INVOLVEMENT OF SMC COMPLEXES IN T CELL DEVELOPMENT

Both the Cohesin and Condensin II complexes have been implicated in T cell development. The Merkenschlager lab generated a conditional knockout allele of Rad21, the α-kleisin protein of the Cohesin complex (Table 2). Using a CD4-*Cre* recombinase they depleted Rad21 from non-dividing DP thymocytes (Seitan et al., 2011). As expected, Rad21 deficiency resulted in chromosome segregation defects when these DP thymocytes were induced to divide *in vitro*. Interestingly, both CD4+ and CD8+ SP cells were able to develop in the absence of Rad21; however, their rate of development was slowed. Further analysis showed that the Cohesin complex was enriched at the *Tcra* locus in DP cells, when α chain rearrangements are still occurring (Table 1). Importantly, the deficiency in Rad21 prevented the recruitment of Rag recombinases to the *Tcra* locus, resulting in impairment of V-J gene segment rearrangements, presumably the culprit for the slowed development. It should be noted that possible roles for the Cohesin complex in V(D)J rearrangements of the *Tcrb* locus could not be determined because Rad21 proteins levels were not depleted until the DP stage. By then, *Tcrb* rearrangements have already occurred (Table 1). Following up on these studies, the Merkenschlager lab investigated the genome-wide phenotype of Rad21 depletion on noncycling thymocytes using chromosome conformation capture. Depletion of Rad21 led to the deregulation of expression of over 1000 genes, almost all of which reside in "open" chromatin compartments (e.g., accessible, actively transcribed regions) within the nucleus (Lieberman-Aiden et al., 2009; Seitan et al., 2013). Interestingly, there was no effect on the status of the compartments themselves in Rad21 depleted cells (i.e., whether a locus is in an "open" or "closed" compartment). Instead, the Cohesin complex appears to be regulating gene expression through interactions with CTCF (described in more detail later) and by facilitating the spatial clustering of enhancer elements (Ing-Simmons et al., 2015).

The roles of the Condensin II complex in T cell development are based on the analysis of the *nessy* mouse. The *nessy* mouse harbors a point mutation in the Kleisin-β subunit of the Condensin II complex. The mutation, a T to A transversion within exon 1 of the long splice variant of the gene, results in a predicted Ile to Asn substitution at position 15 in the amino acid sequence. This mouse was generated in a large-scale ENU mutagenesis screen for T cell developmental or homeostatic defects (Gosling et al., 2007). While they appear healthy and normal, these mice have a striking thymocyte development defect in that the animals have much smaller thymuses with thymocyte cellularities that are approximately one-tenth that of wild-type mice. The reduction in cell numbers is manifested primarily in the DP stage of development where a 100-fold difference in cell numbers between wild type and mutant was observed. These data could suggest that a larger than normal number of cells in *nessy* mice fail +/− selection resulting in a high level of apoptosis. Indeed, a higher proportion of DP cells from *nessy* mice express the apoptotic marker annexin V. Importantly, Gosling et al. (2007) demonstrated that the reduction in DP cell numbers was not due to impairment of TCR rearrangement, as expression of a prerearranged TCR did not rescue the apoptotic phenotype. Importantly, the phenotypes associated with the *nessy* allele are T cell intrinsic, as the phenotype was recapitulated by transplant of *nessy/nessy* bone marrow into wild-type mice and rescued in *nessy* mice by retroviral expression of a wild-type allele (Gosling et al., 2007). Interestingly, Gosling et al. (2007) did not observe any of these phenotypes in B cells, suggesting that the Condensin II complex has a specific function in T cell development. This observation is particularly striking as Kleisin-β is ubiquitously expressed.

Given the known general function of the Condensin II complex in chromatin compaction and the specific T cell phenotype of the *nessy* mouse, we hypothesized that these mice may show defects in their ability to condense chromatin during T cell development, resulting in the increased apoptosis observed during the DP stage as a consequence of altered +/− selection. We developed a novel flow cytometry assay to measure chromatin condensation, allowing us to analyze chromatin status in subpopulations of T cells (Bingham, Lee, & Rawlings, 2015). We applied this technique to the *nessy* mouse, comparing chromatin status to wild-type littermate controls during thymocyte development (Fig. 1). At the DN1 stage, the chromatin condensation state in *nessy* T cells is indistinguishable from wild-type cells. As with wild-type cells, chromatin decondenses in *nessy* T cells at the DN2 stage, albeit to a greater extent. As chromatin condenses as the cells progress through the DN3, DN4, and DP stages in wild-type mice, the chromatin in T cells from *nessy* mice seems to lag behind. The result for *nessy* T cells is that they undergo +/− selection with chromatin that is more accessible than what is seen in wild-type cells. Given that TCR signal strength determines the outcome of selection, having a more open chromatin configuration could result in *nessy* T cells misinterpreting TCR signals during selection as too strong, resulting in the enhanced apoptosis seen at the DP stage. It should be noted that there are no gross autoimmune or immunodeficient phenotypes reported for the *nessy* mouse, suggesting that although the number of T cells that successfully complete +/− selection is reduced, the fidelity of +/− selection has not been compromised in these animals.

In addition to affecting selection, the failure to condense chromatin at the DP stage may also have effects on proliferation. In wild-type mice, as thymocytes transition to the DP stage, they become quiescent, and remain quiescent in the periphery until they are activated by presentation of their cognate antigen. Therefore, one might expect that *nessy* DP thymocytes may fail to become quiescent. In support of this hypothesis, it was shown that DP thymocytes of *nessy* mice have higher rates of proliferation than wild-type thymocytes. Furthermore, peripheral T cells from *nessy* mice appear to be proliferating in the absence of antigen as described later (Gosling et al., 2007).

6. ROLES OF SMC COMPLEXES IN PERIPHERAL T CELL BIOLOGY

Since both the Cohesin and Condensin II complexes have roles in T cell development, it is logical to question whether these roles extend to mature T cells in the periphery. With regard to the Cohesin complex, Seitan et al. (2011) did not report any peripheral T cell phenotypes when they depleted Rad21 from quiescent DP thymocytes. This is not surprising as it is likely that some successful *Tcra* rearrangements occurred before the depletion of Rad21 was complete. Indeed, Seitan et al. were forced to measure secondary *Tcra* rearrangements due to this technical limitation of their system. Interestingly, the authors did test whether DP thymocytes could respond to in vitro activation and found that activation stimuli induced apoptosis in Rad21 depleted thymocytes. Furthermore, they noted abnormal mitotic figures and defects in chromosome segregation, consistent with the known role of the Cohesin complex in cell division (Seitan et al., 2011). Presumably, peripheral T cells would respond in similar fashion.

Compared to Cohesin, the peripheral T cell phenotype of the Condensin II mutant is dramatic. The *nessy* mouse was recovered in the ENU screen described earlier because all peripheral T cells from *nessy* mice express high levels of CD44, a T cell activation marker (Gosling et al., 2007). Our own analysis of these mice indicates that a higher than normal proportion of T cells also expresses the CD25 activation marker (Rawlings et al., 2011). Strikingly, while *nessy* mice have greatly reduced thymocyte numbers as described earlier, they seem to recover in the periphery: a 2.7-fold reduction in T cell numbers in periphery vs an overall 10-fold reduction in the thymus (Gosling et al., 2007). Our analysis of peripheral T cells from *nessy* mice revealed that they have a more open chromatin configuration than WT mice as assessed by flow cytometry and by transmission electron microscopy (Fig. 3). These observations about chromatin status and expression of activation markers suggest that *nessy* peripheral T cells undergo homeostatic expansion once they leave the thymus (i.e., they are not quiescent). Functionally, *nessy* peripheral T cells are able to respond to antigenic challenge and control bacterial load; however, they are slower to upregulate the CD69 activation marker. Furthermore, their ability to stimulate B cells to produce antibody

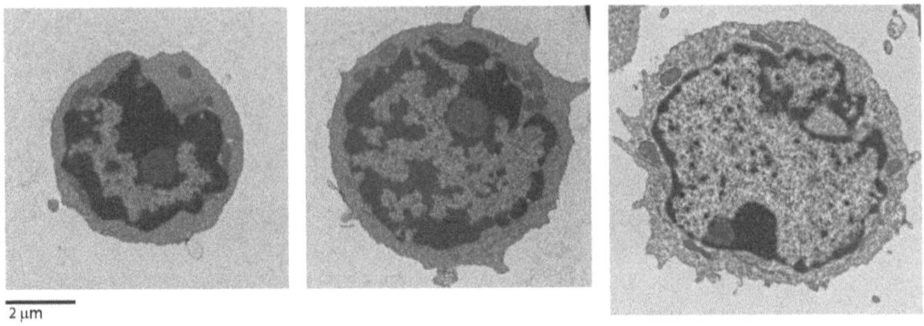

Fig. 3 Peripheral T cells from *nessy* mice have a variable, open chromatin configuration. Transmission electron microscopy (TEM) was used to visualize the chromatin in CD25–peripheral T cells from *nessy* mice. Cells were prepared for TEM as in Fig. 2 and representative images are shown. *Images taken with permission from Rawlings, J. S., Gatzka, M., Thomas, P. G., & Ihle, J. N. (2011). Chromatin condensation via the condensin II complex is required for peripheral T-cell quiescence.* EMBO Journal, 30*(2), 263–276. http://dx.doi.org/10.1038/emboj.2010.314.*

is also diminished (Gosling, Goodnow, Verma, & Fahrer, 2008). It should be noted that the *nessy* mutation is a point mutation resulting in a single-amino acid substitution and is likely a hypomorphic allele and not a complete loss of function. Indeed, we observed a variable expressivity and penetrance in the chromatin phenotype as assessed by transmission electron microscopy (Fig. 3). Therefore, it is possible that the Condensin II complex may have additional critical or even essential functions during T cell development and/or function, only to be revealed by a complete loss-of-function allele. Clearly, T cells are the most sensitive to perturbations in Condensin II complex function as no other phenotypes in other cells have been reported for the *nessy* mouse.

7. REGULATION OF GENE EXPRESSION IN T CELLS BY SMC COMPLEXES

While the roles of SMC complexes are primarily described at the scale of whole chromosome compaction, it is apparent that they may also play roles in regulating specific gene expression. In particular, Cohesin has been shown to be recruited to certain DNase hypersensitive sites by the DNA-binding protein CTCF (CCCTC-binding factor) and can mediate long-range intrachromosomal interactions to regulate gene expression, presumably by blocking enhancers or by preventing heterochromatin from spreading into genes (Labrador & Corces, 2002; Parelho et al., 2008; Stedman et al., 2008; Wendt et al., 2008; West, Huang, Gaszner,

Litt, & Felsenfeld, 2004). In one study, it was demonstrated that this interaction led to the regulation of expression of the cytokine interferon-γ, which is encoded by *IFNG* (Hadjur et al., 2009). Regulation of *IFNG* expression is crucial for proper differentiation of the T_H1 and T_H2 subsets of T helper (CD4+) cells. In order for a T cell to differentiate into a T_H1 cell, *IFNG* expression must be induced, whereas differentiation into a T_H2 cell requires silencing of *IFNG* expression (Ansel, Lee, & Rao, 2003). The inducible expression of *IFNG* in T_H1 cells is dependent on the emergence of DNase hypersensitive sites (Hatton et al., 2006; Lee, Avni, Chen, & Rao, 2004; Schoenborn et al., 2007). CTCF recruits Cohesin to these sites and together they regulate *IFNG* expression (Hadjur et al., 2009). Cohesin may also regulate expression of the T_H2 cytokines IL-4, IL-5, and IL-13, presumably by a similar mechanism (Ribeiro de Almeida et al., 2009). Interestingly, CTCF also has been shown to have a role in regulating *Pdcd1* gene expression in murine CD8+ T cells (Austin, Lu, Majumder, Ahmed, & Boss, 2014). The *Pdcd1* gene encodes for PD-1 (programmed death-1), which negatively regulates CD8+ T cell proliferation and effector functions (Zajac et al., 1998). CTCF has also been shown to regulate expression of all of the major histocompatibility complex (MHC) class II genes. The MHC class II locus is a tight cluster of genes whose expression is required for antigen presentation to T_H cells. The locus contains a total of 10 CTCF-binding sites that can form multiple distinct architectures to control expression of the locus (Majumder & Boss, 2010). It should also be noted that Cohesin has also been shown to interact with transcriptional mediators and regulate gene expression in a tissue-specific manner that is independent of CTCF (Kagey et al., 2010; Schmidt et al., 2010).

8. REGULATION OF SMC COMPLEXES

As SMC complexes play roles in key aspects of T cell development and function, their regulation is critical. Much is known about the regulation of the Cohesin complex in a number of model systems, most notably yeast and in *Xenopus* egg fertilization. The Smc heterodimer of the Cohesin complex, along with the α-kleisin subunit forms a ring-like structure that is large enough to entrap DNA within it. Several mechanisms have been proposed for the dissociation of the Cohesin complex from DNA. In mitotic prophase, a protein called Wapl (wings apart-like), Cohesin release factor has been shown to remove Cohesin from chromosomes (Gandhi, Gillespie, & Hirano, 2006; Kueng et al., 2006). It has been postulated that this is achieved by opening the Cohesin complex ring structure between Smc3 and Scc1 (Buheitel & Stemmann, 2013; Chan et al., 2012; Peters et al., 2008). In addition, as cells transition from metaphase to anaphase, Separase is activated by the anaphase-promoting complex/cyclosome (APC/C). Activated Separase cleaves the α-Kleisin subunit, releasing the DNA from the Cohesin complex, permitting sister chromatid disjunction. In non-dividing cells, Separase is sequestered by another protein called Securin. During cell division, the APC/C which is a ubiquitin ligase, initiates the proteosomal degradation of Securin, releasing Separase (Nasmyth, 2002; Peters et al., 2008; Sullivan & Morgan, 2007; Thornton & Toczyski, 2006). Interestingly, Wapl has been shown to prevent Separase-mediated cleavage of Cohesin (Tedeschi et al., 2013). Another mechanism of control is achieved via the regulation of APC/C-dependent Securin ubiquitination by Cdk1 and Cdc14. It has been shown the Cdk1-dependent phosphorylation of Securin inhibits its ubiquitination by the APC/C; furthermore, the phosphatase Cdc14 can reverse this phosphorylation, promoting proteosomal degradation (Holt, Krutchinsky, & Morgan, 2008).

The human Securin homolog is pituitary tumor transforming gene (PTTG). As its name implies, PTTG is highly expressed in pituitary tumors, and its expression correlates with tumor invasiveness (Saez et al., 1999; Zhang, Horwitz, Heaney, et al., 1999; Zhang, Horwitz, Prezant, et al., 1999). PTTG is also overexpressed in hematopoietic tumors (Dominguez et al., 1998). While PTTG is not expressed in naïve T cells, upon T cell activation with mitogens, the gene is upregulated to levels seen in tumor cells (Stoika & Melmed, 2002). The gene encoding PTTG has been deleted in mice. Surprisingly, the mice are viable, but show very specific phenotypes. In addition to cell cycle defects, mice lacking PTTG exhibit testicular, splenic, pituitary, and pancreatic β-cell hypoplasia. Additionally, these mice have thymic hyperplasia (Chesnokova, Kovacs, Castro, Zonis, & Melmed, 2005; Wang, Moro, Kovacs, Yu, & Melmed, 2003; Wang, Yu, & Melmed, 2001). While these phenotypes suggest defects in T cell development and/or function, this has not been tested directly. However, based on these phenotypes, it is tempting to speculate that PTTG could regulate Cohesin function in a similar fashion as described for Securin.

Less is known about the regulation of the Condensin II complex. While it has a Kleisin subunit (Kleisin-β/CAP-H2, Table 2), it has not been shown to be regulated by Separase or a Separase-like mechanism. Instead, it has recently been shown that the Condensin II complex is regulated, at least in part, by regulating protein levels of the CAP-H2 subunit. Two studies using *Drosophila* as a model system have implicated distinct mechanisms for regulating CAP-H2 levels. First, the SCF^{Slimb} ubiquitin ligase complex has been shown to target CAP-H2 for proteosomal degradation. The inhibition of SCF^{Slimb} results in the reorganization of chromosomes into dense structures, preventing homologous chromosome pairing (Buster et al., 2013). More recently, Casein kinase I alpha (CK1α) has also shown to regulate the CAP-H2 subunit. When CK1α is depleted in *Drosophila*, CAP-H2 protein levels are stabilized and it was shown to accumulate on chromosomes. The resulting phenotypes are reminiscent of what was observed with SCF^{Slimb} (Nguyen et al., 2015). In addition to CAP-H2-mediated regulation, a recent report using chicken DT40 cells indicates that loading of the Condensin II complex during mitosis is mediated by the phosphorylation status of a conserved threonine on CAP-D3 (Bakhrebah, Zhang, Mann, Kalitsis, & Hudson, 2015). Additionally, the protein Mrg15 (MORF4-related gene on chromosome 15) has been shown to interact with the Condensin II complex and is required for its Condensin II-mediated chromatin compaction and homolog pairing in *Drosophila* (Wallace, Klebba, Kusch, Rogers, & Bosco, 2015). Finally, Mps1 has been shown to associate with and phosphorylate CAP-H2. This phosphorylation is required for loading of the Condensin II complex onto chromatin during mitosis (Kagami et al., 2014). It remains to be seen if any of these mechanisms regulate Condensin II in T cells.

9. CONCLUSIONS AND OUTLOOK

Traditionally, the roles of SMC complexes were confined to regulating the large-scale changes in chromosome architecture observed during mitotic and meiotic cell division. However, it has become clear that these complexes play critical roles in interphase biology as well. Indeed, there is a growing body of literature that strongly suggests that the chromosomal architecture of the nucleus can regulate gene expression. For example, it has been suggested that gene expression occurs at discrete locations, termed "transcription factories" and that the physical location of a gene within the nucleus influences its ability to be expressed (Bickmore & van Steensel, 2013; Geyer, Vitalini, & Wallrath, 2011; Osborne et al., 2004). In addition, the physical nature of chromatin itself has been shown to regulate cell functions in a nongenetic fashion (Bustin & Misteli, 2016).

It is clear that a critical part of thymocyte development is the reorganization and condensation of chromatin prior to +/− selection. This reorganization is not only required for proper selection, it also serves to globally restrict gene expression in order to establish the quiescent state in the mature naïve T cell. Although SMC complexes have been implicated in the large-scale chromatin remodeling observed during T cell development; however, many key questions remain unanswered. It is not clear how SMC complex-mediated chromosome condensation is initiated during thymocyte development or how the functions of these complexes are coordinated with the recombination machinery and the onset of selection events. In yeast, it has been shown that Cohesin can be loaded onto chromosomes via Scc2 (NIPBL in humans) (for review, see Nasmyth & Haering, 2009). NIPBL has recently been shown to be involved in regulating class switch recombination of Ig loci during B lymphocyte (B cell) development (Enervald et al., 2013). Indeed, Cohesin binding to chromatin dramatically changes as a function of B cell activation, specifically at class switch regions (Gunal-Sadik et al., 2014). It remains to be seen if and how Cohesin is loaded onto chromosomes in developing T cells. However, it should be noted that mutations in NIPBL (as well as mutations in Smc1A and Smc3) are associated with Cornelia de Lange syndrome (Deardorff et al., 2007; Krantz et al., 2004; Musio et al., 2006). While individuals with this syndrome have a number of developmental phenotypes, it has recently been shown that these phenotypes include decreased ratios of certain T cell subsets in addition to antibody deficiency syndrome. Presumably the latter phenotype is due to faulty class switch recombination (Jyonouchi, Orange, Sullivan, Krantz, & Deardorff, 2013).

In the periphery, decondensation of chromatin is essential for competence to respond to cytokine stimulation, and thus serves as a key mechanism regulating clonal proliferation of T cells during an immune response. As in the thymus, the mechanism of chromatin reorganization leading to decondensation is not well understood. Given the very low metabolic capacity of a naïve peripheral T cell, the fidelity of clonal proliferation, and the kinetics of chromatin decondensation, one could hypothesize that the mechanism would have to be simple (e.g., not requiring protein synthesis), rapid, and easily regulated via the T cell receptor. Our more recent work indicates that the mobilization of calcium from intracellular stores is both sufficient and required to initiate the decondensation of chromatin in peripheral T cells, fitting these criteria (Lee, Bingham, Mitchell, Meredith, & Rawlings, 2015). Two important questions remain: how can calcium initiate the decondensation of chromatin and how do SMC complexes fit into this mechanism? Several studies have implicated calcium in the regulation of Cohesin by at least two mechanisms. First, it has been shown that Calpain-1, a calcium-dependent cysteine protease, can cleave Rad21 to promote sister chromatid disjunction during mitosis in an immortalized T cell line (Panigrahi, Zhang, Mao, & Pati, 2011). Second, the APC/C complex that regulates the release of Separase from Securin (see earlier) is itself been shown to be regulated by calcium via Calmodulin-dependent Kinase II (Liu & Maller, 2005; Rauh, Schmidt, Bormann, Nigg, & Mayer, 2005). It should be noted that our studies were confined to the initiation of decondensation, how chromatin is held open once decondensed remains to be determined. Ultimately, it will be of great interest to learn how TCR signaling-induced chromatin decondensation during T cell activation affects SMC complex localization and function.

The nucleus of a T cell undergoes dramatic reorganization that can restrict or permit the expression of gene programs that are required for proper development in the thymus and activation in the periphery. Even though SMC complexes are ubiquitously expressed and broadly required for cellular function, the fact that mutations in SMC complex subunits (particularly Cohesin and Condensin II) have specific T cell phenotypes suggests that T cells have a unique chromatin architecture or that these cells are uniquely sensitive to disruptions in the functions of SMC complexes. Nonetheless, it is clear that the SMC complexes are a part of a milieu of proteins and other molecules that contribute to the three-dimensional organization of the eukaryotic genome within the nucleus. For example, it has been shown that the SWI/SNF-like BAF complex has been shown to bind chromatin following TCR stimulation (Zhao et al., 1998). How SMC complexes, along with other chromatin remodeling complexes, work together to regulate T cell development and function remains to be fully understood.

REFERENCES

Anderson, D. E., Losada, A., Erickson, H. P., & Hirano, T. (2002). Condensin and cohesin display different arm conformations with characteristic hinge angles. *The Journal of Cell Biology*, *156*(3), 419–424. http://dx.doi.org/10.1083/jcb.200111002.

Ansel, K. M., Lee, D. U., & Rao, A. (2003). An epigenetic view of helper T cell differentiation. *Nature Immunology*, *4*(7), 616–623. http://dx.doi.org/10.1038/ni0703-616.

Aragon, L., Martinez-Perez, E., & Merkenschlager, M. (2013). Condensin, cohesin and the control of chromatin states. *Current Opinion in Genetics & Development*, *23*(2), 204–211. http://dx.doi.org/10.1016/j.gde.2012.11.004.

Austin, J. W., Lu, P., Majumder, P., Ahmed, R., & Boss, J. M. (2014). STAT3, STAT4, NFATc1, and CTCF regulate PD-1 through multiple novel regulatory regions in murine T cells. *The Journal of Immunology*, *192*(10), 4876–4886. http://dx.doi.org/10.4049/jimmunol.1302750.

Bakhrebah, M., Zhang, T., Mann, J. R., Kalitsis, P., & Hudson, D. F. (2015). Disruption of a conserved CAP-D3 threonine alters condensin loading on mitotic chromosomes leading to chromosome hypercondensation. *The Journal of Biological Chemistry*, *290*(10), 6156–6167. http://dx.doi.org/10.1074/jbc.M114.627109.

Barski, A., Jothi, R., Cuddapah, S., Cui, K., Roh, T. Y., Schones, D. E., et al. (2009). Chromatin poises miRNA- and protein-coding genes for expression. *Genome Research*, *19*(10), 1742–1751. http://dx.doi.org/10.1101/gr.090951.109.

Bickmore, W. A., & van Steensel, B. (2013). Genome architecture: Domain organization of interphase chromosomes. *Cell*, *152*(6), 1270–1284. http://dx.doi.org/10.1016/j.cell.2013.02.001.

Bingham, K. N., Lee, M. D., & Rawlings, J. S. (2015). The use of flow cytometry to assess the state of chromatin in T cells. *Journal of Visualized Experiments*, *106*, e53533. http://dx.doi.org/10.3791/53533.

Buheitel, J., & Stemmann, O. (2013). Prophase pathway-dependent removal of cohesin from human chromosomes requires opening of the Smc3-Scc1 gate. *EMBO Journal*, *32*(5), 666–676. http://dx.doi.org/10.1038/emboj.2013.7.

Buster, D. W., Daniel, S. G., Nguyen, H. Q., Windler, S. L., Skwarek, L. C., Peterson, M., et al. (2013). SCFSlimb ubiquitin ligase suppresses condensin II-mediated nuclear reorganization by degrading Cap-H2. *The Journal of Cell Biology*, *201*(1), 49–63. http://dx.doi.org/10.1083/jcb.201207183.

Bustin, M., & Misteli, T. (2016). Nongenetic functions of the genome. *Science*, *352*(6286), aad6933. http://dx.doi.org/10.1126/science.aad6933.

Chan, K. L., Roig, M. B., Hu, B., Beckouet, F., Metson, J., & Nasmyth, K. (2012). Cohesin's DNA exit gate is distinct from its entrance gate and is regulated by acetylation. *Cell*, *150*(5), 961–974. http://dx.doi.org/10.1016/j.cell.2012.07.028.

Chesnokova, V., Kovacs, K., Castro, A. V., Zonis, S., & Melmed, S. (2005). Pituitary hypoplasia in Pttg-/- mice is protective for Rb+/- pituitary tumorigenesis. *Molecular Endocrinology*, *19*(9), 2371–2379. http://dx.doi.org/10.1210/me.2005-0137.

Chong, M. M., Cornish, A. L., Darwiche, R., Stanley, E. G., Purton, J. F., Godfrey, D. I., et al. (2003). Suppressor of cytokine signaling-1 is a critical regulator of interleukin-7-dependent CD8+ T cell differentiation. *Immunity*, *18*(4), 475–487.

Chuang, P. T., Albertson, D. G., & Meyer, B. J. (1994). DPY-27: A chromosome condensation protein homolog that regulates C. elegans dosage compensation through association with the X chromosome. *Cell*, *79*(3), 459–474.

Csankovszki, G., Collette, K., Spahl, K., Carey, J., Snyder, M., Petty, E., et al. (2009). Three distinct condensin complexes control C. elegans chromosome dynamics. *Current Biology: CB*, *19*(1), 9–19. http://dx.doi.org/10.1016/j.cub.2008.12.006.

Deardorff, M. A., Kaur, M., Yaeger, D., Rampuria, A., Korolev, S., Pie, J., et al. (2007). Mutations in cohesin complex members SMC3 and SMC1A cause a mild variant of cornelia de Lange syndrome with predominant mental retardation. *The American Journal of Human Genetics*, *80*(3), 485–494. http://dx.doi.org/10.1086/511888.

Dominguez, A., Ramos-Morales, F., Romero, F., Rios, R. M., Dreyfus, F., Tortolero, M., et al. (1998). hpttg, a human homologue of rat pttg, is overexpressed in hematopoietic neoplasms. Evidence for a transcriptional activation function of hPTTG. *Oncogene*, *17*(17), 2187–2193. http://dx.doi.org/10.1038/sj.onc.1202140.

Enervald, E., Du, L., Visnes, T., Bjorkman, A., Lindgren, E., Wincent, J., et al. (2013). A regulatory role for the cohesin loader NIPBL in nonhomologous end joining during immunoglobulin class switch recombination. *The Journal of Experimental Medicine*, *210*(12), 2503–2513. http://dx.doi.org/10.1084/jem.20130168.

Ercan, S., & Lieb, J. D. (2009). C. elegans dosage compensation: A window into mechanisms of domain-scale gene regulation. *Chromosome Research*, *17*(2), 215–227. http://dx.doi.org/10.1007/s10577-008-9011-0.

Fitzpatrick, D. R., Shirley, K. M., McDonald, L. E., Bielefeldt-Ohmann, H., Kay, G. F., & Kelso, A. (1998). Distinct methylation of the interferon gamma (IFN-gamma) and interleukin 3 (IL-3) genes in newly activated primary CD8+ T lymphocytes: Regional IFN-gamma promoter demethylation and mRNA expression are heritable in CD44 (high)CD8+ T cells. *The Journal of Experimental Medicine*, *188*(1), 103–117.

Fletcher, J., & Starr, R. (2005). The role of suppressors of cytokine signalling in thymopoiesis and T cell activation. *International Journal of Biochemistry and Cell Biology*, *37*(9), 1774–1786. http://dx.doi.org/10.1016/j.biocel.2005.04.005.

Frauwirth, K. A., & Thompson, C. B. (2004). Regulation of T lymphocyte metabolism. *The Journal of Immunology*, *172*(8), 4661–4665.

Gandhi, R., Gillespie, P. J., & Hirano, T. (2006). Human Wapl is a cohesin-binding protein that promotes sister-chromatid resolution in mitotic prophase. *Current Biology: CB*, *16*(24), 2406–2417. http://dx.doi.org/10.1016/j.cub.2006.10.061.

Gatzka, M., Piekorz, R., Moriggl, R., Rawlings, J., & Ihle, J. N. (2006). A role for STAT5A/B in protection of peripheral T-lymphocytes from postactivation apoptosis: Insights from gene expression profiling. *Cytokine*, *34*(3–4), 143–154. http://dx.doi.org/10.1016/j.cyto.2006.04.003.

Geyer, P. K., Vitalini, M. W., & Wallrath, L. L. (2011). Nuclear organization: Taking a position on gene expression. *Current Opinion in Cell Biology*, *23*(3), 354–359. http://dx.doi.org/10.1016/j.ceb.2011.03.002.

Gosling, K. M., Goodnow, C. C., Verma, N. K., & Fahrer, A. M. (2008). Defective T-cell function leading to reduced antibody production in a kleisin-beta mutant mouse. *Immunology*, *125*(2), 208–217. http://dx.doi.org/10.1111/j.1365-2567.2008.02831.x.

Gosling, K. M., Makaroff, L. E., Theodoratos, A., Kim, Y. H., Whittle, B., Rui, L., et al. (2007). A mutation in a chromosome condensin II subunit, kleisin beta, specifically disrupts T cell development. *Proceedings of the National Academy of Sciences of the United States of America*, *104*(30), 12445–12450. http://dx.doi.org/10.1073/pnas.0704870104.

Gunal-Sadik, G., Paszkowski-Rogacz, M., Singaravelu, K., Beyer, A., Buchholz, F., & Jessberger, R. (2014). Stage-specific binding profiles of cohesin in resting and activated B lymphocytes suggest a role for cohesin in immunoglobulin class switching and maturation. *PLoS One*, *9*(11), e111748. http://dx.doi.org/10.1371/journal.pone.0111748.

Hadjur, S., Williams, L. M., Ryan, N. K., Cobb, B. S., Sexton, T., Fraser, P., et al. (2009). Cohesins form chromosomal cis-interactions at the developmentally regulated IFNG locus. *Nature*, *460*(7253), 410–413. http://dx.doi.org/10.1038/nature08079.

Hatton, R. D., Harrington, L. E., Luther, R. J., Wakefield, T., Janowski, K. M., Oliver, J. R., et al. (2006). A distal conserved sequence element controls Ifng gene expression by T cells and NK cells. *Immunity*, *25*(5), 717–729. http://dx.doi.org/10.1016/j.immuni.2006.09.007.

Heard, E., & Disteche, C. M. (2006). Dosage compensation in mammals: Fine-tuning the expression of the X chromosome. *Genes and Development*, *20*(14), 1848–1867. http://dx.doi.org/10.1101/gad.1422906.

Hirano, T. (2005). Condensins: Organizing and segregating the genome. *Current Biology: CB*, *15*(7), R265–R275. http://dx.doi.org/10.1016/j.cub.2005.03.037.

Hirano, T., Kobayashi, R., & Hirano, M. (1997). Condensins, chromosome condensation protein complexes containing XCAP-C, XCAP-E and a Xenopus homolog of the Drosophila Barren protein. *Cell*, *89*(4), 511–521.

Hirota, T., Gerlich, D., Koch, B., Ellenberg, J., & Peters, J. M. (2004). Distinct functions of condensin I and II in mitotic chromosome assembly. *Journal of Cell Science*, *117*(Pt. 26), 6435–6445. http://dx.doi.org/10.1242/jcs.01604.

Holt, L. J., Krutchinsky, A. N., & Morgan, D. O. (2008). Positive feedback sharpens the anaphase switch. *Nature*, *454*(7202), 353–357. http://dx.doi.org/10.1038/nature07050.

Ing-Simmons, E., Seitan, V. C., Faure, A. J., Flicek, P., Carroll, T., Dekker, J., et al. (2015). Spatial enhancer clustering and regulation of enhancer-proximal genes by cohesin. *Genome Research*, *25*(4), 504–513. http://dx.doi.org/10.1101/gr.184986.114.

Jaehning, J. A., Stewart, C. C., & Roeder, R. G. (1975). DNA-dependent RNA polymerase levels during the response of human peripheral lymphocytes to phytohemagglutinin. *Cell*, *4*(1), 51–57.

Jenuwein, T., & Allis, C. D. (2001). Translating the histone code. *Science*, *293*(5532), 1074–1080. http://dx.doi.org/10.1126/science.1063127.

Jeppsson, K., Kanno, T., Shirahige, K., & Sjogren, C. (2014). The maintenance of chromosome structure: Positioning and functioning of SMC complexes. *Nature Reviews Molecular Cell Biology*, *15*(9), 601–614. http://dx.doi.org/10.1038/nrm3857.

Jyonouchi, S., Orange, J., Sullivan, K. E., Krantz, I., & Deardorff, M. (2013). Immunologic features of Cornelia de Lange syndrome. *Pediatrics*, *132*(2), e484–e489. http://dx.doi.org/10.1542/peds.2012-3815.

Kagami, Y., Nihira, K., Wada, S., Ono, M., Honda, M., & Yoshida, K. (2014). Mps1 phosphorylation of condensin II controls chromosome condensation at the onset of mitosis. *The Journal of Cell Biology*, *205*(6), 781–790. http://dx.doi.org/10.1083/jcb.201308172.

Kagey, M. H., Newman, J. J., Bilodeau, S., Zhan, Y., Orlando, D. A., van Berkum, N. L., et al. (2010). Mediator and cohesin connect gene expression and chromatin architecture. *Nature*, *467*(7314), 430–435. http://dx.doi.org/10.1038/nature09380.

Komori, H. K., Hart, T., LaMere, S. A., Chew, P. V., & Salomon, D. R. (2015). Defining CD4 T cell memory by the epigenetic landscape of CpG DNA methylation. *The Journal of Immunology*, *194*(4), 1565–1579. http://dx.doi.org/10.4049/jimmunol.1401162.

Kouzarides, T. (2007). Chromatin modifications and their function. *Cell*, *128*(4), 693–705. http://dx.doi.org/10.1016/j.cell.2007.02.005. S0092-8674(07)00184-5 [pii].

Krantz, I. D., McCallum, J., DeScipio, C., Kaur, M., Gillis, L. A., Yaeger, D., et al. (2004). Cornelia de Lange syndrome is caused by mutations in NIPBL, the human homolog of Drosophila melanogaster Nipped-B. *Nature Genetics*, *36*(6), 631–635. http://dx.doi.org/10.1038/ng1364.

Kueng, S., Hegemann, B., Peters, B. H., Lipp, J. J., Schleiffer, A., Mechtler, K., et al. (2006). Wapl controls the dynamic association of cohesin with chromatin. *Cell*, *127*(5), 955–967. http://dx.doi.org/10.1016/j.cell.2006.09.040.

Labrador, M., & Corces, V. G. (2002). Setting the boundaries of chromatin domains and nuclear organization. *Cell*, *111*(2), 151–154.

Lee, D. U., Agarwal, S., & Rao, A. (2002). Th2 lineage commitment and efficient IL-4 production involves extended demethylation of the IL-4 gene. *Immunity*, *16*(5), 649–660.

Lee, D. U., Avni, O., Chen, L., & Rao, A. (2004). A distal enhancer in the interferon-gamma (IFN-gamma) locus revealed by genome sequence comparison. *The Journal of Biological Chemistry*, *279*(6), 4802–4810. http://dx.doi.org/10.1074/jbc.M307904200.

Lee, M. D., Bingham, K. N., Mitchell, T. Y., Meredith, J. L., & Rawlings, J. S. (2015). Calcium mobilization is both required and sufficient for initiating chromatin decondensation during activation of peripheral T-cells. *Molecular Immunology*, *63*(2), 540–549. http://dx.doi.org/10.1016/j.molimm.2014.10.015.

Lehmann, A. R. (2005). The role of SMC proteins in the responses to DNA damage. *DNA Repair (Amst)*, *4*(3), 309–314. http://dx.doi.org/10.1016/j.dnarep.2004.07.009.

Lehmann, A. R., Walicka, M., Griffiths, D. J., Murray, J. M., Watts, F. Z., McCready, S., et al. (1995). The rad18 gene of Schizosaccharomyces pombe defines a new subgroup of the SMC superfamily involved in DNA repair. *Molecular and Cellular Biology*, *15*(12), 7067–7080.

Lewis, C. D., & Laemmli, U. K. (1982). Higher order metaphase chromosome structure: Evidence for metalloprotein interactions. *Cell*, *29*(1), 171–181.

Li, Y., Chen, G., Ma, L., Ohms, S. J., Sun, C., Shannon, M. F., et al. (2012). Plasticity of DNA methylation in mouse T cell activation and differentiation. *BMC Molecular Biology, 13*, 16. http://dx.doi.org/10.1186/1471-2199-13-16.

Lieb, J. D., Albrecht, M. R., Chuang, P. T., & Meyer, B. J. (1998). MIX-1: An essential component of the C. elegans mitotic machinery executes X chromosome dosage compensation. *Cell, 92*(2), 265–277.

Lieb, J. D., Capowski, E. E., Meneely, P., & Meyer, B. J. (1996). DPY-26, a link between dosage compensation and meiotic chromosome segregation in the nematode. *Science, 274*(5293), 1732–1736.

Lieberman-Aiden, E., van Berkum, N. L., Williams, L., Imakaev, M., Ragoczy, T., Telling, A., et al. (2009). Comprehensive mapping of long-range interactions reveals folding principles of the human genome. *Science, 326*(5950), 289–293. http://dx.doi.org/10.1126/science.1181369.

Lim, P. S., Hardy, K., Bunting, K. L., Ma, L., Peng, K., Chen, X., et al. (2009). Defining the chromatin signature of inducible genes in T cells. *Genome Biology, 10*(10), R107. http://dx.doi.org/10.1186/gb-2009-10-10-r107.

Lim, P. S., Shannon, M. F., & Hardy, K. (2010). Epigenetic control of inducible gene expression in the immune system. *Epigenomics, 2*(6), 775–795. http://dx.doi.org/10.2217/epi.10.55.

Lin, J. X., & Leonard, W. J. (1997). Signaling from the IL-2 receptor to the nucleus (Review). *Cytokine and Growth Factor Reviews, 8*(4), 313–332.

Liu, J., & Maller, J. L. (2005). Calcium elevation at fertilization coordinates phosphorylation of XErp1/Emi2 by Plx1 and CaMK II to release metaphase arrest by cytostatic factor. *Current Biology: CB, 15*(16), 1458–1468. http://dx.doi.org/10.1016/j.cub.2005.07.030.

Losada, A., Hirano, M., & Hirano, T. (1998). Identification of Xenopus SMC protein complexes required for sister chromatid cohesion. *Genes and Development, 12*(13), 1986–1997.

Majumder, P., & Boss, J. M. (2010). CTCF controls expression and chromatin architecture of the human major histocompatibility complex class II locus. *Molecular and Cellular Biology, 30*(17), 4211–4223. http://dx.doi.org/10.1128/MCB.00327-10.

Marine, J. C., Topham, D. J., McKay, C., Wang, D., Parganas, E., Stravopodis, D., et al. (1999). SOCS1 deficiency causes a lymphocyte-dependent perinatal lethality. *Cell, 98*(5), 609–616.

Melby, T. E., Ciampaglio, C. N., Briscoe, G., & Erickson, H. P. (1998). The symmetrical structure of structural maintenance of chromosomes (SMC) and MukB proteins: Long, antiparallel coiled coils, folded at a flexible hinge. *The Journal of Cell Biology, 142*(6), 1595–1604.

Michaelis, C., Ciosk, R., & Nasmyth, K. (1997). Cohesins: Chromosomal proteins that prevent premature separation of sister chromatids. *Cell, 91*(1), 35–45.

Moriggl, R., Sexl, V., Piekorz, R., Topham, D., & Ihle, J. N. (1999). Stat5 activation is uniquely associated with cytokine signaling in peripheral T cells. *Immunity, 11*(2), 225–230.

Moriggl, R., Topham, D. J., Teglund, S., Sexl, V., McKay, C., Wang, D., et al. (1999). Stat5 is required for IL-2-induced cell cycle progression of peripheral T cells. *Immunity, 10*(2), 249–259.

Morley, S. J., Rau, M., Kay, J. E., & Pain, V. M. (1993). Increased phosphorylation of eukaryotic initiation factor 4 alpha during early activation of T lymphocytes correlates with increased initiation factor 4 F complex formation. *European Journal of Biochemistry, 218*(1), 39–48.

Musio, A., Selicorni, A., Focarelli, M. L., Gervasini, C., Milani, D., Russo, S., et al. (2006). X-linked Cornelia de Lange syndrome owing to SMC1L1 mutations. *Nature Genetics, 38*(5), 528–530. http://dx.doi.org/10.1038/ng1779.

Nasim, A., & Smith, B. P. (1975). Genetic control of radiation sensitivity in Schizosaccharomyces pombe. *Genetics*, *79*(4), 573–582.
Nasmyth, K. (2002). Segregating sister genomes: The molecular biology of chromosome separation. *Science*, *297*(5581), 559–565. http://dx.doi.org/10.1126/science.1074757.
Nasmyth, K., & Haering, C. H. (2009). Cohesin: Its roles and mechanisms. *Annual Review of Genetics*, *43*, 525–558. http://dx.doi.org/10.1146/annurev-genet-102108-134233.
Nguyen, H. Q., Nye, J., Buster, D. W., Klebba, J. E., Rogers, G. C., & Bosco, G. (2015). Drosophila casein kinase I alpha regulates homolog pairing and genome organization by modulating condensin II subunit Cap-H2 levels. *PLoS Genetics*, *11*(2), e1005014. http://dx.doi.org/10.1371/journal.pgen.1005014.
Ono, T., Fang, Y., Spector, D. L., & Hirano, T. (2004). Spatial and temporal regulation of Condensins I and II in mitotic chromosome assembly in human cells. *Molecular Biology of the Cell*, *15*(7), 3296–3308. http://dx.doi.org/10.1091/mbc.E04-03-0242.
Ono, T., Losada, A., Hirano, M., Myers, M. P., Neuwald, A. F., & Hirano, T. (2003). Differential contributions of condensin I and condensin II to mitotic chromosome architecture in vertebrate cells. *Cell*, *115*(1), 109–121.
Osborne, C. S., Chakalova, L., Brown, K. E., Carter, D., Horton, A., Debrand, E., et al. (2004). Active genes dynamically colocalize to shared sites of ongoing transcription. *Nature Genetics*, *36*(10), 1065–1071. http://dx.doi.org/10.1038/ng1423.
Panigrahi, A. K., Zhang, N., Mao, Q., & Pati, D. (2011). Calpain-1 cleaves Rad21 to promote sister chromatid separation. *Molecular and Cellular Biology*, *31*(21), 4335–4347. http://dx.doi.org/10.1128/MCB.06075-11.
Parelho, V., Hadjur, S., Spivakov, M., Leleu, M., Sauer, S., Gregson, H. C., et al. (2008). Cohesins functionally associate with CTCF on mammalian chromosome arms. *Cell*, *132*(3), 422–433. http://dx.doi.org/10.1016/j.cell.2008.01.011.
Peters, J. M., Tedeschi, A., & Schmitz, J. (2008). The cohesin complex and its roles in chromosome biology. *Genes and Development*, *22*(22), 3089–3114. http://dx.doi.org/10.1101/gad.1724308.
Phipps, J., Nasim, A., & Miller, D. R. (1985). Recovery, repair, and mutagenesis in Schizosaccharomyces pombe. *Advances in Genetics*, *23*, 1–72.
Rauh, N. R., Schmidt, A., Bormann, J., Nigg, E. A., & Mayer, T. U. (2005). Calcium triggers exit from meiosis II by targeting the APC/C inhibitor XErp1 for degradation. *Nature*, *437*(7061), 1048–1052. http://dx.doi.org/10.1038/nature04093.
Rawlings, J. S., Gatzka, M., Thomas, P. G., & Ihle, J. N. (2011). Chromatin condensation via the condensin II complex is required for peripheral T-cell quiescence. *EMBO Journal*, *30*(2), 263–276. http://dx.doi.org/10.1038/emboj.2010.314.
Rawlings, J. S., Rosler, K. M., & Harrison, D. A. (2004). The JAK/STAT signaling pathway. *Journal of Cell Science*, *117*(Pt. 8), 1281–1283. http://dx.doi.org/10.1242/jcs.00963.
Ribeiro de Almeida, C., Heath, H., Krpic, S., Dingjan, G. M., van Hamburg, J. P., Bergen, I., et al. (2009). Critical role for the transcription regulator CCCTC-binding factor in the control of Th2 cytokine expression. *The Journal of Immunology*, *182*(2), 999–1010.
Rodriguez, R. M., Suarez-Alvarez, B., Mosen-Ansorena, D., Garcia-Peydro, M., Fuentes, P., Garcia-Leon, M. J., et al. (2015). Regulation of the transcriptional program by DNA methylation during human alphabeta T-cell development. *Nucleic Acids Research*, *43*(2), 760–774. http://dx.doi.org/10.1093/nar/gku1340.
Saez, C., Japon, M. A., Ramos-Morales, F., Romero, F., Segura, D. I., Tortolero, M., et al. (1999). hpttg is over-expressed in pituitary adenomas and other primary epithelial neoplasias. *Oncogene*, *18*(39), 5473–5476. http://dx.doi.org/10.1038/sj.onc.1202914.
Saitoh, N., Goldberg, I. G., Wood, E. R., & Earnshaw, W. C. (1994). ScII: An abundant chromosome scaffold protein is a member of a family of putative ATPases with an unusual predicted tertiary structure. *The Journal of Cell Biology*, *127*(2), 303–318.

Scharer, C. D., Barwick, B. G., Youngblood, B. A., Ahmed, R., & Boss, J. M. (2013). Global DNA methylation remodeling accompanies CD8 T cell effector function. *The Journal of Immunology, 191*(6), 3419–3429. http://dx.doi.org/10.4049/jimmunol.1301395.

Schmidt, D., Schwalie, P. C., Ross-Innes, C. S., Hurtado, A., Brown, G. D., Carroll, J. S., et al. (2010). A CTCF-independent role for cohesin in tissue-specific transcription. *Genome Research, 20*(5), 578–588. http://dx.doi.org/10.1101/gr.100479.109.

Schoenborn, J. R., Dorschner, M. O., Sekimata, M., Santer, D. M., Shnyreva, M., Fitzpatrick, D. R., et al. (2007). Comprehensive epigenetic profiling identifies multiple distal regulatory elements directing transcription of the gene encoding interferon-gamma. *Nature Immunology, 8*(7), 732–742. http://dx.doi.org/10.1038/ni1474.

Schones, D. E., & Zhao, K. (2008). Genome-wide approaches to studying chromatin modifications. *Nature Reviews Genetics, 9*(3), 179–191. http://dx.doi.org/10.1038/nrg2270.

Seitan, V. C., Faure, A. J., Zhan, Y., McCord, R. P., Lajoie, B. R., Ing-Simmons, E., et al. (2013). Cohesin-based chromatin interactions enable regulated gene expression within preexisting architectural compartments. *Genome Research, 23*(12), 2066–2077. http://dx.doi.org/10.1101/gr.161620.113.

Seitan, V. C., Hao, B., Tachibana-Konwalski, K., Lavagnolli, T., Mira-Bontenbal, H., Brown, K. E., et al. (2011). A role for cohesin in T-cell-receptor rearrangement and thymocyte differentiation. *Nature, 476*(7361), 467–471. http://dx.doi.org/10.1038/nature10312.

Sjogren, C., & Nasmyth, K. (2001). Sister chromatid cohesion is required for postreplicative double-strand break repair in Saccharomyces cerevisiae. *Current Biology: CB, 11*(12), 991–995.

Stedman, W., Kang, H., Lin, S., Kissil, J. L., Bartolomei, M. S., & Lieberman, P. M. (2008). Cohesins localize with CTCF at the KSHV latency control region and at cellular c-myc and H19/Igf2 insulators. *EMBO Journal, 27*(4), 654–666. http://dx.doi.org/10.1038/emboj.2008.1.

Stoika, R., & Melmed, S. (2002). Expression and function of pituitary tumour transforming gene for T-lymphocyte activation. *British Journal of Haematology, 119*(4), 1070–1074.

Strom, L., & Sjogren, C. (2007). Chromosome segregation and double-strand break repair—A complex connection. *Current Opinion in Cell Biology, 19*(3), 344–349. http://dx.doi.org/10.1016/j.ceb.2007.04.003.

Strunnikov, A. V., Larionov, V. L., & Koshland, D. (1993). SMC1: An essential yeast gene encoding a putative head-rod-tail protein is required for nuclear division and defines a new ubiquitous protein family. *The Journal of Cell Biology, 123*(6 Pt. 2), 1635–1648.

Sullivan, M., & Morgan, D. O. (2007). Finishing mitosis, one step at a time. *Nature Reviews Molecular Cell Biology, 8*(11), 894–903. http://dx.doi.org/10.1038/nrm2276.

Sumara, I., Vorlaufer, E., Gieffers, C., Peters, B. H., & Peters, J. M. (2000). Characterization of vertebrate cohesin complexes and their regulation in prophase. *The Journal of Cell Biology, 151*(4), 749–762.

Tapia-Alveal, C., Lin, S. J., & O'Connell, M. J. (2014). Functional interplay between cohesin and Smc5/6 complexes. *Chromosoma, 123*(5), 437–445. http://dx.doi.org/10.1007/s00412-014-0474-9.

Tedeschi, A., Wutz, G., Huet, S., Jaritz, M., Wuensche, A., Schirghuber, E., et al. (2013). Wapl is an essential regulator of chromatin structure and chromosome segregation. *Nature, 501*(7468), 564–568. http://dx.doi.org/10.1038/nature12471.

Thornton, B. R., & Toczyski, D. P. (2006). Precise destruction: An emerging picture of the APC. *Genes and Development, 20*(22), 3069–3078. http://dx.doi.org/10.1101/gad.1478306.

Toth, A., Ciosk, R., Uhlmann, F., Galova, M., Schleiffer, A., & Nasmyth, K. (1999). Yeast cohesin complex requires a conserved protein, Eco1p(Ctf7), to establish cohesion between sister chromatids during DNA replication. *Genes and Development, 13*(3), 320–333.

Tsai, C. J., Mets, D. G., Albrecht, M. R., Nix, P., Chan, A., & Meyer, B. J. (2008). Meiotic crossover number and distribution are regulated by a dosage compensation protein that resembles a condensin subunit. *Genes and Development*, *22*(2), 194–211. http://dx.doi.org/10.1101/gad.1618508.

Verkade, H. M., Bugg, S. J., Lindsay, H. D., Carr, A. M., & O'Connell, M. J. (1999). Rad18 is required for DNA repair and checkpoint responses in fission yeast. *Molecular Biology of the Cell*, *10*(9), 2905–2918.

von Freeden-Jeffry, U., Solvason, N., Howard, M., & Murray, R. (1997). The earliest T lineage-committed cells depend on IL-7 for Bcl-2 expression and normal cell cycle progression. *Immunity*, 7(1), 147–154.

Wallace, H. A., Klebba, J. E., Kusch, T., Rogers, G. C., & Bosco, G. (2015). Condensin II regulates interphase chromatin organization through the Mrg-binding motif of Cap-H2. *G3 (Bethesda)*, *5*(5), 803–817. http://dx.doi.org/10.1534/g3.115.016634.

Wang, Z., Moro, E., Kovacs, K., Yu, R., & Melmed, S. (2003). Pituitary tumor transforming gene-null male mice exhibit impaired pancreatic beta cell proliferation and diabetes. *Proceedings of the National Academy of Sciences of the United States of America*, *100*(6), 3428–3432. http://dx.doi.org/10.1073/pnas.0638052100.

Wang, Z., Yu, R., & Melmed, S. (2001). Mice lacking pituitary tumor transforming gene show testicular and splenic hypoplasia, thymic hyperplasia, thrombocytopenia, aberrant cell cycle progression, and premature centromere division. *Molecular Endocrinology*, *15*(11), 1870–1879. http://dx.doi.org/10.1210/mend.15.11.0729.

Wendt, K. S., Yoshida, K., Itoh, T., Bando, M., Koch, B., Schirghuber, E., et al. (2008). Cohesin mediates transcriptional insulation by CCCTC-binding factor. *Nature*, *451*(7180), 796–801. http://dx.doi.org/10.1038/nature06634.

West, A. G., Huang, S., Gaszner, M., Litt, M. D., & Felsenfeld, G. (2004). Recruitment of histone modifications by USF proteins at a vertebrate barrier element. *Molecular Cell*, *16*(3), 453–463. http://dx.doi.org/10.1016/j.molcel.2004.10.005.

Yeong, F. M., Hombauer, H., Wendt, K. S., Hirota, T., Mudrak, I., Mechtler, K., et al. (2003). Identification of a subunit of a novel Kleisin-beta/SMC complex as a potential substrate of protein phosphatase 2A. *Current Biology: CB*, *13*(23), 2058–2064.

Zajac, A. J., Blattman, J. N., Murali-Krishna, K., Sourdive, D. J., Suresh, M., Altman, J. D., et al. (1998). Viral immune evasion due to persistence of activated T cells without effector function. *The Journal of Experimental Medicine*, *188*(12), 2205–2213.

Zhang, X., Horwitz, G. A., Heaney, A. P., Nakashima, M., Prezant, T. R., Bronstein, M. D., et al. (1999a). Pituitary tumor transforming gene (PTTG) expression in pituitary adenomas. *The Journal of Clinical Endocrinology and Metabolism*, *84*(2), 761–767. http://dx.doi.org/10.1210/jcem.84.2.5432.

Zhang, X., Horwitz, G. A., Prezant, T. R., Valentini, A., Nakashima, M., Bronstein, M. D., et al. (1999b). Structure, expression, and function of human pituitary tumor-transforming gene (PTTG). *Molecular Endocrinology*, *13*(1), 156–166. http://dx.doi.org/10.1210/mend.13.1.0225.

Zhao, K., Wang, W., Rando, O. J., Xue, Y., Swiderek, K., Kuo, A., et al. (1998). Rapid and phosphoinositol-dependent binding of the SWI/SNF-like BAF complex to chromatin after T lymphocyte receptor signaling. *Cell*, *95*(5), 625–636.

CHAPTER THREE

The Role of Epigenetic Regulation in Transcriptional Memory in the Immune System

A.M. Woodworth, A.F. Holloway[1]
School of Medicine, University of Tasmania, Hobart, TAS, Australia
[1]Corresponding author: e-mail address: a.f.holloway@utas.edu.au

Contents

Abstract

The immune system is exquisitely poised to identify, respond to, and eradicate pathogens from the body, as well as to produce a more rapid and augmented response to a subsequent encounter with the pathogen. These cellular responses rely on the highly coordinated and rapid activation of gene expression programs as well as the ability of the cell to retain a memory of the initial gene response. It is clear that chromatin structure and epigenetic mechanisms play a crucial role in determining these gene responses, and in fact the immune system has proved an instructive model for investigating the multifaceted mechanisms through which the chromatin landscape contributes to gene expression programs. These mechanisms include modifications to the DNA and histone proteins, the positioning, composition, and remodeling of nucleosomes, as well as the formation of higher-order chromatin structures. Moreover, it is now apparent that epigenetic mechanisms also provide an instrument by which cells can retain memory of the initial transcriptional response, "priming" the genome so that

Advances in Protein Chemistry and Structural Biology, Volume 106
ISSN 1876-1623
http://dx.doi.org/10.1016/bs.apcsb.2016.09.002

it can respond more quickly to subsequent exposure to the signal. Here, we use the immune system as a model to demonstrate the complex interplay between transcription factors and the chromatin landscape required to orchestrate precise gene responses to external stimuli and further to demonstrate how these interactions can establish memory of past transcriptional events. We focus on what we have learnt from the immune system and how this can inform our understanding of other cellular systems.

1. INTRODUCTION

Cellular identity is defined by the specific gene expression profile of a cell, and this is determined by the repertoire of transcription factors that operate within the particular cellular context. In addition, the chromatin landscape and epigenetic mechanisms play an important role in modulating interactions between transcription factors and the genome and therefore also shaping cell identity. The composition and structure of the chromatin into which the DNA is assembled as well as modifications to the DNA and histone proteins provides an additional layer of regulatory control of gene expression programs (Lim, Li, Holloway, & Rao, 2013). Preservation of gene expression programs following cell division therefore relies on maintenance of the transcription factor repertoire as well as the chromatin landscape, and this provides the basis for a form of cellular memory. In addition, it is now apparent that chromatin structure and epigenetic modifications can provide a mechanism to enable a cell to retain memory of a transcriptional response to an environmental cue so that it can respond more quickly and in an augmented fashion upon reencountering that cue. The adaptive immune system is an exemplar of this type of response, with memory T cells able to retain "transcriptional memory" of an initial response in order to respond with a more rapid and effective transcriptional response on reencountering the same antigen (Dunn, McCuaig, Tu, Hardy, & Rao, 2015). However, it is becoming apparent that this type of "transcriptional memory" is not unique to the adaptive immune system, or even multicellular organisms and a picture is now emerging of how the chromatin environment can provide memory of past transcriptional events. Herein, we will focus on the multilayered mechanisms that contribute to epigenetic and transcriptional memory in the immune system, keeping in mind that the lessons that we have learnt from the immune system are informing our understanding of other biological systems and also disease processes.

2. CHROMATIN STRUCTURE AND FUNCTION

2.1 Chromatin Organization

Chromatin is a highly organized DNA:protein structure which enables the approximately 2 m of DNA contained in each human cell to be packaged into the nucleus. The fundamental unit of chromatin is the nucleosome, which consists of 147 base pairs of DNA wrapped around a histone octamer (Luger, Mäder, Richmond, Sargent, & Richmond, 1997). The nucleosome is comprised of eight histone proteins, two each of histones H2A, H2B, H3, and H4 (Luger et al., 1997), and is connected to adjacent nucleosomes by short segments of linker DNA. Variants of the canonical histones are assembled into nucleosomes in certain genomic or cellular contexts (Li & Fang, 2015) and themselves affect chromatin structure. The linker histones, H1 and H5, also interact with the nucleosome and have a role in stabilizing chromatin (Carruthers, Bednar, Woodcock, & Hansen, 1998).

Depending on nucleosome composition and posttranslational modifications, DNA is then condensed further as neighboring nucleosomes interact with each other to form higher-order structures. There is evidence that chromatin can form a 30-nm diameter fiber in vitro (Dorigo et al., 2004), although the existence and configuration of this structure in the nucleus remains a matter of debate (Fussner, Ching, & Bazett-Jones, 2011), and it is clear that there is a significant level of plasticity in the formation of chromatin into higher-order structures (Luger, Dechassa, & Tremethick, 2012). Higher-order chromatin structures are then organized into topologically associating domains of approximately 0.1–1 megabase pairs (Mb) in size (Dixon et al., 2012; Lieberman-Aiden et al., 2009). Further, within the nucleus itself, chromosomes occupy specific regions, referred to as chromosome territories (Gerlich et al., 2003). The exact mechanisms driving the formation of chromosome territories are as yet unknown, but the interaction of chromatin with the nuclear membrane as well as condensin-mediated chromosomal interactions (Iwasaki, Corcoran, & Noma, 2015) is thought to contribute to these spatial patterns.

While these hierarchical levels of chromatin organization ensure the orderly packaging of the genomic material into the cell nucleus, they have an equally important role in regulating genome function (Gonzalez-Sandoval & Gasser, 2016; Khorasanizadeh, 2004). Gene expression can be regulated at every level of chromatin organization, from the positioning of nucleosomes on the DNA to the localization of chromosome territories within the nucleus. At the most fundamental level, the chromatin structure influences gene expression by restricting access of transcription factors and the transcription machinery to the DNA (Hu, Chen, Li, & Tyler, 2014). DNA exists in two general forms within the nucleus: heterochromatin and euchromatin. At one extreme, heterochromatin is nucleosome dense, highly condensed and tends to be transcriptionally repressive due to the exclusion of transcriptional regulators from the DNA, while at the other extreme euchromatin is less nucleosome dense, less compacted, and more accessible to transcriptional regulators and conducive to transcription. However, the chromatin structure is highly dynamic and can be converted between hetero- and euchromatin states, or an endless permutation of intermediate states, in response to intracellular or extracellular signals. Such changes in chromatin state have the potential to expose or obscure gene regulatory elements to the nuclear transcriptional machinery, thus facilitating alterations in gene expression (Li, Carey, & Workman, 2007). These dynamic changes in chromatin structure involve complex interdependent and multilayered processes driven in concert by transcription factors and epigenetic modifiers.

2.2 Mechanisms for Altering Chromatin Structure and Function

Chromatin state at particular loci is defined by chemical modifications to the DNA and histones, the presence and positioning of nucleosomes, and the incorporation of histone variants into these nucleosomes in exchange for the canonical histones. Changes to the chromatin state are brought about by an array of enzymes, often loosely grouped under the term "epigenetic enzymes," which can affect chromatin function either directly through the modifications they catalyze or through the interpretation of these modifications by other proteins. Generally speaking, these enzymes can be broadly classified as either DNA and histone-modifying enzymes or ATP-dependent chromatin remodelers. These enzymes function in a coordinated manner, together with transcription factors, to affect changes in chromatin structure and function.

2.2.1 DNA Methylation

Methylation of DNA is the unequivocal epigenetic modification and the most widely studied. In mammalian cells, methylation occurs predominantly as 5-methylcytosine (5mC) in cytosine–guanine dinucleotides (CpGs) throughout the genome, although non-CpG methylation has been described in some instances, for example, in embryonic stem cells

(Ramsahoye et al., 2000). CpG dinucleotides are generally depleted in the genome, but are found in clusters of high density associated with over 50% of gene promoters, referred to as CpG islands. The majority of CpG dinucleotides dispersed throughout the genome are methylated, while CpG islands tend to be demethylated in normal somatic cells (Deaton & Bird, 2011). Methylation of CpG islands overlapping promoters has long been associated with gene repression, acting in some instances to inhibit the binding of transcription factors (Wiench et al., 2011), but also contributing to stable gene silencing through the recruitment of a family of methyl-CpG-binding proteins. The methyl-CpG-binding proteins complex with epigenetic enzymes that direct proximal histone modifications, contributing to transcriptional repression (Nan et al., 1998). There is now a large body of literature documenting the association of promoter methylation with gene silencing. For example, the differentiation-dependent methylation of individual gene promoters coincident with gene silencing has been well described during the maturation and commitment of T cells, as reviewed in Rothenberg and Zhang (2012). Further, CpG islands are predisposed to hypermethylation in cancer cells and there is a plethora of studies describing the aberrant methylation and associated silencing of individual tumor-suppressor gene promoters in cancer (Taberlay & Jones, 2011).

While CpG methylation has become synonymous with gene silencing and the association of CpG island methylation with promoter silencing is well described, more recent studies suggest that the effects of DNA methylation are more complex than originally thought (Jones, 2012). Methylation of the regions surrounding CpG islands, referred to as shelves and shores, tends to be more variable than CpG islands (Irizarry et al., 2009), and the regulatory role of DNA methylation in these regions is less well understood. DNA methylation also occurs within the transcribed regions of genes and in these regions its occurrence is positively correlated with gene expression (Lee, Choi, Lee, & Kim, 2015; Yang et al., 2014). Although the role of gene body methylation is still unclear, it has in some cases been found to influence alternative splicing mechanisms (Lee et al., 2015).

DNA methylation is catalyzed by a family of methylating enzymes, including the de novo methylases, DNMT3A and DNMT3B, as well as maintenance methylase DNMT1, which recognizes and methylates hemimethylated DNA, ensuring that DNA methylation patterns are propagated following cell division (Hendrich & Bird, 1999). For this reason, together with the inability of researchers for many years to identify enzymes capable of demethylating DNA, 5mC was long regarded as a stable modification. However, enigmatic to this is the well-described phenomena of genome-wide demethylation during fertilization and embryonic development (Messerschmidt, Knowles, & Solter, 2014). Originally it was thought that this was a result of passive demethylation, but more recent studies have demonstrated that the demethylation occurs without prior DNA replication. In addition to this are isolated reports of CpG demethylation at specific loci without cell division, such as described for the human IL-2 promoter following T cell activation (Bruniquel & Schwartz, 2003). While it is clear that demethylation can and does occur passively during DNA replication, considerable insight into the mechanisms of DNA demethylation was provided by the discovery of an additional cytosine modification, 5-hydroxymethylcytosine (5hmC) in 2009 (Kriaucionis & Heintz, 2009; Tahiliani et al., 2009). This modification is generated by oxidation of 5mC by the ten eleven translocation (TET) enzymes (Ito et al., 2011; Tahiliani et al., 2009). The TET enzymes (of which there are three: TET1, TET2, and TET3) can then further oxidize 5hmC to 5-formylcytosine and 5-carboxylcytosine. This provides an additional mechanism for demethylation of the genome, either by the passive removal of the 5hmC modification following DNA methylation, as this modification is inefficiently recognized by DNMT1, or actively through mechanisms including the action of the base excision repair machinery, as reviewed previously (Delatte, Deplus, & Fuks, 2014).

2.2.2 Histone Modifications

Histone proteins are subject to extensive posttranslational modifications, with dozens of different modifications now described (Tan et al., 2011). These modifications occur extensively not only throughout the histone tails but also to residues within the globular histone folds and include acetylation, methylation, phosphorylation, ubiquitination, and sumoylation (Kouzarides, 2007; Lawrence, Daujat, & Schneider, 2016). Histone modifications can affect all genomic functions, including gene transcription, DNA repair, and replication. However, the effect of the modification depends on both the functional group added and the location at which it occurs, and while the function of some of these modifications is now well understood, the role of others is unclear. In some instances, the histone modification alters the chemical structure of the histone core, changing the affinity of the core for the negatively charged DNA molecule and impacting histone–histone interactions (Tessarz & Kouzarides, 2014). In other instances, the functional outcome is mediated by proteins that recognize and bind to the modifications. Acetylation of N-tail lysine residues is generally associated with gene activation, with acetylation neutralizing the positively charged lysine residues, and thus proposed to impact higher-order chromatin stability (Wang & Hayes, 2008). However, lysine acetylation can also act to recruit other proteins, which influence chromatin function, such as chromatin-remodeling complexes (Chatterjee et al., 2011). The effect of histone methylation, which can occur at either lysine or arginine residues, depends on the particular residue modified and also the number of methyl groups added, due to their recruitment of different protein complexes (Gardner, Allis, & Strahl, 2011). For example, histone 3 lysine 4 trimethylation (H3K4me3) is generally found associated with active promoters (Ucar & Bayarsaihan, 2015). In contrast, monomethylation of the same lysine residue tends to be associated with both promoters as well as other regulatory elements, such as enhancer regions (Rada-Iglesias et al., 2011; Taberlay et al., 2011). Both active and inactive, or "poised enhancers," are marked by H3K4me1 modifications, while active enhancers tend to also be marked with acetylation at H3K27 (Rada-Iglesias et al., 2011). In contrast, H3K27me3 is a repressive modification, generally associated with repressed enhancers and promoters (see Fig. 1). This is the basis of the "histone code" as proposed by Allis and colleagues, in which modifications can be "written" or "erased" by epigenetic enzymes, and these modifications can then be interpreted by proteins which recognize these modifications and bring about functional changes at the chromatin (Wang et al., 2004).

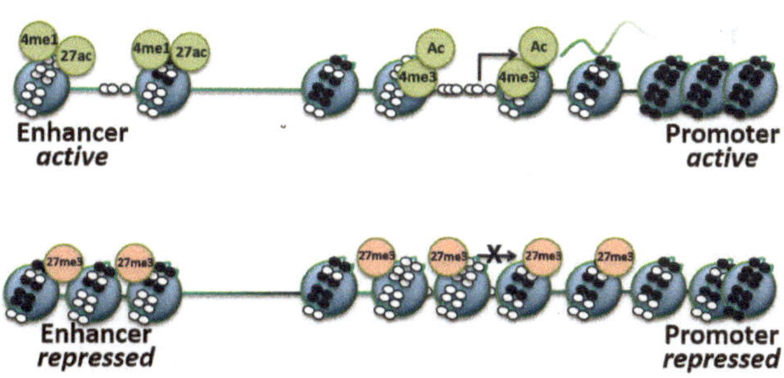

Fig. 1 Epigenetic signatures of regulatory elements. Genome-wide studies are deciphering some of the epigenetic signatures that define particular transcriptional states. Nucleosomes are depicted as *blue circles*, with methylated (*black circles*) and unmethylated (*white circles*) CpGs. H3K27me3 is depicted associated with transcriptionally repressed enhancers and promoters, while H3K27me3 defines active promoters. Further, H3K4me1 is associated with both active and poised enhancers, while active enhancers are also marked by H3K27 acetylation.

Extensive genome-wide analysis has described histone modification marks that are associated with particular regulatory DNA features and particular functional outcomes, and T cell models proved especially informative, particularly in the early studies during the development of these methodologies (Barski et al., 2007; Wang et al., 2008). However, while the generation of genome-wide epigenetic maps has proved valuable in identification and annotation of regulatory regions, and understanding how the epigenetic signature of a cell changes during development and differentiation, in many cases it is still unclear how the modifications relate to the functional outcomes, that is, whether they are drivers or a consequence of the functional outcome, and it is likely that this is specific to the particular modification. The best described and understood examples of modifications that direct a functional change at chromatin are probably those that provide mechanisms for gene silencing, for example, the H3K9me3 modification, which is associated with heterochromatic regions and gene silencing (Nakayama, Rice, Strahl, Allis, & Grewal, 2001). This modification is generated by the histone methyltransferases G9A and GLP (Krishnan, Horowitz, & Trievel, 2011), and the modification then recruits heterochromatin protein 1 (HP1), which facilitates the formation of heterochromatin (Ayyanathan et al., 2003). Similarly, the polycomb family of proteins have a well-described role in chromatin remodeling and gene silencing that is distinct from that of DNA methylation. These proteins form complexes which can be categorized into two major families: Polycomb repressive complexes (PRC) 1 and 2. While both groups work to silence transcription, the families are functionally distinct. PRC1 complexes have E3 ubiquitin ligase activity and mainly target monoubiquitinated histone H2A at lysine 119 (Di Croce & Helin, 2013). PRC2 complexes di- or trimethylate histone H3 on lysine 27 (Di Croce & Helin, 2013), and this is associated with chromatin condensation and gene silencing. The mechanism by which Polycomb complexes are recruited to specific DNA loci is still unclear, although like most chromatin-modifying enzymes this is thought to involve recruitment by DNA binding proteins such as transcription factors and epigenetic modifications (Aranda, Mas, & Di Croce, 2015). Once recruited to DNA, the PRC influences the chromatin platform by modifying histone proteins and noncatalytically inducing chromatin condensation, through a mechanism that is not fully understood (Francis, Kingston, & Woodcock, 2004). As well documented for DNA methylation, gene silencing by polycomb complexes plays an important role in specifying gene expression programs during cell differentiation and development. For example, PRC complexes have been demonstrated to play a role in T cell differentiation, including the differentiation and function of particular T cell subsets (Raaphorst et al., 2001; Yang et al., 2015).

2.2.3 Histone Variants

In addition to the four canonical histone proteins involved in the formation of the nucleosome core and the linker histone H1 and H5, additional histone variants exist. Variants differ from canonical histones by substitutions or changes in their amino acid sequence, and their incorporation into nucleosomes therefore has the capacity to alter the chemical environment of the nucleosome. Alterations lead to a variety of effects depending on the specific variant involved, allowing the variants to perform specialized functions (Li & Fang, 2015), and localized incorporation of some of these variants is specifically associated with transcriptional regulation or transcriptional competency. The histone H2A, variant H2A.Z, has been associated with a number of functions, including DNA repair. The histone variant is deposited at sites of double-strand DNA breaks, ensuring that the chromatin environment remains open and accessible to enzymes required for the repair process (Xu et al., 2012). However, a role in transcriptional regulation has also been described, with H2A.Z associated with both transcriptional activation and repression. In eukaryotes, a nucleosome-free region (NFR) is commonly found at the transcription start site (TSS) of active genes and this is associated with a strongly positioned nucleosome immediately downstream of the TSS, referred to as the +1 nucleosome. This nucleosome is generally marked by the incorporation of H2A.Z, as is the nucleosome immediately upstream of the TSS, and in higher eukaryotes, this is associated with genes that are either poised or active. There is some evidence that H2A.Z incorporation into these nucleosomes has a direct role in recruiting RNA polymerase II (Adam, Robert, Larochelle, & Gaudreau, 2001), as H2A.Z has been found to be recruited to promoter regions in human cells prior to RNA polymerase II binding (Hardy et al., 2009). Another histone variant, histone H3.3, is also often located at transcriptionally active regions and has been associated with increased transcription during T cell activation. Changes in chromatin accessibility at the promoters of several highly inducible genes during T cell activation were found to involve loss of H2A.Z and deposition of H3.3 (Sutcliffe et al., 2009). This change in nucleosome composition leads to alteration of histone modifications and restructuring of the chromatin. Further, genome-wide analysis has demonstrated that the nucleosomes containing both H2A.Z and H3.3 are enriched at NFRs and active promoters and enhancers, and the instability of these nucleosomes is thought important in facilitating access of transcription factors to these regions (Jin et al., 2009).

2.2.4 Chromatin-Remodeling Machinery

Chromatin structure can also be modified by a group of complexes collectively referred to as chromatin-remodeling complexes. These complexes are distinct from the DNA and histone-modifying proteins described earlier and are defined by a core subunit within the complex that has the ability to harness energy generated by the hydrolysis of ATP to alter the chromatin structure. Additional to the ATPase subunit, these complexes contain up to 20 noncatalytic subunits, which regulate activity and targeting of the complex. While the chromatin-remodeling complexes may contain subunits that interact with DNA, they lack specificity in their DNA binding ability and therefore must be recruited to specific genomic regions by other elements such as histone modifications and transcription factors. ATP-dependent remodeling enzymes assist in chromatin assembly, nucleosome folding, unfolding and displacement, histone replacement, and the sliding of nucleosome cores along the DNA molecule (Wurster & Pazin, 2011).

The chromatin-remodeling complexes are generally grouped according to their ATPase subunit into the SWI/SNF, ISWI, CHD, and INO80 families. The first identified and most well studied of these is the SWI/SNF complexes, components of which were first discovered in studies of yeast signaling molecules (Neigeborn & Carlson, 1984; Stern, Jensen, & Herskowitz, 1984). The mammalian homologue of this complex, BAF, comprises 15 subunits encoded by 29 genes, all with unique functional roles (Kadoch & Crabtree, 2015). The function of the BAF complex depends on the subunits involved, and its composition varies in different cell types. The function of these complexes was initially thought to be relatively straightforward, acting following recruitment to DNA by sequence-specific transcription factors where it then targets and remodels nucleosomes to facilitate binding of additional transcriptional regulators and subsequent gene expression (Cosma, Tanaka, & Nasmyth, 1999). However, more recent evidence indicates that BAF complexes are able to bind DNA and activate genes in the absence of transcription factors (Kadoch & Crabtree, 2013), suggesting that the remodeling mechanism is more complex than originally suspected.

The function of chromatin-remodeling complexes has been illustrated clearly in the immune system including in T lymphocytes. These cells spend long periods in a dormant state, with very low levels of transcriptional activity and DNA stored in a highly condensed heterochromatin state. However, in order for these cells to effectively respond to antigens, they must be able to quickly reinstate expression of specific genes. Chromatin-remodeling complexes such as BAF have a role in facilitating this, as they are able to rapidly associate with lymphocytic nuclei upon antigenic stimulation. Stimulation by an antigen causes a membrane signaling event in lymphocytes, which alters cellular phosphatidyl inositol 4,5-bisphosphate (PIP2) levels. PIP2, in turn, influences the strength of BAF association in the nucleus, promoting chromatin remodeling and increased gene expression (Zhao et al., 1998). The ATPase component of the BAF complex, BRG1, has also been demonstrated to play an important role in both T cell differentiation and activation, with genome-wide analysis demonstrating binding of BRG1 to gene promoters and enhancers in association with gene activation (De et al., 2011). More recent analysis found that a set of differentiation genes are bivalently marked with H3K4me3 and H3K27me3 in hemopoietic stem and progenitor cells, but are also primed for subsequent expression by association of BRG1 and RNA polymerase II (Abraham, Cui, Tang, & Zhao, 2013).

3. EPIGENETIC MECHANISMS AS A BASIS FOR CELLULAR MEMORY

3.1 A Word About Epigenetics

The term *epigenetics* was originally coined in the 1940s by Conrad Waddington to describe the relationship between genotype and phenotype during development (Waddington, 2012) and has since come to describe changes to DNA that result in a heritable phenotype that does not involve changes in DNA sequence (Berger, Kouzarides, Shiekhattar, & Shilatifard, 2009). The term *epigenetics* is now widely and broadly used within the scientific literature, and while often used to encompass all the mechanisms by which chromatin structure and function can be altered, the use of the term in this way is contentious given the ambiguity around the heritability of some chromatin features. However, given the interrelated nature of these various mechanisms, they all contribute to the generation of the epigenetic landscape of the cell, even though some of the features, such as particular histone modifications, may not be truly epigenetic in nature themselves.

3.2 Establishing the Epigenetic Landscape

The interplay between transcription factors and the chromatin environment is complex and bidirectional. The effect of chromatin state, in particular nucleosome positioning, on the binding of transcription factors to regulatory elements is well established (Lelli, Slattery, & Mann, 2012). It is also well documented that transcription factors are themselves able to recruit epigenetic modifiers and chromatin-remodeling complexes to the DNA in order to direct transient changes to the chromatin environment and epigenetic signatures (Kadoch & Crabtree, 2015). Further, the role of transcription factors in establishing the epigenetic profile of a cell during developmental processes and cell differentiation is now well described and formative studies characterizing these processes during immune cell differentiation have been particularly instructive, as reviewed extensively elsewhere (Obier & Bonifer, 2016; Rothenberg, 2014). Certain transcription factors, called "pioneer" factors, are unique in that they possess the ability to bind to their DNA recognition sequences even within relatively inaccessible chromatin. The binding of pioneer transcription factors was first described in mouse liver cells (Gualdi et al., 1996) where members of the GATA family were found to bind to regulatory sites in the serum albumin gene prior to differentiation (Bossard & Zaret, 1998). Indeed, pioneer factors are distinguished from other transcription factors by several criteria. First, the factor must bind to the regulatory region before the gene is transcriptionally active. Additionally, it must bind prior to other transcription factors and bind to its target sites, while the chromatin is still in its condensed heterochromatin state. Finally, the transcription factor must have the capacity to remodel the chromatin structure so as to expose regulatory elements for transcriptional activation (Zaret & Carroll, 2011). Pioneer factors therefore play an important role in initiating chromatin-remodeling events that are required during development, cellular reprogramming, and activation of inducible genes (Zaret, Lerner, & Iwafuchi-Doi, 2016).

The best studied of the pioneer transcription factors are the FOX1 family, which have been shown to bind condensed chromatin through a winged-helix domain, consisting of a recognition helix with polypeptide "wings" on either side which interact with DNA. This domain resembles the binding region of linker histones and is therefore capable of targeting DNA in nucleosomes (Cirillo et al., 1998). A hierarchical model of transcription factor binding has been proposed in which pioneer factors are able to bind to nucleosomal DNA and direct remodeling of chromatin, allowing so-called settler transcription factors to bind. A third class of transcription factors termed "migrant" factors are also able to bind once the chromatin has been remodeled, but these factors also require cooperative interactions with other transcription factors or cofactors to do so (Sherwood et al., 2014). In keeping with this model, FoxA1 has been shown to bind to its cognate recognition sites in nucleosome-assembled enhancers, facilitating opening of the chromatin (Belikov, Astrand, & Wrange, 2009). This remodeling is then proposed to enable access of other transcription factors that require open chromatin to bind. These remodeling events are also associated with modifications to the DNA and histones with genome-wide analysis of FOXA1 binding demonstrating that this coincides with DNA demethylation and H3K4 methylation (Serandour et al., 2011). These modifications were proposed to stabilize FOXA1 binding and also facilitate subsequent transcription factor binding. Similarly, genome-wide binding of the PU.1 transcription factor, which functions as a pioneer factor in B cell and macrophage lineage specification, is associated with chromatin remodeling and the deposition of H3K4me1 at enhancers (Heinz et al., 2010).

We are therefore gaining an understanding of the transcription factor hierarchy and cooperative interactions required to generate cell-specific chromatin landscapes and epigenetic signatures, particularly those that are established during differentiation and development. However, it is clear that transcription factors do not occupy all their potential DNA binding sites throughout the genome, and we still have much need to learn about this selectivity for DNA binding and how the existing chromatin state and also the surrounding DNA sequence influences this.

3.3 Perpetuating Epigenetic Signatures to Maintain Cellular Identity

The identity of a cell is determined by its unique gene expression programs, which are dictated by the interactions between the specific set of transcription factors expressed within the cell and the distinctive chromatin landscape into which the genome is assembled. Replication of the distinct set of transcription factors and also the specific chromatin landscape must occur in daughter cells, in order for these cells to inherit the transcriptional programs. This can be achieved through either the maintenance of the chromatin state through mitosis or the reestablishment of the chromatin state in the daughter cells following cell division. In the first scenario, the epigenetic environment would provide a form of cellular memory to ensure replication of transcriptional programs after cell division, while in the second, transcription regulators are also required to reestablish the chromatin state.

Maintenance of the epigenetic signature of a cell through cell division is complicated by the lack of epigenetic marks, such as 5mC, on the newly synthesized DNA strand, and the removal of modified histone proteins during the replication process. Thus tightly controlled cellular mechanisms are required to reestablish these marks in the daughter cells and preserve the gene expression programs. Maintenance of DNA methylation from one cellular generation to another is relatively straightforward, as the template DNA strand retains its methylation patterns through the replication process. Hemimethylated CpG sites on the newly synthesized DNA recruit the maintenance DNA methyltransferase DNMT1, which converts new cytosine residues into 5mC (Hermann, Goyal, & Jeltsch, 2004). Hence, DNA methylation patterns can be faithfully replicated following cell division and are a mechanism by which an epigenetic signature can be remembered and gene expression patterns perpetuated in daughter cells. How faithfully histone modification patterns at specific genomic loci are replicated following cell division is less clear. Most modified histones are removed from the DNA during replication and a number of models have been proposed to explain how the histone modification state is "remembered." Initially, it was thought that the tetrameric histone component, formed from two H3–H4 subunit pairs, was split into dimers with each one binding to a daughter DNA molecule (Weintraub, Worcel, & Alberts, 1976). However, evidence has emerged to suggest that the subunits remain as a tetramer during replication. Another, more likely, mechanism is that components of the parent histone tetramers are able to remain associated with daughter DNA strands and reestablish modifications in the newly added histone components (Lowe, Hostager, & Kikyo, 2016). In this scenario, the parental H3–H4 tetramer and both H2A–H2B dimers are randomly allocated to each new strand, in which case, it is important that the modified parental histone components are accurately directed to the appropriate loci to act as templates for the reestablishment of histone marks. This is achieved by a number of histone chaperons working to direct the parental histone components to new DNA and to recruit newly synthesized histones for incorporation into the nucleosome. Once reassembled, the parental histones recruit the appropriate epigenetic writers that then "copy" the parental histone modifications on the new histone components. There is evidence to support such a scenario, with examples reported in which parental histone methylation modifications associated with active genes were preserved through mitosis with the transcription-dependent modification of the newly incorporated histones, resulting in replication of the histone modification state (Kouskouti & Talianidis, 2005).

While mechanisms exist to ensure that certain modifications to DNA and histones can be replicated during cell division, there is also evidence that transcriptional proteins can act as so-called bookmarking factors which facilitate the reestablishment of the cellular gene expression program after cell division (Zaidi et al., 2010). Bookmarking factors remain associated with particular genes during mitosis, thus allowing rapid reexpression of the genes following cell division. During DNA replication, histone H3 is phosphorylated which facilitates chromatin condensation (Hans & Dimitrov, 2001; Nowak & Corces, 2004; Van Hooser, Goodrich, Allis, Brinkley, & Mancini, 1998), preventing binding of transcription factors and transcriptional repression of most genes. However, some transcriptional proteins can remain associated with the condensed chromatin. For example, elegant studies have demonstrated that the HNF1β transcription factor remains associated with chromatin at target genes during mitosis, despite these genes acquiring the modifications associated with heterochromatin (Verdeguer et al., 2010). A similar study examining bookmarking in hemopoietic cells demonstrated retention of the GATA1 transcription factor at some key hemopoietic genes during mitosis, ensuring their subsequent rapid reactivation (Kadauke et al., 2012). Chromatin-modifying proteins have also been reported to act as bookmarking factors. The trithorax protein MLL, which methylates H3K4, has also been shown to bookmark active genes during mitosis. Interestingly, MLL was shown to relocate from the sites where it is deposited during interphase to a different set of genes, becoming associated during mitosis with genes that are normally highly expressed (Blobel et al., 2009). Finally, there is also evidence that histone variants can bookmark active genes through mitosis. For example, the histone variant H3.3 has been found to associate with actively transcribed genes, as outlined earlier, and there is evidence that this association can persist through cell division (Chow et al., 2005).

The repositioning of nucleosomes themselves has also been shown to provide a mechanism to enable genes to be repressed during mitosis and then rapidly reactivated afterward (Kelly et al., 2010). Genome-wide analysis demonstrated the shifting of the +1 nucleosome to cover the TSS of active genes during mitosis, reducing the NFR and physically repressing gene expression during mitosis. Interestingly, the +1 nucleosome contains H2A.Z which may destabilize the nucleosome facilitating this movement during cell division. The nucleosome is also marked by H3K4me3, which unlike some other histone modifications is maintained during mitosis, and may enable the rapid reexpression of the gene following exit from mitosis (Kelly et al., 2010). In contrast, silenced genes that are marked by the presence of DNA methylation do not have H2A.Z-containing nucleosomes in the +1 position and remain silenced following mitosis (Kelly et al., 2010).

4. EPIGENETIC MECHANISMS AS A BASIS FOR TRANSCRIPTIONAL MEMORY

4.1 Transcriptional Memory in the Adaptive Immune Response

Not only has the extensive analysis of immune cell differentiation and development provided elegant examples of the contribution of chromatin to cellular identity, as described earlier, but it also provides dramatic evidence of the role of the chromatin landscape in providing memory of past transcriptional events. Immunological memory is the ability of the immune system to produce a more rapid and augmented response to a pathogen that has been encountered previously. T lymphocytes play a central role in this adaptive immune response. Upon encountering antigen naïve T cells undergo clonal expansion generating effector T cells. While most of these cells are removed by apoptosis, a small number of so-called memory T cells remain. These cells are able to respond with a more rapid and effective transcriptional response on reencountering the same antigen, ensuring that gene expression programs required to orchestrate an immune response are quickly activated. While the molecular basis of these responses remains to be fully elucidated, it is clear that changes to chromatin structure and epigenetic signatures are critical for activation of T cells during the orchestration of an immune response, as reviewed previously (Lim et al., 2013), and there is now evidence that the maintenance of these remodeled states underlies immunological memory.

T cell activation provided some of the earliest evidence for the requirement of chromatin remodeling for gene activation in mammalian cells. Early studies demonstrated the appearance of inducible DNase I hypersensitivity (DH) sites associated with cytokine gene activation (Cockerill, Shannon, Bert, Ryan, & Vadas, 1993) and subsequent studies have demonstrated that these inducible DH sites represent the highly targeted remodeling of one or two nucleosomes at promoters and enhancers of cytokine genes such as *IL-2*, *IL-3*, and *GM-CSF* (Chen, Wang, Woltring, Gerondakis, &

Shannon, 2005; Holloway, Rao, Chen, & Shannon, 2003; Johnson, Bert, Ryan, Condina, & Cockerill, 2004; Rao, Procko, & Shannon, 2001). Studies in cell lines have revealed that these remodeling events are driven by inducible transcription factors (Brettingham-Moore, Rao, Juelich, Shannon, & Holloway, 2005; Rao, Gerondakis, Woltring, & Shannon, 2003). These remodeling events have the potential to provide memory of the initial activation event as once this remodeling had occurred, cytokine genes can respond more quickly and effectively to restimulation with the same signal (Brettingham-Moore et al., 2005; Rao et al., 2003), but not once the evicted histones have been redeposited (Poke et al., 2012). This type of scenario in which maintenance of the remodeled state allows a more rapid and augmented response to subsequent stimulus underlies the cellular memory that is a feature of memory T cells and the adaptive immune response.

The genes that respond with an augmented and more rapid transcriptional response also appear to be epigenetically primed in memory T cells. Increased histone H3 acetylation is associated with these genes when they are activated, and there is evidence that they retain this modification status in memory T cells (Fann et al., 2006; Northrop, Thomas, Wells, & Shen, 2006). Studies of human peripheral T cells have demonstrated that in naïve T cells the *IL-3/GM-CSF* cytokine locus is maintained in an inactive chromatin state. Upon stimulation, chromatin-remodeling events occur, which involve the appearance of DH sites upstream of the IL-3 gene, and this remodeled state is maintained in memory T cells (Mirabella, Baxter, Boissinot, James, & Cockerill, 2010). A further study extended this to examine changes in chromatin structure on a genome-wide scale in memory T cells (Bevington et al., 2016) and demonstrated the appearance of a large number of DH sites following stimulation of naïve T cells. These DH sites were associated with active histone marks (H3K4me2 and H3K27ac) and often encompassed inducible enhancers. These DH sites persisted after the activating signal had dissipated and were maintained in memory T cells. While inducible transcription factors (NFAT and AP1) were required for the appearance of these sites and the reactivation of these genes in memory T cells, the evidence suggests that the sites were maintained in memory T cells by occupancy with the constitutive transcription factors ETS-1 and RUNX1 (Bevington et al., 2016). This binding maintains the gene in a primed state allowing it to respond more rapidly to subsequent stimulation, which among other things activates the AP-1 and NFAT transcription factors. The maintenance of the remodeled state therefore provides memory of the original transcriptional event.

A number of studies have now demonstrated a clear association between the loss of the repressive H3K27me3 modification and the retention or acquisition of the activating H3K4me3 modification during transcriptional activation of genes in effector T cells (Araki et al., 2009). Further, there is evidence that the H3K4me3 modification persists in memory T cells (Araki et al., 2009). However, there is also evidence that H3K4me2 is a specific mark of those genes that are held in a primed state, ready for rapid activation in response to stimulation, as indicated by its association with the remodeled chromatin in memory T cells earlier (Bevington et al., 2016), as depicted in Fig. 2. Further, a recent study analyzing epigenetic changes

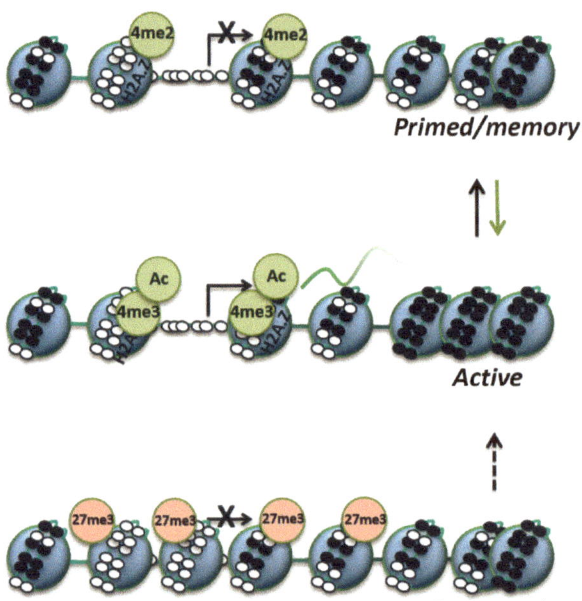

Fig. 2 A model for transcriptional memory. A model is emerging from a range of cellular contexts of the epigenetic changes that occur as genes become transcriptionally active and those that are associated with the retention of a memory of the transcriptional event. Transcriptionally repressed genes commonly display nucleosome-assembled promoters associated with the repressive H3K27me3 modification. Nucleosomes are depicted as *blue circles*, with methylated (*black circles*) and unmethylated (*white circles*) CpGs. In response to an activating signal, chromatin-remodeling events result in the appearance of a nucleosome-depleted region at the gene promoter, the depletion of the H3K27me3 modification and acquisition of H3K4me3. Studies in a number of models suggest that a primed or memory state is associated with retention of the remodeled state, retention of H2A.Z, depletion of H3K4me3 but retention of H3K4me2.

associated with cytotoxic T lymphocyte differentiation found that upon differentiation of naïve to effector cells, acquisition of H3K4me3 and loss of H3K27me3 were associated with the establishment of a permissive transcriptional state (Fig. 2). However, H3K4me2 was associated with a small subset of genes that were primed for rapid activation upon stimulation (Russ et al., 2014).

4.2 Transcriptional Memory in the Innate Immune System and Beyond

While the adaptive immune response provides an elegant example of how chromatin changes can generate cellular memory, there is accumulating evidence that other cells also have the ability to retain memory of a previous stimulatory event. For example, the innate immune response also involves a similar type of cellular memory which has been termed "trained immunity" (Netea et al., 2016). Similar to the adaptive immune response, epigenetic changes have been shown to underlie this ability of innate immune cells such as monocytes, macrophages, and natural killer cells, to orchestrate a more rapid and augmented transcriptional response to repeated stimulation. This type of response differs from the adaptive immune response in that the cells become "primed" by an initial response which allows them to respond in an augmented fashion to a second signal. The initial "training" signal induces chromatin remodeling and changes in epigenetic signatures, which persist for a number of days (Saeed et al., 2014). Acquisition of H3K4me1 and H3K27ac was observed coincident with the remodeled chromatin, with H3K4me1 persisting after transcription had ceased.

The idea of the genome being primed following an initial stimulus to respond more rapidly to a subsequent signal also extends to other cell types and in fact to lower organisms. In yeast the *INO1* and *GAL1* genes were found to localize to the nuclear periphery upon activation, but remained localized to this region upon repression and remained there through several cell divisions (Brickner et al., 2007). H2A.Z incorporation into the nucleosomes and demethylation of H3K4 was required to maintain the repressed genes at the nuclear periphery and to prime them for rapid reactivation. More recently, the importance of the H3K4me2 modification in generating the transcriptional memory state in yeast has also been demonstrated. While the active INO1 gene is hyperacetylated and marked by both H3K4me2 and H3K4me3, upon repression and retention in the nuclear periphery *INO1* is depleted of histone acetylation and H3K4me3 but retains the H3K4me2 modification and poised RNA polymerase II (D'Urso et al., 2016), as depicted in Fig. 2. In keeping with the idea that the acquisition of the H3K4me2 modification appears to be a more general mark of transcriptional memory and/or the priming of genes for rapid activation, this modification has been observed at primed genes in other cell models also. A study in HeLa cells demonstrated the association of recently activated IFNγ responsive genes with the nuclear pore where they retain the memory of the transcriptional event for several days. With remarkable similarity to the yeast studies, these genes were also marked with H3K4me2 and a poised RNA polymerase II (Light & Brickner, 2013).

5. CONCLUSIONS

In the last decade, we have learnt much about the mechanisms by which epigenetic modifications and chromatin structure, in cooperation with transcription factors, establish and maintain gene expression programs within a cell. Immune cells provide elegant and tractable models in which to examine transcriptional programs and have been enormously valuable in deciphering the "language" of the DNA and histone modifications. We therefore now have a good understanding of not only the signatures that mark repressed or active genes and how these signatures change during development and differentiation to define cell lineages, but are also starting to understand the nuances that allow genes to "remember" past events and remain primed to respond rapidly to future activating signals. Future challenges will be to understand how disruption of epigenetic memory may contribute to disease states. While the obvious avenue of inquiry is in immune diseases such as asthma and autoimmune disease, the emergence of evidence of epigenetic memory in a range of cellular contexts suggests that disruption of these processes may contribute to other disease states also.

ACKNOWLEDGMENTS

The authors thank Phillippa Taberlay for insightful comments on the manuscript and figures.

REFERENCES

Abraham, B. J., Cui, K., Tang, Q., & Zhao, K. (2013). Dynamic regulation of epigenomic landscapes during hematopoiesis. *BMC Genomics*, *14*, 193. http://dx.doi.org/10.1186/1471-2164-14-193.

Adam, M., Robert, F., Larochelle, M., & Gaudreau, L. (2001). H2A. Z is required for global chromatin integrity and for recruitment of RNA polymerase II under specific conditions. *Molecular and Cellular Biology*, *21*(18), 6270–6279.

Araki, Y., Wang, Z., Zang, C., Wood, W. H., 3rd, Schones, D., Cui, K., ... Weng, N. P. (2009). Genome-wide analysis of histone methylation reveals chromatin state-based regulation of gene transcription and function of memory CD8+ T cells. *Immunity*, *30*(6), 912–925. http://dx.doi.org/10.1016/j.immuni.2009.05.006.

Aranda, S., Mas, G., & Di Croce, L. (2015). Regulation of gene transcription by Polycomb proteins. *Science Advances*, *1*(11), e1500737.

Ayyanathan, K., Lechner, M. S., Bell, P., Maul, G. G., Schultz, D. C., Yamada, Y., ... Rauscher, F. J. (2003). Regulated recruitment of HP1 to a euchromatic gene induces mitotically heritable, epigenetic gene silencing: A mammalian cell culture model of gene variegation. *Genes & Development*, *17*(15), 1855–1869.

Barski, A., Cuddapah, S., Cui, K., Roh, T.-Y., Schones, D. E., Wang, Z., ... Zhao, K. (2007). High-resolution profiling of histone methylations in the human genome. *Cell*, *129*(4), 823–837.

Belikov, S., Astrand, C., & Wrange, O. (2009). FoxA1 binding directs chromatin structure and the functional response of a glucocorticoid receptor-regulated promoter. *Molecular and Cellular Biology*, *29*(20), 5413–5425. http://dx.doi.org/10.1128/MCB.00368-09.

Berger, S. L., Kouzarides, T., Shiekhattar, R., & Shilatifard, A. (2009). An operational definition of epigenetics. *Genes & Development*, *23*(7), 781–783. http://dx.doi.org/10.1101/gad.1787609.

Bevington, S. L., Cauchy, P., Piper, J., Bertrand, E., Lalli, N., Jarvis, R. C., ... Cockerill, P. N. (2016). Inducible chromatin priming is associated with the establishment of immunological memory in T cells. *The EMBO Journal*, *35*, 515–535. http://dx.doi.org/10.15252/embj.201592534.

Blobel, G. A., Kadauke, S., Wang, E., Lau, A. W., Zuber, J., Chou, M. M., & Vakoc, C. R. (2009). A reconfigured pattern of MLL occupancy within mitotic chromatin promotes rapid transcriptional reactivation following mitotic exit. *Molecular Cell*, *36*(6), 970–983.

Bossard, P., & Zaret, K. S. (1998). GATA transcription factors as potentiators of gut endoderm differentiation. *Development*, *125*(24), 4909–4917.

Brettingham-Moore, K. H., Rao, S., Juelich, T., Shannon, M. F., & Holloway, A. F. (2005). GM-CSF promoter chromatin remodelling and gene transcription display distinct signal and transcription factor requirements. *Nucleic Acids Research*, *33*(1), 225–234.

Brickner, D. G., Cajigas, I., Fondufe-Mittendorf, Y., Ahmed, S., Lee, P. C., Widom, J., & Brickner, J. H. (2007). H2A.Z-mediated localization of genes at the nuclear periphery confers epigenetic memory of previous transcriptional state. *PLoS Biology*, *5*(4), e81. http://dx.doi.org/10.1371/journal.pbio.0050081.

Bruniquel, D., & Schwartz, R. H. (2003). Selective, stable demethylation of the interleukin-2 gene enhances transcription by an active process. *Nature Immunology*, *4*(3), 235–240.

Carruthers, L. M., Bednar, J., Woodcock, C. L., & Hansen, J. C. (1998). Linker histones stabilize the intrinsic salt-dependent folding of nucleosomal arrays: Mechanistic ramifications for higher-order chromatin folding. *Biochemistry*, *37*(42), 14776–14787.

Chatterjee, N., Sinha, D., Lemma-Dechassa, M., Tan, S., Shogren-Knaak, M. A., & Bartholomew, B. (2011). Histone H3 tail acetylation modulates ATP-dependent remodeling through multiple mechanisms. *Nucleic Acids Research*, *39*(19), 8378–8391.

Chen, X., Wang, J., Woltring, D., Gerondakis, S., & Shannon, M. F. (2005). Histone dynamics on the interleukin-2 gene in response to T-cell activation. *Molecular and Cellular Biology*, *25*(8), 3209–3219.

Chow, C. M., Georgiou, A., Szutorisz, H., Maia e Silva, A., Pombo, A., Barahona, I., ... Dillon, N. (2005). Variant histone H3.3 marks promoters of transcriptionally active genes during mammalian cell division. *EMBO Reports*, *6*(4), 354–360. http://dx.doi.org/10.1038/sj.embor.7400366.

Cirillo, L. A., McPherson, C. E., Bossard, P., Stevens, K., Cherian, S., Shim, E. Y., … Zaret, K. S. (1998). Binding of the winged-helix transcription factor HNF3 to a linker histone site on the nucleosome. *The EMBO Journal*, *17*(1), 244–254.

Cockerill, P. N., Shannon, M. F., Bert, A. G., Ryan, G. R., & Vadas, M. A. (1993). The granulocyte-macrophage colony-stimulating factor/interleukin 3 locus is regulated by an inducible cyclosporin A-sensitive enhancer. *Proceedings of the National Academy of Sciences*, *90*(6), 2466–2470.

Cosma, M. P., Tanaka, T., & Nasmyth, K. (1999). Ordered recruitment of transcription and chromatin remodeling factors to a cell cycle–and developmentally regulated promoter. *Cell*, *97*(3), 299–311.

De, S., Wurster, A. L., Precht, P., Wood, W. H., 3rd, Becker, K. G., & Pazin, M. J. (2011). Dynamic BRG1 recruitment during T helper differentiation and activation reveals distal regulatory elements. *Molecular and Cellular Biology*, *31*(7), 1512–1527. http://dx.doi.org/10.1128/MCB.00920-10.

Deaton, A. M., & Bird, A. (2011). CpG islands and the regulation of transcription. *Genes & Development*, *25*(10), 1010–1022.

Delatte, B., Deplus, R., & Fuks, F. (2014). Playing TETris with DNA modifications. *The EMBO Journal*, *33*(11), 1198–1211.

Di Croce, L., & Helin, K. (2013). Transcriptional regulation by Polycomb group proteins. *Nature Structural & Molecular Biology*, *20*(10), 1147–1155.

Dixon, J. R., Selvaraj, S., Yue, F., Kim, A., Li, Y., Shen, Y., … Ren, B. (2012). Topological domains in mammalian genomes identified by analysis of chromatin interactions. *Nature*, *485*(7398), 376–380.

Dorigo, B., Schalch, T., Kulangara, A., Duda, S., Schroeder, R. R., & Richmond, T. J. (2004). Nucleosome arrays reveal the two-start organization of the chromatin fiber. *Science*, *306*(5701), 1571–1573.

Dunn, J., McCuaig, R., Tu, W. J., Hardy, K., & Rao, S. (2015). Multi-layered epigenetic mechanisms contribute to transcriptional memory in T lymphocytes. *BMC Immunology*, *16*(1), 1.

D'Urso, A., Takahashi, Y. H., Xiong, B., Marone, J., Coukos, R., Randise-Hinchliff, C., … Brickner, J. H. (2016). Set1/COMPASS and mediator are repurposed to promote epigenetic transcriptional memory. *eLife*, *5*, e16691. http://dx.doi.org/10.7554/eLife.16691.

Fann, M., Godlove, J. M., Catalfamo, M., Wood, W. H., 3rd, Chrest, F. J., Chun, N., … Weng, N. P. (2006). Histone acetylation is associated with differential gene expression in the rapid and robust memory CD8(+) T-cell response. *Blood*, *108*(10), 3363–3370. http://dx.doi.org/10.1182/blood-2006-02-005520.

Francis, N. J., Kingston, R. E., & Woodcock, C. L. (2004). Chromatin compaction by a polycomb group protein complex. *Science*, *306*(5701), 1574–1577.

Fussner, E., Ching, R. W., & Bazett-Jones, D. P. (2011). Living without 30 nm chromatin fibers. *Trends in Biochemical Sciences*, *36*(1), 1–6.

Gardner, K. E., Allis, C. D., & Strahl, B. D. (2011). Operating on chromatin, a colorful language where context matters. *Journal of Molecular Biology*, *409*(1), 36–46.

Gerlich, D., Beaudouin, J., Kalbfuss, B., Daigle, N., Eils, R., & Ellenberg, J. (2003). Global chromosome positions are transmitted through mitosis in mammalian cells. *Cell*, *112*(6), 751–764.

Gonzalez-Sandoval, A., & Gasser, S. M. (2016). On TADs and LADs: Spatial control over gene expression. *Trends in Genetics*, *32*, 485–495.

Gualdi, R., Bossard, P., Zheng, M., Hamada, Y., Coleman, J. R., & Zaret, K. S. (1996). Hepatic specification of the gut endoderm in vitro: Cell signaling and transcriptional control. *Genes & Development*, *10*(13), 1670–1682.

Hans, F., & Dimitrov, S. (2001). Histone H3 phosphorylation and cell division. *Oncogene, 20*(24), 3021–3027.

Hardy, S., Jacques, P.-É., Gévry, N., Forest, A., Fortin, M.-È., Laflamme, L., ... Robert, F. (2009). The euchromatic and heterochromatic landscapes are shaped by antagonizing effects of transcription on H2A. Z deposition. *PLoS Genetics, 5*(10), e1000687.

Heinz, S., Benner, C., Spann, N., Bertolino, E., Lin, Y. C., Laslo, P., ... Glass, C. K. (2010). Simple combinations of lineage-determining transcription factors prime cis-regulatory elements required for macrophage and B cell identities. *Molecular Cell, 38*(4), 576–589. http://dx.doi.org/10.1016/j.molcel.2010.05.004.

Hendrich, B., & Bird, A. (1999). Mammalian methyltransferases and methyl-CpG-binding domains: Proteins involved in DNA methylation. *Current Topics in Microbiology and Immunology, 249*, 55–74.

Hermann, A., Goyal, R., & Jeltsch, A. (2004). The Dnmt1 DNA-(cytosine-C5)-methyltransferase methylates DNA processively with high preference for hemimethylated target sites. *Journal of Biological Chemistry, 279*(46), 48350–48359.

Holloway, A. F., Rao, S., Chen, X., & Shannon, M. F. (2003). Changes in chromatin accessibility across the GM-CSF promoter upon T cell activation are dependent on nuclear factor kB proteins. *The Journal of Experimental Medicine, 197*(4), 413–423.

Hu, Z., Chen, K., Li, W., & Tyler, J. K. (2014). A matter of access: Nucleosome disassembly from gene promoters is the central goal of transcriptional activators. *Transcription, 5*(2), e29355.

Irizarry, R. A., Ladd-Acosta, C., Wen, B., Wu, Z., Montano, C., Onyango, P., ... Webster, M. (2009). Genome-wide methylation analysis of human colon cancer reveals similar hypo-and hypermethylation at conserved tissue-specific CpG island shores. *Nature Genetics, 41*(2), 178.

Ito, S., Shen, L., Dai, Q., Wu, S. C., Collins, L. B., Swenberg, J. A., ... Zhang, Y. (2011). Tet proteins can convert 5-methylcytosine to 5-formylcytosine and 5-carboxylcytosine. *Science, 333*(6047), 1300–1303.

Iwasaki, O., Corcoran, C. J., & Noma, K.-i. (2016). Involvement of condensin-directed gene associations in the organization and regulation of chromosome territories during the cell cycle. *Nucleic Acids Research, 44*, 3618–3628. http://dx.doi.org/10.1093/nar/gkv1502.

Jin, C., Zang, C., Wei, G., Cui, K., Peng, W., Zhao, K., & Felsenfeld, G. (2009). H3.3/H2A.Z double variant-containing nucleosomes mark 'nucleosome-free regions' of active promoters and other regulatory regions. *Nature Genetics, 41*(8), 941–945. http://dx.doi.org/10.1038/ng.409.

Johnson, B. V., Bert, A. G., Ryan, G. R., Condina, A., & Cockerill, P. N. (2004). Granulocyte-macrophage colony-stimulating factor enhancer activation requires cooperation between NFAT and AP-1 elements and is associated with extensive nucleosome reorganization. *Molecular and Cellular Biology, 24*(18), 7914–7930.

Jones, P. A. (2012). Functions of DNA methylation: Islands, start sites, gene bodies and beyond. *Nature Reviews. Genetics, 13*(7), 484–492.

Kadauke, S., Udugama, M. I., Pawlicki, J. M., Achtman, J. C., Jain, D. P., Cheng, Y., ... Blobel, G. A. (2012). Tissue-specific mitotic bookmarking by hematopoietic transcription factor GATA1. *Cell, 150*(4), 725–737. http://dx.doi.org/10.1016/j.cell.2012.06.038.

Kadoch, C., & Crabtree, G. R. (2013). Reversible disruption of mSWI/SNF (BAF) complexes by the SS18-SSX oncogenic fusion in synovial sarcoma. *Cell, 153*(1), 71–85.

Kadoch, C., & Crabtree, G. R. (2015). Mammalian SWI/SNF chromatin remodeling complexes and cancer: Mechanistic insights gained from human genomics. *Science Advances, 1*(5), e1500447.

Kelly, T. K., Miranda, T. B., Liang, G., Berman, B. P., Lin, J. C., Tanay, A., & Jones, P. A. (2010). H2A.Z maintenance during mitosis reveals nucleosome shifting on mitotically silenced genes. *Molecular Cell*, *39*(6), 901–911.

Khorasanizadeh, S. (2004). The nucleosome: From genomic organization to genomic regulation. *Cell*, *116*(2), 259–272.

Kouskouti, A., & Talianidis, I. (2005). Histone modifications defining active genes persist after transcriptional and mitotic inactivation. *The EMBO Journal*, *24*(2), 347–357. http://dx.doi.org/10.1038/sj.emboj.7600516.

Kouzarides, T. (2007). Chromatin modifications and their function. *Cell*, *128*(4), 693–705.

Kriaucionis, S., & Heintz, N. (2009). The nuclear DNA base 5-hydroxymethylcytosine is present in Purkinje neurons and the brain. *Science*, *324*(5929), 929–930.

Krishnan, S., Horowitz, S., & Trievel, R. C. (2011). Structure and function of histone H3 lysine 9 methyltransferases and demethylases. *Chembiochem*, *12*(2), 254–263.

Lawrence, M., Daujat, S., & Schneider, R. (2016). Lateral thinking: How histone modifications regulate gene expression. *Trends in Genetics*, *32*(1), 42–56.

Lee, S.-M., Choi, W.-Y., Lee, J., & Kim, Y.-J. (2015). The regulatory mechanisms of intragenic DNA methylation. *Epigenomics*, 7(4), 527–531.

Lelli, K. M., Slattery, M., & Mann, R. S. (2012). Disentangling the many layers of eukaryotic transcriptional regulation. *Annual Review of Genetics*, *46*, 43.

Li, B., Carey, M., & Workman, J. L. (2007). The role of chromatin during transcription. *Cell*, *128*(4), 707–719.

Li, M., & Fang, Y. (2015). Histone variants: The artists of eukaryotic chromatin. *Science China Life Sciences*, *58*(3), 232–239.

Lieberman-Aiden, E., Van Berkum, N. L., Williams, L., Imakaev, M., Ragoczy, T., Telling, A., … Dorschner, M. O. (2009). Comprehensive mapping of long-range interactions reveals folding principles of the human genome. *Science*, *326*(5950), 289–293.

Light, W. H., & Brickner, J. H. (2013). Nuclear pore proteins regulate chromatin structure and transcriptional memory by a conserved mechanism. *Nucleus*, *4*(5), 357–360. http://dx.doi.org/10.4161/nucl.26209.

Lim, P. S., Li, J., Holloway, A. F., & Rao, S. (2013). Epigenetic regulation of inducible gene expression in the immune system. *Immunology*, *139*(3), 285–293.

Lowe, M., Hostager, R., & Kikyo, N. (2016). Preservation of epigenetic memory during DNA replication. *Journal of Stem Cell Research & Therapeutics*, *1*(1), 00007.

Luger, K., Dechassa, M. L., & Tremethick, D. J. (2012). New insights into nucleosome and chromatin structure: An ordered state or a disordered affair? *Nature Reviews. Molecular Cell Biology*, *13*(7), 436–447.

Luger, K., Mäder, A. W., Richmond, R. K., Sargent, D. F., & Richmond, T. J. (1997). Crystal structure of the nucleosome core particle at 2.8 Å resolution. *Nature*, *389*(6648), 251–260.

Messerschmidt, D. M., Knowles, B. B., & Solter, D. (2014). DNA methylation dynamics during epigenetic reprogramming in the germline and preimplantation embryos. *Genes & Development*, *28*(8), 812–828.

Mirabella, F., Baxter, E. W., Boissinot, M., James, S. R., & Cockerill, P. N. (2010). The human IL-3/granulocyte-macrophage colony-stimulating factor locus is epigenetically silent in immature thymocytes and is progressively activated during T cell development. *The Journal of Immunology*, *184*(6), 3043–3054.

Nakayama, J.-i, Rice, J. C., Strahl, B. D., Allis, C. D., & Grewal, S. I. (2001). Role of histone H3 lysine 9 methylation in epigenetic control of heterochromatin assembly. *Science*, *292*(5514), 110–113.

Nan, X., Ng, H.-H., Johnson, C. A., Laherty, C. D., Turner, B. M., Eisenman, R. N., & Bird, A. (1998). Transcriptional repression by the methyl-CpG-binding protein MeCP2 involves a histone deacetylase complex. *Nature*, *393*(6683), 386–389.

Neigeborn, L., & Carlson, M. (1984). Genes affecting the regulation of SUC2 gene expression by glucose repression in Saccharomyces cerevisiae. *Genetics*, *108*(4), 845–858.

Netea, M. G., Joosten, L. A., Latz, E., Mills, K. H., Natoli, G., Stunnenberg, H. G., … Xavier, R. J. (2016). Trained immunity: A program of innate immune memory in health and disease. *Science*, *352*(6284), aaf1098. http://dx.doi.org/10.1126/science.aaf1098.

Northrop, J. K., Thomas, R. M., Wells, A. D., & Shen, H. (2006). Epigenetic remodeling of the IL-2 and IFN-gamma loci in memory CD8 T cells is influenced by CD4 T cells. *Journal of Immunology*, *177*(2), 1062–1069.

Nowak, S. J., & Corces, V. G. (2004). Phosphorylation of histone H3: A balancing act between chromosome condensation and transcriptional activation. *Trends in Genetics*, *20*(4), 214–220.

Obier, N., & Bonifer, C. (2016). Chromatin programming by developmentally regulated transcription factors: Lessons from the study of haematopoietic stem cell specification and differentiation. *FEBS Letters*. http://dx.doi.org/10.1002/1873-3468.12343.

Poke, F. S., Upcher, W. R., Sprod, O. R., Young, A., Brettingham-Moore, K. H., & Holloway, A. F. (2012). Depletion of c-Rel from cytokine gene promoters is required for chromatin reassembly and termination of gene responses to T cell activation. *PLoS One*, 7(7), e41734.

Raaphorst, F. M., Otte, A. P., van Kemenade, F. J., Blokzijl, T., Fieret, E., Hamer, K. M., … Meijer, C. J. (2001). Distinct BMI-1 and EZH2 expression patterns in thymocytes and mature T cells suggest a role for Polycomb genes in human T cell differentiation. *Journal of Immunology*, *166*(10), 5925–5934.

Rada-Iglesias, A., Bajpai, R., Swigut, T., Brugmann, S. A., Flynn, R. A., & Wysocka, J. (2011). A unique chromatin signature uncovers early developmental enhancers in humans. *Nature*, *470*(7333), 279–283.

Ramsahoye, B. H., Biniszkiewicz, D., Lyko, F., Clark, V., Bird, A. P., & Jaenisch, R. (2000). Non-CpG methylation is prevalent in embryonic stem cells and may be mediated by DNA methyltransferase 3a. *Proceedings of the National Academy of Sciences of the United States of America*, *97*(10), 5237–5242.

Rao, S., Gerondakis, S., Woltring, D., & Shannon, M. F. (2003). c-Rel is required for chromatin remodeling across the IL-2 gene promoter. *The Journal of Immunology*, *170*(7), 3724–3731.

Rao, S., Procko, E., & Shannon, M. F. (2001). Chromatin remodeling, measured by a novel real-time polymerase chain reaction assay, across the proximal promoter region of the IL-2 gene. *The Journal of Immunology*, *167*(8), 4494–4503.

Rothenberg, E. V. (2014). The chromatin landscape and transcription factors in T cell programming. *Trends in Immunology*, *35*(5), 195–204. http://dx.doi.org/10.1016/j.it.2014.03.001.

Rothenberg, E. V., & Zhang, J. A. (2012). T-cell identity and epigenetic memory. *Current Topics in Microbiology and Immunology*, *356*, 117–143.

Russ, B. E., Olshanksy, M., Smallwood, H. S., Li, J., Denton, A. E., Prier, J. E., … Turner, S. J. (2014). Distinct epigenetic signatures delineate transcriptional programs during virus-specific CD8(+) T cell differentiation. *Immunity*, *41*(5), 853–865. http://dx.doi.org/10.1016/j.immuni.2014.11.001.

Saeed, S., Quintin, J., Kerstens, H. H., Rao, N. A., Aghajanirefah, A., Matarese, F., … Stunnenberg, H. G. (2014). Epigenetic programming of monocyte-to-macrophage differentiation and trained innate immunity. *Science*, *345*(6204), 1251086. http://dx.doi.org/10.1126/science.1251086.

Serandour, A. A., Avner, S., Percevault, F., Demay, F., Bizot, M., Lucchetti-Miganeh, C., ... Eeckhoute, J. (2011). Epigenetic switch involved in activation of pioneer factor FOXA1-dependent enhancers. *Genome Research, 21*(4), 555–565. http://dx.doi.org/10.1101/gr.111534.110.

Sherwood, R. I., Hashimoto, T., O'Donnell, C. W., Lewis, S., Barkal, A. A., van Hoff, J. P., ... Gifford, D. K. (2014). Discovery of directional and nondirectional pioneer transcription factors by modeling DNase profile magnitude and shape. *Nature Biotechnology, 32*(2), 171–178.

Stern, M., Jensen, R., & Herskowitz, I. (1984). Five SWI genes are required for expression of the HO gene in yeast. *Journal of Molecular Biology, 178*(4), 853–868.

Sutcliffe, E. L., Parish, I. A., He, Y. Q., Juelich, T., Tierney, M. L., Rangasamy, D., ... Rao, S. (2009). Dynamic histone variant exchange accompanies gene induction in T cells. *Molecular and Cellular Biology, 29*(7), 1972–1986.

Taberlay, P. C., & Jones, P. A. (2011). DNA methylation and cancer. *Progress in Drug Research, 67*, 1–23.

Taberlay, P. C., Kelly, T. K., Liu, C.-C., You, J. S., De Carvalho, D. D., Miranda, T. B., ... Jones, P. A. (2011). Polycomb-repressed genes have permissive enhancers that initiate reprogramming. *Cell, 147*(6), 1283–1294.

Tahiliani, M., Koh, K. P., Shen, Y., Pastor, W. A., Bandukwala, H., Brudno, Y., ... Aravind, L. (2009). Conversion of 5-methylcytosine to 5-hydroxymethylcytosine in mammalian DNA by MLL partner TET1. *Science, 324*(5929), 930–935.

Tan, M., Luo, H., Lee, S., Jin, F., Yang, J. S., Montellier, E., ... Rajagopal, N. (2011). Identification of 67 histone marks and histone lysine crotonylation as a new type of histone modification. *Cell, 146*(6), 1016–1028.

Tessarz, P., & Kouzarides, T. (2014). Histone core modifications regulating nucleosome structure and dynamics. *Nature Reviews. Molecular Cell Biology, 15*(11), 703–708.

Ucar, D., & Bayarsaihan, D. (2015). Cell-specific gene promoters are marked by broader spans of H3K4me3 and are associated with robust gene expression patterns. *Epigenomics*, 7(2), 129–131.

Van Hooser, A., Goodrich, D. W., Allis, C. D., Brinkley, B., & Mancini, M. A. (1998). Histone H3 phosphorylation is required for the initiation, but not maintenance, of mammalian chromosome condensation. *Journal of Cell Science, 111*(23), 3497–3506.

Verdeguer, F., Le Corre, S., Fischer, E., Callens, C., Garbay, S., Doyen, A., ... Pontoglio, M. (2010). A mitotic transcriptional switch in polycystic kidney disease. *Nature Medicine, 16*(1), 106–110. http://dx.doi.org/10.1038/nm.2068.

Waddington, C. H. (2012). The epigenotype. 1942. *International Journal of Epidemiology, 41*(1), 10–13. http://dx.doi.org/10.1093/ije/dyr184.

Wang, Y., Fischle, W., Cheung, W., Jacobs, S., Khorasanizadeh, S., & Allis, C. D. (2004). Beyond the double helix: Writing and reading the histone code. *Novartis Foundation Symposium, 259*, 3–17. discussion 17–21, 163–169.

Wang, X., & Hayes, J. J. (2008). Acetylation mimics within individual core histone tail domains indicate distinct roles in regulating the stability of higher-order chromatin structure. *Molecular and Cellular Biology, 28*(1), 227–236.

Wang, Z., Zang, C., Rosenfeld, J. A., Schones, D. E., Barski, A., Cuddapah, S., ... Zhang, M. Q. (2008). Combinatorial patterns of histone acetylations and methylations in the human genome. *Nature Genetics, 40*(7), 897–903.

Weintraub, H., Worcel, A., & Alberts, B. (1976). A model for chromatin based upon two symmetrically paired half-nucleosomes. *Cell, 9*(3), 409–417.

Wiench, M., John, S., Baek, S., Johnson, T. A., Sung, M. H., Escobar, T., ... Sabo, P. J. (2011). DNA methylation status predicts cell type-specific enhancer activity. *The EMBO Journal, 30*(15), 3028–3039.

Wurster, A. L., & Pazin, M. J. (2011). ATP-dependent chromatin remodeling in T cells. *Biochemistry and Cell Biology*, *90*(1), 1–13.

Xu, Y., Ayrapetov, M. K., Xu, C., Gursoy-Yuzugullu, O., Hu, Y., & Price, B. D. (2012). Histone H2A. Z controls a critical chromatin remodeling step required for DNA double-strand break repair. *Molecular Cell*, *48*(5), 723–733.

Yang, X., Han, H., De Carvalho, D. D., Lay, F. D., Jones, P. A., & Liang, G. (2014). Gene body methylation can alter gene expression and is a therapeutic target in cancer. *Cancer Cell*, *26*(4), 577–590.

Yang, X. P., Jiang, K., Hirahara, K., Vahedi, G., Afzali, B., Sciume, G., … Laurence, A. (2015). EZH2 is crucial for both differentiation of regulatory T cells and T effector cell expansion. *Scientific Reports*, *5*, 10643. http://dx.doi.org/10.1038/srep10643.

Zaidi, S. K., Young, D. W., Montecino, M. A., Lian, J. B., van Wijnen, A. J., Stein, J. L., & Stein, G. S. (2010). Mitotic bookmarking of genes: A novel dimension to epigenetic control. *Nature Reviews. Genetics*, *11*(8), 583–589. http://dx.doi.org/10.1038/nrg2827.

Zaret, K. S., & Carroll, J. S. (2011). Pioneer transcription factors: Establishing competence for gene expression. *Genes & Development*, *25*(21), 2227–2241.

Zaret, K. S., Lerner, J., & Iwafuchi-Doi, M. (2016). Chromatin scanning by dynamic binding of pioneer factors. *Molecular Cell*, *62*(5), 665–667.

Zhao, K., Wang, W., Rando, O. J., Xue, Y., Swiderek, K., Kuo, A., & Crabtree, G. R. (1998). Rapid and phosphoinositol-dependent binding of the SWI/SNF-like BAF complex to chromatin after T lymphocyte receptor signaling. *Cell*, *95*(5), 625–636.

CHAPTER FOUR

Histone Acetylation and the Regulation of Major Histocompatibility Class II Gene Expression

K. Suzuki[1], Y. Luo
Faculty of Medical Technology, Teikyo University, Itabashi, Japan
[1]Corresponding author: e-mail address: koichis0923@med.teikyo-u.ac.jp

Contents

Advances in Protein Chemistry and Structural Biology, Volume 106
ISSN 1876-1623
http://dx.doi.org/10.1016/bs.apcsb.2016.08.002

Abstract

Major histocompatibility complex (MHC) class II molecules are essential for processing and presenting exogenous pathogen antigens to activate $CD4^+$ T cells. Given their central role in adaptive immune responses, MHC class II genes are tightly regulated in a tissue- and activation-specific manner. The regulation of MHC class II gene expression involves various transcription factors that interact with conserved proximal *cis*-acting regulatory promoter elements, as well as MHC class II transactivator that interacts with a variety of chromatin remodeling machineries. Recent studies also identified distal regulatory elements within MHC class II gene locus that provide enormous insight into the long-range coordination of MHC class II gene expression. Novel therapeutic modalities that can modify MHC class II genes at the epigenetic level are emerging and are currently in preclinical and clinical trials. This review will focus on the role of chromatin remodeling, particularly remodeling that involves histone acetylation, in the constitutive and inducible regulation of MHC class II gene expression.

1. INTRODUCTION

1.1 Major Histocompatibility Complex Class II Molecules

Major histocompatibility complex (MHC) class II molecules are a family of αβ-heterodimer cell surface receptors that are generally expressed only on professional antigen-presenting cells (APCs) such as dendritic cells, mononuclear phagocytes, thymic epithelial cells, and B cells (Janeway, Travers, Walport, & Shlomchik, 2001). Human MHC class II molecules, known as human leukocyte antigens (HLAs), include three classical isotypes: HLA-DR, DP, and DQ that each consists of two subunit chains, α and β (Fig. 1). Each subunit contains two extracellular domains, a membrane-spanning domain and a C-terminal cytoplasmic domain, and each are anchored in the cell membrane. The extracellular N-terminal domain of both chains, termed the α1 and β1 domains, form an α-helix cleft (peptide-binding groove) that can accommodate peptides of 10–30 residues (Fig. 1). The C-terminal cytoplasmic domains interact with each other to form a β-sheet under the peptide-binding groove. During synthesis of MHC class II molecules in the endoplasmic reticulum (ER) of professional APCs, α and β subunits are assembled and then complex with a special polypeptide known as the class II invariant chain peptide (CLIP) that occupies the peptide-binding groove of the newly synthesized MHC class II proteins and prevents this groove from binding to self-antigens. During transport of occupied MHC class II molecules from the ER to the cell surface,

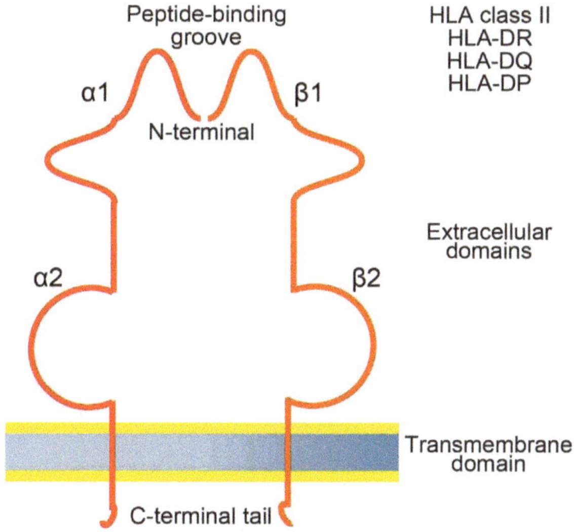

Fig. 1 Schematic representation of a HLA class II molecule. Each HLA class II molecule consists of two subunit chains, α and β. Each subunit anchors in the cell membrane and contains two extracellular domains, a membrane-spanning domain, and a C-terminal cytoplasmic domain.

exocytic vesicles that carry the MHC class II molecules can fuse with endosomal vesicles containing peptides derived from ingested extracellular proteins, e.g., bacterial antigens. These endosomal vesicles also carry a MHC class II-like protein called HLA-DM that functions to remove the CLIP structure from the MHC class II molecule and unmask the peptide-binding groove to accept peptides. When the MHC class II molecules bind peptides derived from ingested proteins, the complex is stabilized. Antigen-bound MHC class II molecules are then delivered to the cell surface, where they activate $CD4^+$ T helper cells through T cell receptors with the help of costimulatory molecules to initiate an appropriate immune response (Janeway et al., 2001). Indeed, $CD4^+$ T cells can only respond to the peptides presented by MHC class II molecules, which make MHC class II proteins central players in initiating vigorous responses to foreign pathogens while minimizing the risk of presenting self-antigens. Given their essential role in immunity, the expression of MHC class II molecules is tightly regulated at both the transcriptional and epigenetic level. Although the constitutive expression of MHC class II molecules is restricted to professional APCs, their expression can be induced upon exposure to certain cytokines with interferon γ (IFNγ) being the most potent stimulator of MHC class II induction (Ting & Trowsdale, 2002).

1.2 Human HLA Class II Locus

The HLA class II supergene locus, spanning nearly 700 kb, is located on the short arm of human chromosome 6 at 6p21.31. The locus contains a dense cluster of highly homologous genes, including *HLA-DR* (*HLA-DRA1* encodes HLA-DR α chain and *DRB1*, *DRB3*, and *DRB5* encode the HLA-DR β chain), *HLA-DQ* (*HLA-DQA1* and *-DRQ1* encode HLA-DQ α and β chains, respectively), *HLA-DP* (*HLA-DPA1* and *-DRP1* encode HLA-DP α and β chains, respectively), *HLA-DM* (*HLA-DMA* and *-DMB* encode HLA-DM α and β chains, respectively), and the accessory molecule *HLA-DO* (*HLA-DOA* and *-DOB* encode HLA-DO α and β chains, respectively) (Fig. 2) (Shiina, Hosomichi, Inoko, & Kulski, 2009). In addition, nine CCCTC-binding factor (CTCF)-associated insulator elements punctuate the HLA class II supergene locus and separate the HLA class II gene subregions (Fig. 2). The mammalian protein CTCF demarcates and insulates regions of regulatory activity within the genome by functioning as an enhancer blocker or by preventing the spread of heterochromatin into active genes (Cuddapah et al., 2009). In vertebrates, all known insulator elements are associated with CTCF function (Shiina et al., 2009). CTCF has been putatively shown to regulate the expression of HLA class II genes by modifying their spatial relationships and topological architecture (see later).

1.3 Histone Tail Modification: A Key to "Open" Chromatin

In eukaryotic cells, DNA strands are wrapped around an octamer of core histones to form the basic chromatin unit, the nucleosome core particle (Fig. 3) (Bradbury, 2002). The histone octamer is composed of a central heterotetramer formed by histones H3 and H4 that is flanked by two heterodimers of histones H2A and H2B. Each nucleosome is separated by 10–60 bp of "linker" DNA, and the resulting nucleosomal array constitutes a chromatin fiber that is 10 nm in diameter (Fig. 3) (Bradbury, 2002). This

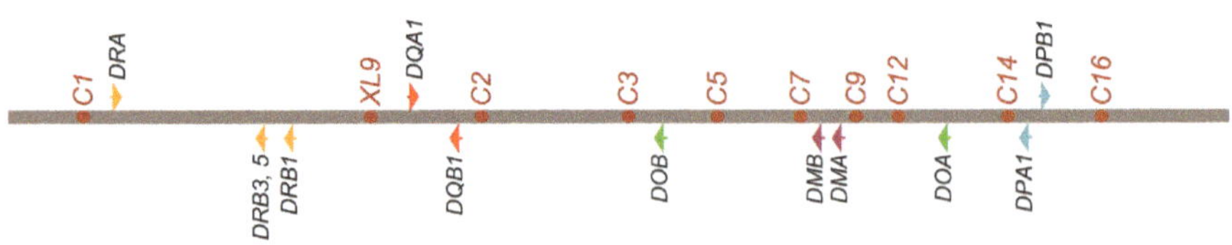

Fig. 2 Schematic representation of the human HLA class II supergene locus. The HLA class II supergene locus spans nearly 700 kb and contains a dense cluster of highly homologous genes that are separated by CTCF-associated insulator elements (*red ovals*). The *arrowheads* indicate transcription orientation.

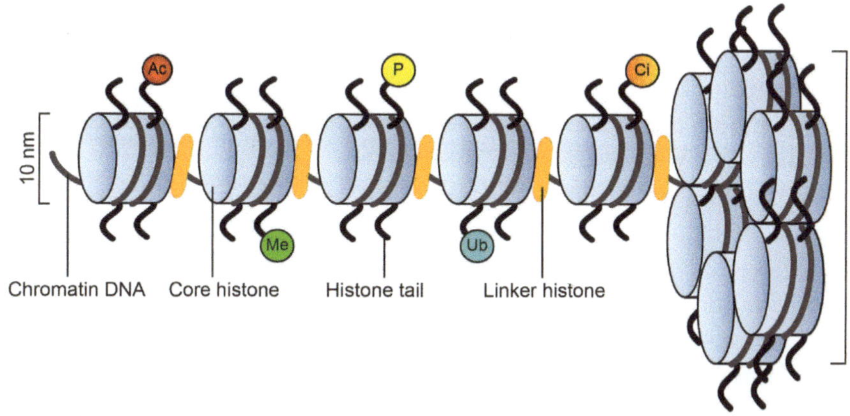

Fig. 3 Schematic representation of the "bead-on-a-string" chromatin structure. A nucleosomal array constitutes a chromatin fiber of 10 nm in diameter. This arrangement further folds into more condensed 30 nm thick fibers that are stabilized by the binding of a linker histone to each nucleosome core. Histone tails extend from the nucleosome surface where they can be modified through acetylation (Ac), phosphorylation (P), methylation (Me), ubiquitination (Ub), and citrullination (Ci).

simple "beads-on-a-string" arrangement is then folded into 30 nm thick fibers that are stabilized by binding of a linker histone, histone H1, to each nucleosome core (Fig. 3) (Bradbury, 2002). The 30 nm fibers can be further condensed in vivo to form 100–400 nm thick interphase fibers or the more tightly compacted metaphase chromosome structures (Bradbury, 2002). This organization of DNA into chromatin fibers hinders its accessibility to proteins, e.g., transcriptional factors, which must "read" the nucleotide base sequence to initiate transcription. Thus, in order for gene transcription to occur, chromatin structures must be dynamically regulated to transit between folded and unfolded conformations (Bradbury, 2002).

Nucleosomal histones can be chemically modified to dynamically regulate chromatin folding conformations (Bradbury, 2002). Each core histone has a domain that mediates histone–histone interactions within the octamer, and also organizes the wrapping of nucleosomal DNA. Additionally, each histone harbors a terminal 20–35 residue segment known as the histone tail that is rich in basic amino acids and extends from the nucleosome surface, where it can undergo posttranslational modifications such as acetylation, phosphorylation, methylation, and ubiquitination (Fig. 3). These histone tails may not contribute significantly to the structure or stability of individual nucleosomes, but they do play an essential role in mediating the folding of nucleosomal arrays, which make them crucial for the regulation of gene transcription. Indeed, removal of the histone tails in vitro produces nucleosomal arrays that cannot unfold beyond the "bead-on-a-string" 10 nm fiber (Peterson & Laniel, 2004). Several studies also showed that site-specific histone modifications correlate with particular biological functions (Hadnagy, Beaulieu, & Balicki, 2008).

Among the various histone tail modifications, acetylation/deacetylation has generated the most interest given the relationship with transcription activity (Sterner & Berger, 2000). The rationale for chromatin remodeling mediated by histone acetylation is that acetylation neutralizes the positively charged lysine residues in the histone N-termini that in turn decreases the DNA affinity of histones and displaces the N-termini from the nucleosome. Following this displacement, the nucleosomes can unfold and the accessibility of DNA is increased. In the past two decades, enormous advances have been made in our understanding of epigenetic changes, particularly by the identification and cloning of enzymes that can modify histones, e.g., histone acetyltransferase (HAT) and histone deacetylase (HDAC), which catalyze histone (as well as nonhistone protein) acetylation and deacetylation, respectively (Ogryzko, Schiltz, Russanova, Howard, & Nakatani, 1996; Sterner & Berger, 2000). Specific combinations of histone modifications are thought to establish various "histone code" or recognition platforms that enable the regulation of transcription at an epigenetic level (Jenuwein & Allis, 2001). MHC class II has become one of the most extensively studied systems that clarified the association of histone modifications with the regulation of specific gene expression. This review will focus on our current level of knowledge and recent advances in understanding how MHC class II gene expression is regulated by chromatin remodeling, in particular remodeling that occurs following histone acetylation.

1.4 Proximal Promoters of MHC Class II Genes

Until recently, control of MHC class II molecule transcription was thought to occur primarily through the action of its proximal promoter: a highly conserved, combinatorial *cis*-regulatory sequence located approximately 100–200 bp upstream of the transcription start site of each MHC class II gene (Fig. 4). The proximal promoter S-Y module consists of S (also known as W, S/W), X1, X2, and Y box elements, in addition to an octamer element (Oct) and TATA box (Fig. 4). The Y box, which contains a CCAAT sequence that was identified decades ago as a canonical regulatory element, is bound by the nuclear factor (NF)-Y complex that comprises NF-YA,

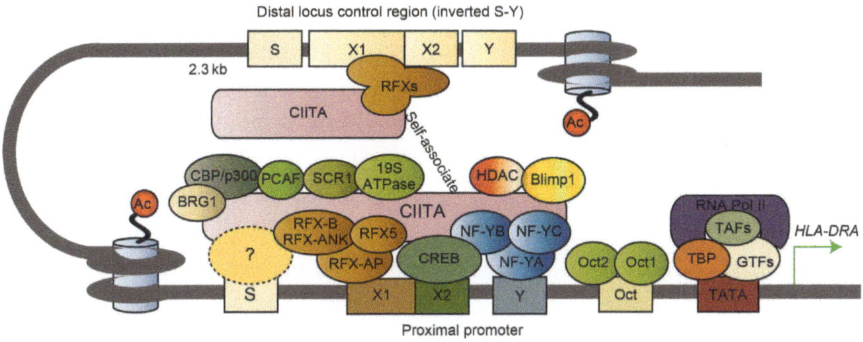

Fig. 4 Schematic representation of promoter proximal and distal locus control of MHC class II genes (HLA-DRA as an example). The highly conserved proximal promoter S-Y module is bound by an array of transcription factors such as RFX, CREB, and NF-Y. These transcription factors cannot induce MHC class II gene transcription on their own but instead form a multiprotein complex to provide a landing pad for CIITA, the master regulator of MHC class II transcription, to recognize and bind. CIITA interacts with a variety of chromatin remodeling factors, including CBP/p300, PCAF, SCR1, 19S ATPase, BRG1, HDACs, and Blimp1. A locus control region consisting of an inverted S-Y module was identified 2.3 kb upstream of the proximal promoter that also interacts with RFX and CIITA. Strong histone acetylation (Ac) was found upstream of both the HLA-DRA proximal promoter and the distal control regions. The ability of CIITA to self-associate supports the notion that chromatin interactions occur between proximal promoters with distal control elements.

NF-YB, and NF-YC (Jabrane-Ferrat, Nekrep, Tosi, Esserman, & Peterlin, 2002; Louis-Plence, Moreno, & Boss, 1997). The NF-Y complex interacts with the regulatory factor-X (RFX) (Ronchi, Bellorini, Mongelli, & Mantovani, 1995) complex that cooperatively binds the X box adjacent to the CCAAT sequence. Moreover, NF-Y is known to distort and compact DNA structure to facilitate combinatorial transcription factor binding (Guerra, Imperadori, Mantovani, Dunlap, & Finzi, 2007) that stabilizes the promoter–enhanceosome interaction, which is critical for MHC class II transcription. Meanwhile, the X2 box is bound by cyclic adenosine monophosphate (cAMP) response element binding protein (CREB) (Moreno, Beresford, Louis-Plence, Morris, & Boss, 1999). Although chromatin immunoprecipitation (ChIP) assays detected the presence of phosphorylated CREB at the *HLA-DRA* promoter, and phos-CREB was shown to mildly enhance the transcription of a MHC class II reporter plasmid, CREB is dispensable for MHC class II gene transcription (Lochamy, Rogers, & Boss, 2007). On the other hand, CREB can physically interact with RFX5 and class II transactivator (CIITA) (Lochamy et al., 2007). Thus, for regulating the MHC class II gene system, the role of CREB as a stabilizing partner of

X1 box binding proteins at the promoter is considered to be a more important function than its phosphorylation-related activity. The RFX complex, which binds to the X1 box, is a multiprotein complex that comprises RFX5, RFX-AP, and RFX-B/RFXANK. RFX components interact directly with CREB and NF-Y, forming a combinatorial enhanceosome over the X-Y box region. An intact RFX complex is crucial for MHC class II gene transcription and enhanceosome assembly, as elucidated by studies using cells derived from bare lymphocyte syndrome (BLS) patients, who are deficient in MHC class II gene expression due to defects in RFX or CIITA (Durand et al., 1997; Garvie & Boss, 2008; Masternak et al., 1998; Nagarajan et al., 1999; Steimle et al., 1995). In contrast to the well-studied X-Y box region, the role of the S region remains controversial. Early reports suggested that the RFX complex binds to the S box, and NF-Y reinforces the binding of RFX to the S box (Jabrane-Ferrat, Fontes, Boss, & Peterlin, 1996). However, recent studies showed that the S box is a critical determinant for tethering CIITA to the MHC class II enhanceosome complex (Muhlethaler-Mottet et al., 2004). Interaction of CIITA with the enhanceosome requires both an intact S box and remarkably stringent spacing between the S and X regions (Muhlethaler-Mottet et al., 2004). Even a 1–2-bp change in the native S-X spacing distance negatively affects CIITA recruitment and normal promoter function (Muhlethaler-Mottet et al., 2004). In contrast, the binding of RFX to a putative binding site in the S box is not essential for either CIITA recruitment or promoter activity (Muhlethaler-Mottet et al., 2004).

Although these transcription factors are ubiquitous and are expressed constitutively, they cannot induce MHC class II gene transcription on their own. Instead, they form the multiprotein MHC class II enhanceosome that coordinately binds the S-Y module in a stereospecific and cooperative manner. The MHC class II enhanceosome serves as a landing pad for CIITA to recognize and interact with. CIITA is a well-known master regulator of MHC class II gene expression, as CIITA overexpression induced MHC class II gene transcription in virtually all cell types tested (Mori-Aoki et al., 2000). In addition to the S-Y module located proximal to the *HLA-DRA* transcription start site, an inverted S-Y module approximately 2.3 kb upstream of *HLA-DRA* may be a distal locus control region (Dorn et al., 1988) (Fig. 4). An inverted S-Y module also interacts with RFX and CIITA in vivo, and is correlated with strong acetylation of histones H3 and H4, as well as recruitment of basal transcription machinery (Masternak, Peyraud, Krawczyk, Barras, & Reith, 2003).

2. HISTONE ACETYLATION CORRELATES WITH ACTIVE MHC CLASS II TRANSCRIPTION

Acetylation of N-terminal lysines in histones H3 and H4 is associated with transcription processes at multiple levels (Torok & Grant, 2004). MHC class II is one of the best examples to demonstrate the association between histone acetylation and active gene transcription. ChIP assays that compared wild-type B cell lymphoma Raji cells and MHC class II-deficient BLS cell lines (e.g., RJ2.2.5, Ramia, and SJO cells, defective for *CIITA*, *RFE-B*, or *RFX5*, respectively) showed strong acetylation of H3 and H4 histones at the *HLA-DRA* promoter of Raji cells, but acetylation was barely detectable in BLS cells (Beresford & Boss, 2001), indicating that acetylation is a potential histone code for active *HLA-DRA* transcription. Histone acetylation at the *HLA-DRA* promoter is also highly dynamic and inducible by IFNγ, which gradually increases histone H3 and H4 acetylation at the *HLA-DRA* promoter (Beresford & Boss, 2001). This acetylation preceded the increase in HLA-DRA mRNA expression levels in HeLa cells, while acetylation levels rapidly returned to the baseline after IFNγ removal (Beresford & Boss, 2001). Acetylation of histones H3 and H4 is detectable over a broad range (~8 kb) upstream of the proximal promoter region in Raji cells but not in BLS cells, and the amount of acetylation peaks just upstream of the proximal S-Y module (Beresford & Boss, 2001; Masternak et al., 2003). Intriguingly, another histone acetylation peak was detected adjacent to the distal inverted S-Y modules, which also supports a role for this distal region in MHC class II transcription (Masternak et al., 2003). Additionally, the distal inverted S-Y module interacts with RFX and CIITA as well as basal transcription machinery (Masternak et al., 2003), and thus this module has functions similar to that of the proximal promoter.

The vital role for histone acetylation in MHC class II transcription is also strongly supported by the remarkable effect of HDAC inhibitors, e.g., trichostatin A (TSA), to induce MHC class II expression in a CIITA-independent manner. TSA treatment enhanced histone H3 and H4 acetylation at the *HLA-DRA* promoter that was correlated with significantly induced *HLA-DRA* expression in CIITA-deficient RJ2.2.5 cells (Chou & Tomasi, 2008). Similarly, TSA treatment strongly induced MHC class II expression in SM9-1 and Colon 26 cell lines in the absence of CIITA expression (Chou, Khan, Magner, & Tomasi, 2005; Chou & Tomasi, 2008). MHC class II expression was also induced by TSA treatment in primary culture cells derived from CIITA, interferon regulatory factor 1 (IRF1), or signal transducer and activator of transcription 1 (STAT1) knockout mice (Chou et al., 2005). Together these results suggest that histone acetylation induced by HDAC inhibitors can bypass the requirement for CIITA in MHC class II expression, presumably by directly "opening" chromatin to allow stable binding by the transcription factor complex and subsequent engagement of RNA polymerase II. Indeed, ChIP assays in CIITA-deficient RJ2.2.5 cells showed that TSA treatment significantly enhanced the association of the *HLA-DRA* promoter with transcription factors such as RFX5, NF-YA, NF-YB, and general transcription machinery components, including TATA-binding protein (TBP) and RNA polymerase II (Gialitakis et al., 2006; Zika, Greer, Zhu, & Ting, 2003).

On the other hand, overexpression of wild-type HDAC1, but not a deacetylase-defective mutant HDAC, suppressed IFNγ- and CIITA-inducible MHC class II expression in a dose-dependent manner. This suppression could be rescued by TSA treatment (Kong, Fang, Li, Fang, & Xu, 2009; McKinsey, Kuwahara, Bezprozvannaya, & Olson, 2006; Zika et al., 2003), suggesting that histone deacetylation per se hinders MHC class II transcription presumably by condensing chromatin to impede MHC class II enhanceosome binding. As such, HDAC overexpression antagonizes the transcription activity of CIITA, not by repressing CIITA expression, but rather by hindering the association of CIITA with the MHC class II promoter and rendering the promoter inaccessible to CIITA (Kong et al., 2009; Zika et al., 2003).

3. INTERPLAY BETWEEN CIITA CHROMATIN REMODELING FACTORS

In general, both constitutive and inducible MHC class II expression is globally regulated at the transcriptional level by CIITA. All known MHC class II-related genes containing the conserved S-Y promoter motif are regulated by CIITA (Reith, LeibundGut-Landmann, & Waldburger, 2005). Defects in the CIITA protein (BLS group A) result in fatal immunodeficiency (Dziembowska et al., 2002) because CIITA is the chief control point for the induction of MHC class II genes and other IFNγ-inducible genes. As noted earlier, CIITA does not bind to DNA directly, but instead exerts its transactivating effect through protein–protein interactions with transcription factor complexes that directly bind to the S-Y regulatory module at the MHC class II promoter (Jabrane-Ferrat, Nekrep, Tosi, Esserman, &

Peterlin, 2003; Masternak et al., 2000; Zhu et al., 2000). In vivo genomic footprinting of CIITA-overexpressing G3A cells, which lack endogenous CIITA, revealed that CIITA is required for the assembly of transcription factor complexes at MHC class II gene promoters (Wright et al., 1998). This evidence indicates that CIITA is important for "opening" the chromatin at MHC class II promoters to accept transcription factors and that CIITA promotes the "opening" either through direct effects on chromatin structure or by recruiting certain chromatin remodeling factors. ChIP assays revealed that the association of CIITA with the *HLA-DRA* promoter was correlated with strong acetylation of histones H3 and H4 at the promoter upon IFNγ exposure in HeLa cells (Beresford & Boss, 2001). Meanwhile, IFNγ removal was followed by CIITA dissociation from the *HLA-DRA* promoter and was correlated with histone deacetylation at the *HLA-DRA* promoter (Beresford & Boss, 2001). Raji cells that express endogenous CIITA had histone H3 and H4 acetylation that spanned a broad range upstream of the proximal *HLA-DRA* promoter and the distal inverted S-Y module. In contrast, CIITA-deficient RJ2.2.5 cells show little histone acetylation in the same area (Beresford & Boss, 2001; Masternak et al., 2003). These findings suggest that the induction of histone acetylation is one mechanism by which CIITA mediates chromatin remodeling at the promoter regions of MHC class II genes.

3.1 CIITA Associates with HAT Coactivators

The molecular mechanism by which CIITA mediates histone acetylation to activate MHC class II gene expression is an area of intense interest. CIITA is an 1130-amino acid (aa) protein that contains several functional domains, including an N-terminal transcriptional acidic activation domain (AAD), a proline/serine/threonine-rich domain (P/S/T), a GTP-binding domain (GBD), and a C-terminal leucine-rich repeat (LRR) (Fig. 5). CIITA also has three identified nuclear localization sequences (NLS) that mediate its nuclear translocation (Fig. 5) (Cressman, O'Connor, Greer, Zhu, & Ting, 2001). The N-terminal AAD of CIITA has been mapped to interact with several transcription coactivators: CREB-binding protein (CBP)/p300 (Fontes, Kanazawa, Jean, & Peterlin, 1999; Kretsovali et al., 1998), P300/CBP-associated factor (PCAF) (Spilianakis, Papamatheakis, & Kretsovali, 2000), and steroid receptor coactivator 1 (SRC1) (also known as nuclear receptor coactivator 1) (Tzortzakaki, Spilianakis, Zika, Kretsovali, & Papamatheakis, 2003). The discovery that many coactivators, including

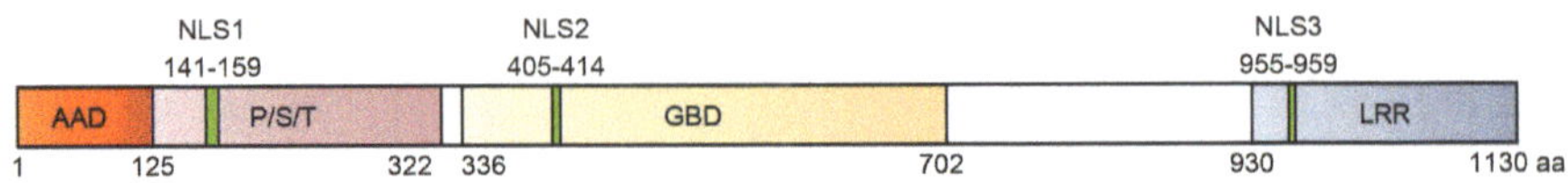

Fig. 5 Schematic representation of the CIITA protein. CIITA is an 1130 amino acid (aa) protein that contains an N-terminal transcriptional acidic activation domain (ADD), a proline/serine/threonine-rich domain (P/S/T), a GTP-binding domain (GBD), and a C-terminal leucine-rich repeat (LRR). Three identified nuclear localization sequences (NLS) are mapped within CIITA. CIITA interacts with HAT coactivators, 19S ATPase, and chromatin remodeling factor BRG1. GBD and LRR have also been shown to be important for CIITA self-association.

CBP/p300, PCAF, and SRC1 possess intrinsic HAT activity (Sterner & Berger, 2000) inspired the hypothesis that CIITA mediates chromatin remodeling indirectly through recruitment of HAT coactivators.

Reporter gene assays showed that overexpression of each of these coactivators enhanced the CIITA- or IFNγ-induced transcription of MHC class II genes (Fontes et al., 1999; Sheppard, Harries, Hussain, Bevan, & Heery, 2001; Spilianakis et al., 2000) and a dominant negative form of CBP inhibited the cell surface expression of MHC class II in B-lymphoblastoid RM3 cells (Fontes et al., 1999). Estrogen [17β-estradiol (17β-E_2)] can downregulate IFNγ-inducible MHC class II expression without affecting CIITA expression levels. Instead, 17β-E_2 affects the interaction between CBP and CIITA, which correlated with attenuated histone H3 and H4 acetylation at the MHC class II promoter (Adamski, Ma, Nozell, & Benveniste, 2004). Similarly, adenoviral oncoprotein E1A exerts a strong inhibitory effect on both IFNγ-inducible and constitutive expression of MHC class II genes that can be attributed to its competitive association with CBP/p300 against CIITA (Kretsovali et al., 1998). These results indicate that these coactivators are indeed important for the synergic activation of MHC class II gene transcription that is modulated by CIITA.

CBP and PCAF have also been shown to acetylate CIITA at the lysine pair K141 and K144 within the N-terminal NLS1. Exogenously expressed PCAF increased the nuclear accumulation of CIITA, whereas a HAT-deficient PCAF mutant was significantly less effective than the wild type (Spilianakis et al., 2000). The CIITA NLS double mutant at K141 and K144 did not undergo nuclear import in response to exogenous PCAF expression, whereas a control NLS mutant that had mutations at a non-acetylated lysine pair translocated to the nucleus in the presence of wild-type PCAF but not HAT-deficient PCAF (Spilianakis et al., 2000). The HDAC inhibitor TSA also induced significant nuclear accumulation of CIITA

(Spilianakis et al., 2000). These results together suggest that acetylation at NLS lysines K141 and K144 mediates the shuttling of CIITA between the nucleus and the cytoplasm. Thus, the critical question is whether acetylation affects the ability of CIITA to transactivate MHC class II expression. A study by Harton et al. analyzed this issue. Using reporter gene assays with COS-7 cells they compared the ability of the CIITA deletion mutant Δ132–209 (residues 132–209 deleted) to activate *HLA-DRA* transcription relative wild-type CIITA (Harton, Zika, & Ting, 2001). Their results showed that residues 132–209, which encompass the N-terminal NLS1, were needed for CIITA transactivity only when exogenous CIITA was stringently expressed (at 0.1 and 0.02 μg). However, this NLS was completely dispensable when CIITA was abundantly expressed (at 1 μg), since Δ132–209 activated *HLA-DRA* transcription to levels that were comparable to wild-type CIITA (Harton et al., 2001). When wild-type CIITA was coexpressed to high levels (at 1 μg) with HAT-defective CBP or/and PCAF mutants, the synergic effect on *HLA-DRA* transcription activation was not significantly affected relative to coexpression with wild-type CBP or/and PCAF (Harton et al., 2001). These results indicate that the HAT activity of CBP or PCAF or even both in combination is not necessary for the observed cooperativity involved in CIITA-mediated MHC class II gene expression.

This HAT independence of CBP and PCAF action could lead to several interpretations: (a) Coactivators, CBP and PCAF, regulate MHC class II expression through mechanisms other than HAT activity, e.g., protein–protein interaction with transcription factors; (b) Other HAT coactivators, e.g., SRC1, TBP-associated factor $(TAF)_{II}250$, may be recruited to MHC class II promoters and exert HAT activity; or (c) CIITA itself possesses intrinsic HAT activity. Indeed, HAT activity reportedly coimmunoprecipitated with exogenously expressed CIITA and this interaction could be abolished by deleting residues 94–135 within the N-terminal AAD of CIITA, which suggests that CIITA indeed has intrinsic HAT activity (Raval et al., 2001). However, HAT coactivators are also known to interact with CIITA through the same domain. As such, the possibility that the HAT activity of these coactivators coimmunoprecipitated with CIITA could not be excluded (Raval et al., 2001) and thus whether CIITA has intrinsic HAT activity remains an open question.

It should be noted that current reporter gene assays can be limited in their ability to evaluate transcription regulation that involves histone modifications or chromatin remodeling. Reporter vectors are typically derived from bacterial plasmids that do not contain histones or a dynamic chromatin structure, although a portion of exogenous plasmid DNA may be complexed with endogenous histones and acquire a nucleosome-like structure upon introduction into mammalian cells (Nakagawa, Bulger, Muramatsu, & Ito, 2001). Moreover, episomal plasmids may acquire a chromatin-like structure after integration into the host chromosome, but such chromatin-like structures may differ significantly from that of the mammalian chromosome (Nakagawa et al., 2001). Therefore, HAT or HDAC activity demonstrated by plasmid-based reporter gene assays (Fontes et al., 1999; Sheppard et al., 2001; Spilianakis et al., 2000; Zika et al., 2003) most likely reflects indirect effects on transcription factors and/or coactivators rather than direct effects on histones or chromatin structure.

3.2 CIITA Associates with HDACs

The role of CIITA-associated HATs in activating MHC class II gene transcription has been extensively studied. Meanwhile, the role of HDACs in this process has also emerged. Mammalian HDACs fall into four classes according to their action and sequence homology with the originally identified yeast enzymes. Class I, II, and IV are considered to be classical HDACs with activities that can be inhibited by TSA. Class I includes the yeast reduced potassium dependency 3 (Rpd3)-like HDACs (HDAC1, 2, 3, 8); Class II includes yeast histone deacetylase 1 (Hda1)-like HDACs (HDAC4, 5, 6, 7, 9); and Class IV includes HDAC11, which is not highly homologous to either yeast Rpd3 or Hda1 (Yang & Seto, 2008). Meanwhile, Class III, known as sirtuin deacetylases, is a separate type of HDAC and has a different action in that the activity is nicotinamide adenine dinucleotide (NAD)-dependent and is not affected by TSA (Imai, Armstrong, Kaeberlein, & Guarente, 2000). As opposed to HAT activity, HDACs primarily remove acetyl groups from ε-*N*-acetyl lysine amino acids in histone tails to increase the positive charge of histones and encourage high-affinity binding between histones and the DNA backbone. This increased binding affinity allows the histones to wrap the DNA strands more tightly, which is generally correlated with transcriptional repression. HDACs also regulate transcription through deacetylation of nonhistone proteins (Ikenoue, Inoki, Zhao, & Guan, 2008) and through protein–protein interactions with multiple transcription factors (Juan et al., 2000).

As mentioned earlier, the role of HDACs in regulating MHC class II expression was also supported by the observation that the HDAC inhibitor

TSA remarkably induced MHC class II gene expression in CIITA-deficient cells (Chou et al., 2005). Additionally, exogenously expressed HDAC1 suppressed IFNγ-induced endogenous MHC class II mRNA expression without inhibiting CIITA expression in HeLa cells (Zika et al., 2003), indicating that HDAC1 inhibits MHC class II gene transcription by antagonizing CIITA activity rather than reducing its expression. Wild-type HDAC1, but not a HDAC1 deacetylase-defective mutant, suppressed CIITA-induced endogenous MHC class II expression, suggesting that intact deacetylase activity is indeed required for HDAC1 to exert its inhibitory effect on MHC class II expression (Zika et al., 2003). ChIP assays revealed that exogenously expressed HDAC2 inhibited the recruitment of CIITA to the *HLA-DRA* promoter in macrophages (Kong et al., 2009). These observations are in agreement with the notion that HDACs primarily condense chromatin to repress transcription.

However, HDACs cannot be regarded as affecting gene transcription only by modifying histones and chromatin structure, although this may be their predominant function. Both HDAC1 and HDAC2 inhibited CIITA-mediated MHC class II transcription in plasmid-based reporter gene assays (Kong et al., 2009; Zika et al., 2003), indicating that nonhistone effects may also be involved in HDAC1/HDAC2-mediated MHC class II suppression. Exogenously expressed HDAC2 deacetylates CIITA, whereas both TSA treatment and overexpression of HDAC2 interfering RNA increased CIITA acetylation (Kong et al., 2009), thus raising the question of how deacetylation affects CIITA activity. Wild-type HDAC2, but not a deacetylase-deficient mutant, promoted proteasome-dependent CIITA degradation (Kong et al., 2009), suggesting that deacetylation of CIITA affects its stability. Intriguingly, coimmunoprecipitation assays showed that both HDAC1 and HDAC2 interact with CIITA in cultured cells (Kong et al., 2009; Zika et al., 2003). Nonetheless, further studies are required to elucidate the mechanism that regulates the balance between CIITA-associated HATs and HDACs.

Although HDACs are not known to possess intrinsic DNA-binding activity, they can be directed to specific gene promoters through interactions with certain transcription factors, thus allowing the target genes to be specifically regulated by HDACs. The transcriptional repressor YY1 can physically associate with human HDAC1, 2, and 3 (Yang, Yao, Sun, Davie, & Seto, 1997). A consensus YY1-binding element is present in the first exon of the *HLA-DRA* sequence from +62 to 72 bp relative to the transcriptional start site (Hehlgans & Strominger, 1995). Both YY1 and HDAC1 overexpression repressed *HLA-DRA* promoter activation in reporter gene assays only when the plasmid YY1-binding element was intact (Osborne, Zhang, Yang, Seto, & Blanck, 2001), indicating that YY1 could direct HDAC1 to the vicinity of the *HLA-DRA* promoter. Sin3A is transcriptional repressor that can tether HDAC1/HDAC2 activity to the MHC class II gene. HDAC1 and HDAC2 stably associate with the Sin3A corepressor (Ayer, 1999), and this complex can be recruited to specific promoters through interactions with multiple sequence-specific transcription factors (Kuzmichev, Zhang, Erdjument-Bromage, Tempst, & Reinberg, 2002; Zhang & Dufau, 2003). Overexpression of mouse Sin3A completely inhibited CIITA-induced MHC class II transcription in reporter gene assays. Meanwhile, overexpression of NcoR, a corepressor known to preferentially associate with HDAC3, did not affect the same MHC class II promoter plasmid (Zika et al., 2003).

In addition to Class I HDACs, exogenously expressed HDAC4/HDAC5 repressed CIITA-induced endogenous *HLA-DRA* in HeLa cells and CIITA-induced *HLA-DRA* transcription in plasmid-based reporter gene assays (McKinsey et al., 2006), indicating the involvement of Class II HDACs in regulating MHC class II transcription. Coimmunoprecipitation assays using mammalian cells showed that HDAC4/HDAC5 interacts with REXANK, a subunit of the RFX complex that directly binds to the X box in MHC class II promoters (McKinsey et al., 2006). Phosphorylation of HDAC4/HADC5 by Ca^{2+}/calmodulin-dependent protein kinase (CaMK) induced nuclear export of the HDAC4/5–REXANK complex as demonstrated by double immunofluorescence staining (McKinsey et al., 2006). However, whether this nuclear export increases or inhibits MHC class II expression awaits further evaluation.

3.3 CIITA Associates with 19S Proteasomal ATPases

The 26S proteasome degrades a majority of cellular proteins, including transcription regulatory proteins. The 26S proteasome comprises a 20S proteolytic core and two 19S regulatory subunits. The 20S proteolytic core, composed of four stacked ring-structure subunits, is hollow and provides an enclosed cavity to receive, and degrades poly-ubiquitinated proteins. Each end of the 20S core associates with a 19S regulatory subunit that contains multiple ATPase active sites and ubiquitin-binding sites. 19S recognizes poly-ubiquitinated proteins and transfers them to 20S core. The 19S subunit consists of a base subunit composed of six ATPase subunits (S4, S6a,

S6B, S7, Sug1, and S10b) and three non-ATPase subunits (S1, S2, and S5a), and a lid subunit that includes eight non-ATPase subunits. Interactions between 19S ATPase subunits and CIITA have been shown to play an important role in CIITA-mediated MHC class II transcriptional activation via a mechanism that involves chromatin remodeling (Bhat, Truax, Brooks, & Greer, 2010; Bhat et al., 2008; Koues et al., 2008). The 19S ATPase subunit Sug1 also associates with CIITA expressed in macrophages as assayed by coimmunoprecipitation (Bhat et al., 2010, 2008). The interaction of Sug1 with CIITA was mapped to aa 132–301 of the P/S/T domain of CIITA (Bhat et al., 2010). Additionally, the 19S ATPase subunits S6a and S7 also interact with wild-type CIITA but not a CIITA with aa 132–301 deleted (Bhat et al., 2010), suggesting that the 19S ATPase associates with CIITA through the P/S/T domain as a hexameric complex. The CIITA P/S/T domain contains a proteolytic signal site recognized as a degron, which typically labels proteins for degradation. Half-life assays showed that CIITA with a deletion of the 19S ATPase binding domain (aa 132–301) was much more stable than wild-type CIITA (Bhat et al., 2010). These data indicate that the 19S ATPase binding domain is required for normal CIITA protein turnover. Although deletion of the 19S ATPase binding domain stabilizes the CIITA protein, it abolished CIITA-induced cell surface MHC class II expression in HeLa cells (Bhat et al., 2010), indicating that the 19S ATPase binding domain is also essential for CIITA transactivity. In HeLa cells, Sug1 is recruited to the MHC class II promoter 4 h after IFNγ treatment, which is correlated with CIITA recruitment to the promoter, as demonstrated in ChIP assays (Bhat et al., 2008). Sug1 knockdown strongly inhibited CIITA and CBP recruitment to the MHC class II promoter in a manner that could not be reconstituted by proteasome inhibition, indicating that the loss CIITA promoter binding was not due to altered protein stability (Bhat et al., 2008). Moreover, Sug1 interacts with acetylated H3 histone at the MHC class II promoter in IFNγ-treated HeLa cells (Koues et al., 2008). In the absence of Sug1, histone H3 acetylation was dramatically decreased at the MHC class II promoter, with a preferential loss of histone H3 K18 acetylation (Koues et al., 2008). In addition to Sug1 knockdown, knockdown of either S6a or S7 also abolished acetylation of histone H3 at the MHC class II promoter in IFNγ-treated HeLa cells (Koues et al., 2008). These results suggest a role for 19S ATPase subunits in modifying histones to mediate MHC class II transcription. Furthermore, proteasome inhibition by either MG132 or lactacystin did not affect acetylation levels of histone H3 at the proximal MHC class II promoter in IFNγ-treated HeLa cells (Koues et al., 2008), indicating that the function of 19S ATPase subunits in regulating MHC class II expression is independent of its proteolytic function.

3.4 CIITA Associates with SWI/SNF Chromatin Remodeling Protein

The SWI/SNF complex is an evolutionarily conserved multisubunit chromatin remodeling complex that possesses DNA-stimulated ATPase activity, and can destabilize histone–DNA interactions in an ATP-dependent manner (van Holde & Yager, 2003). HATs covalently modify histones and promote relaxation of chromatin structure. SWI/SNF uses ATP hydrolysis to slide nucleosomes along the DNA helix, and in turn expose the DNA to transcription factors, although the nature of this structural alteration is unclear (van Holde & Yager, 2003). HATs and SWI/SNF often work in concert to regulate transcription (Roberts & Orkin, 2004). The mammalian SWI/SNF complex is defined by two ATPase subunits: brahma-related gene 1 (BRG1) and human brahma (hBRM) that are present in various forms of the complex. A growing body of evidence supports a model in which gene-specific transcription factors recruit human SWI/SNF to promoters to facilitate transcription (Decristofaro et al., 2001). For example, BRG1 has been shown to be required for CIITA-mediated MHC class II expression. When introduced into BRG1-deficient SW13 cells, CIITA could not activate cellular MHC class II expression (Mudhasani & Fontes, 2002). Meanwhile, reexpression of wild type but not an ATP-binding-defective BRG1 protein in SW13 cells restored the ability of CIITA to activate MHC class II expression (Mudhasani & Fontes, 2002). Dominant-negative BRG1 abolished the ability of CIITA to activate MHC class II in HeLa cells (Mudhasani & Fontes, 2002). These observations suggest a key role for BRG1 in CIITA-mediated MHC class II transcription. However, CIITA activated an episomal plasmid *HLA-DRA* promoter in BRG1-deficient SW13 cells, indicating that transcription factors and coactivators required for CIITA-mediated MHC class II expression are intact in SW13 cells (Mudhasani & Fontes, 2002). These findings also suggested that an episomal *HLA-DRA* promoter may have a chromatin structure that significantly differs from the chromosomal *HLA-DRA* promoter and does not require BRG1 activity. Furthermore, BRG1 has been shown to associate with CIITA as assayed by coimmunoprecipitation (Mudhasani & Fontes, 2002). In addition to CIITA, BRG1 interacts with RFXAP, a subunit of the DNA-binding RFX complex. RFXAP and CIITA bind to different regions of BRG1 with the CIITA N-terminal AAD and central region both interacting with BRG1

(Mudhasani & Fontes, 2005). The RFXAP–BRG1 interaction coincides with a minimally required fragment of RFXAP that is necessary for MHC class II activation (Mudhasani & Fontes, 2002). The RFXAP–BRG1 interaction may provide an explanation for CIITA-independent chromatin remodeling at the MHC class II promoter. Notably, the MHC class II promoter is fully occupied by transcription factors in CIITA-deficient RJ2.2.5 cells. In contrast, cells lacking one of the RFX subunits show no transcription factor binding to MHC class II promoters (Kara & Glimcher, 1991, 1993). These observations indicate a mechanism by which transcription factors, e.g., RFX, can mediate chromatin remodeling independently of CIITA, at least to some extent.

4. REGULATION OF MHC CLASS II GENE EXPRESSION BY LONG-RANGE CHROMATIN REMODELING

The promoter-proximal S-Y modules are certainly essential for the expression of MHC class II genes. Meanwhile, there is accumulating evidence showing that additional distal regulatory elements are required for optimal MHC class II expression. For instance, a distal inverted S-Y module situated 2.3 kb upstream of the human *HLA-DRA* gene interacts with RFX and CIITA, and is correlated with long-range hyperacetylation of histones H3 and H4 in Raji cells (Masternak et al., 2003). Moreover, deletion of upstream sequences of a MHC class II gene in a transgenic mouse resulted in aberrant MHC class II expression, thus indicating a key role for the upstream sequence. This evidence suggests that MHC class II gene expression is regulated by a complex system that involves long-range regulation.

4.1 Distal S′-Y′ Modules

To explore potential broad-ranging regulatory locus regions in MHC class II genes, Krawczyk et al. used a computational approach to search for novel S-Y-like modules (called S′-Y′) based on the similarity with promoter-proximal S-Y modules. They identified six novel S′-Y′ modules, including four sites dispersed at strategic positions within the MHC class II gene locus (Krawczyk et al., 2004). These S′-Y′ modules had similar features to the promoter-proximal S-Y module with respect to protein interactions, histone acetylation, and ability to activate MHC class II genes. S′-Y′ modules are occupied by a full panel of RFX, CREB, and CIITA proteins in Raji cells but not in BLS cell lines as revealed in ChIP assays (Krawczyk et al., 2004). In Me67.8 melanoma cells, recruitment of RFX and CIITA to S′-Y′ modules can also be induced by IFNγ (Krawczyk et al., 2004). Two S′-Y′

modules activated *HLA-DRA* transcription in a RFX- and CIITA-dependent manner in plasmid-based reported gene assays (Krawczyk et al., 2004), supporting a role for these S′-Y′ modules as long-distance enhancers of MHC class II gene expression. Moreover, strong acetylation of histones H3 and H4 at the S′-Y′ modules was detected in Raji cells but such acetylation was present to a significantly smaller degree in BLS cells (Krawczyk et al., 2004). RFX- and CIITA-dependent histone acetylation is not restricted to the proximal S-Y module or distal S′-Y′ modules, but is present over a broad range upstream of MHC class gene promoters and covers the S′-Y′ modules. This global histone acetylation across the MHC class II locus may be essential for the formation of long-distance chromatin loops that can coordinately regulate MHC class II gene expression (see later).

4.2 Distal X Box-Like Sequences

Using computer analysis, Gomez et al. identified sequences that are homologous to the canonical MHC class II gene X box regulatory elements within the HLA-DR locus, which they termed X box-like (XL) sequences (Gomez, Majumder, Nagarajan, & Boss, 2005). Several XL sequences (referred to as active XL sequences) also bound to RFX/CIITA and RNA polymerase II in Raji cells and could activate *HLA-DRA* transcription in a CIITA-dependent manner in plasmid-based reporter gene assays (Gomez et al., 2005). In addition, strong acetylation of histones H3 and H4 is associated with active XL sequences in Raji cells but not CIITA-deficient B cells. Although a remarkable similarity was observed for histone modifications at the *HLA-DRA* proximal promoter region and active XL sequences, differences in the density of histone modifications that are associated with the assembly of general transcription factors, e.g., histone H3 K9 acetylation, distinguish the very active *HLA-DRA* promoter from the active XL regions (Gomez et al., 2005). Moreover, in epithelial cells treated with IFNγ increased histone acetylation in active XL regions coincident with CIITA expression was also observed (Gomez et al., 2005). Intriguingly, chromatin conformation capture assays revealed an interaction between one XL sequence (upstream HLA-DRA) and the HLA-DRA promoter (Gomez et al., 2005). These data suggest that certain XL sequences associated with acetylated histones contribute to a global increase in chromatin accessibility of *HLA-DRA* in wild-type B cells in response to IFNγ stimulation, and support the involvement of long-range XL-mediated regulation of MHC class II gene expression.

4.3 XL9 Insulator Element

The XL element XL9 is located between the *HLA-DRB1* and *HLA-DQA1* genes (Fig. 2) and displays high levels of histone acetylation but does not bind CIITA and RFX (Majumder, Gomez, & Boss, 2006). In plasmid-based reporter gene assays, XL9 did not activate reporter gene transcription, but instead blocked the activity of the SV40 transcriptional enhancer to activate a reporter gene when the XL9 fragment was inserted between SV40 and a luciferase reporter gene. This finding indicates that XL9 functions as a transcriptional insulator (Majumder et al., 2006). Insulators or enhancer-blocking elements are present in a number of organisms and are thought to localize in the nuclear matrix, where they organize chromatin into higher order looped structures (Dunn, Zhao, & Davie, 2003). These chromatin loops can protect the promoter of one gene from the activity of enhancers for another gene or inhibit enhancer–promoter interactions through association with the nuclear matrix (Dunn et al., 2003). Insulators are often associated with strong histone acetylation and with the CTCF that is important for both insulator function and normal gene expression (Phillips & Corces, 2009). Consistent with these insulator features, ChIP assays and in vitro DNA-binding assays revealed that the XL9 sequence shows a peak of histone acetylation that coincided with sequences that bound the CTCF protein in Raji cells (Majumder, Gomez, Chadwick, & Boss, 2008). XL9 also associates with the nuclear matrix (Majumder et al., 2008). Furthermore, chromatin conformation capture assays showed that long-range chromatin interactions form between XL9 and the proximal promoter region of both *HLA-DRB1* and *HLA-DQA1* in Raji cells, but not in BLS cells (Majumder et al., 2008). Similar long-range chromatin interactions were also induced in A431 epithelial cells upon IFNγ treatment, which was correlated with induced, time-dependent expression of CIITA, HLA-DRB1, and HLA-DQA1 (Majumder et al., 2008). CTCF binding to XL9 is indispensable for the formation of chromatin interactions, as suggested by the effect of CTCF knockdown that abolished crosslinking between XL9 and *HLA-DRB1* or *HLA-DQA1*, and is correlated with histone deacetylation in the proximal promoters of *HLA-DRB1* and *HLA-DQA1* (Majumder et al., 2008). Moreover, CTCF knockdown abolished mRNA expression and cell surface expression of HLA-DRB1 and HLA-DQA1 in Raji cells without significantly affecting CIITA or RFX5 expression levels (Majumder et al., 2008), indicating that the formation of CTCF-dependent long-range chromatin interactions is essential for MHC class II gene expression. CTCF may mediate the formation of chromatin loops between XL9 and neighboring MHC class II genes by interacting with CIITA at the proximal promoter regions, which is supported by the finding that CTCF and CIITA could be coimmunoprecipitated in Raji cells (Majumder et al., 2008). Based on these observations, an active state of MHC class II genes may exist, wherein interaction of promoter-associated CIITA with XL9-associated CTCF facilitates the formation of chromatin loop structures within the MHC class II gene locus. Thus, in the presence of XL9 as a putative insulator, *HLA-DRB1* and *HLA-DQA1* might be simultaneously expressed without mutual interference. Indeed, RNA fluorescence in situ hybridizations confirmed that both genes can be expressed simultaneously from the same chromosome (Majumder et al., 2008).

4.4 CTCF-Binding Sites

Further studies revealed that in Raji cells CTCF activity is required for expression of all MHC class II genes, including *HLA-DRA*, *DRB1*, *DQA1*, *DQB1*, *DPA1*, *DPB1*, *DQA2*, and *DQB2*, as shown by knockdown experiments (Majumder & Boss, 2010). CTCF dependency also extends to MHC class II antigen-processing genes such as *HLA-DOA*, *DOB*, *DMA*, and *DMB*. In contrast, CTCF knockdown did not affect the expression of the peptide transporter genes *TAP1* and *TAP2*, proteasome genes *PSMB8* and *PSMB9*, or glyceraldehyde 3-phosphate dehydrogenase expression (Majumder & Boss, 2010). The finding that *HLA-DRA* knockdown did not affect the expression of *HLA-DRB1* or other MHC class II genes demonstrated that inhibition of one gene does not affect the expression of other MHC class II genes. These data suggest that CTCF dependency is specific for optimal expression of MHC class II genes and accessory genes, and that an array of CTCF-associated insulators may be present at the MHC class II locus. Indeed, in addition to XL9, 16 CTCF-binding sites (termed C1–C16) across the HLA-DR locus are known (Fig. 2), and among these, nine sites bind strongly with CTCF as shown by ChIP assays (Majumder & Boss, 2010). Chromatin conformation capture assays further revealed that these CTCF sites crosslink with proximal promoter regions of MHC class II genes in Raji cells, but not in RJ2.2.5 cells (Majumder & Boss, 2010). The interaction of CTCF with MHC class II gene promoters could also be induced by IFNγ treatment in A431 epithelial cells. CTCF knockdown resulted in loss of such chromatin loops across the MHC class II locus, which correlated with reduced MHC class II gene expression in Raji cells (Majumder & Boss, 2010). Thus, CIITA-bound promoter regions of MHC class II genes can likely crosslink with one or several nearby CTCF sites, and that CTCF may orchestrate MHC class II gene expression by changing the spatial relationships and topological architecture of genes (Majumder & Boss, 2010).

4.5 CIITA Self-Association

The presence of high order chromatin loop structures within the MHC class II locus is also supported by CIITA self-association. In addition to its association with transcription factors, coactivators, and chromatin remodeling factors, CIITA self-associates, possibly in a multimeric self-association complex (Linhoff, Harton, Cressman, Martin, & Ting, 2001; Sisk, Nickerson, Kwok, & Chang, 2003; Sisk, Roys, & Chang, 2001; Tosi, Jabrane-Ferrat, & Peterlin, 2002). Multiple regions within the CIITA protein likely engage in this self-association (Fig. 5), including the N-terminal AAD, a central region that includes a GTP-binding domain (Sisk et al., 2001), phosphorylated residues between aa 253–321 (Tosi et al., 2002) and the N-terminal LRR domain. Site-specific mutations (Linhoff et al., 2001; Sisk et al., 2001) or certain posttranslational modifications (Sisk et al., 2003; Tosi et al., 2002) generated within the putative self-associating regions interfered with CIITA self-association, and intriguingly, impaired its ability to activate endogenous MHC class II expression. Whether the inhibited transactivity directly resulted from the inhibited self-association, or was due to other impairments of intrinsic functions awaits further study. Nevertheless, one model proposes that CIITA self-association could facilitate the formation of chromatin loops between the proximal promoters of MHC class II genes and distal locus control regions, e.g., inverted S-Y modules, S′-Y′ modules, or XL sequences (Ting & Trowsdale, 2002), which have been shown to bind RFX and CIITA (Gomez et al., 2005; Masternak et al., 2003). Thus, disruption of CIITA self-association may prevent the formation of higher order chromatin loop structures that are emerging as a hallmark of a transcriptional "on" state of MHC class II genes.

5. EPIGENETIC REGULATION OF CIITA EXPRESSION

The coincidental expression of CIITA in MHC class II-expressing cells and loss of CIITA expression in MHC class II-negative cells indicates that CIITA is a switch to turn on/off MHC class II expression. Studies of the mechanisms by which CIITA expression are regulated or silenced in

MHC class II-expressing cells (e.g., APCs and IFNγ-stimulated cells) and in MHC class II-deficient cells (e.g., plasma cells and trophoblasts) provided important insights for understanding the regulation of MHC class II gene expression. Histone acetylation and deacetylation have also been shown to be closely associated with the "on" and "off" state of CIITA expression. Transcription of *MHC2TA*, the gene encoding CIITA, is mediated through the activity of four independent promoters: CIITA-PI, PII, PIII, and PIV (LeibundGut-Landmann et al., 2004). These promoters function in a cell- and activation-specific manner. CIITA-PIV is predominantly involved in IFNγ-inducible expression, while CIITA-PI and PIII are used for constitutive expression in dendritic cells and B cells, respectively. CIITA-PIII has an important role in activated human T cells. Meanwhile, the promoter function of CIITA-PII remains unclear. The various *MHC2TA* promoters each transcribe a unique first exon and are located within a ~14 kb region (Muhlethaler-Mottet, Otten, Steimle, & Mach, 1997). Several regulatory elements in *MHC2TA* promoters and interacting transcription factors that are important for transcriptional activation have since been identified. Moreover, histone modifications are also essential for regulating the transcription of all *MHC2TA* promoters.

5.1 CIITA-PIV

CIITA-PIV contains three elements that are required for transcription in response to IFNγ, including a gamma-activated sequence (GAS) element that binds STAT1, an E-box that binds the ubiquitous factor upstream stimulatory factor 1 (USF1), and an interferon-stimulated response element (ISRE) that binds IRF1. In IFNγ-treated epithelial cells, histones H3 and H4 are acetylated at CIITA-PIV, which is paralleled by STAT1/USF1 binding to the GAS/E-box element. This binding precedes that of IRF1 to the ISRE element and the accumulation of CIITA transcripts (Morris, Beresford, Mooney, & Boss, 2002). In contrast, neither transcription factor binding nor histone acetylation occurs at CIITA-PIV in trophoblasts, which are deficient in CIITA and MHC class II expression either constitutively or after exposure to IFNγ (Morris et al., 2002). The absence of MHC class II antigens on trophoblasts is thought to be essential in preventing rejection of the fetus by the maternal immune system. Intriguingly, *MHC2TA* is expressed in human choriocarcinoma cells, and the transcription of plasmid CIITA-PIV is IFNγ-inducible in both human and mouse choriocarcinoma cells (Holtz, Choi, Petroff, Piskurich, & Murphy, 2003). These data suggest that the inability of trophoblasts to express endogenous CIITA in response to IFNγ likely involves epigenetic silencing. DNA methylation of the CIITA-PIV gene does not correlate with CIITA silencing in human and mouse trophoblasts (Holtz et al., 2003), while the HDAC inhibitor TSA restored IFNγ-inducible endogenous CIITA expression in human and mouse trophoblasts (Holtz et al., 2003), indicating that CIITA silencing in these cells may be due to histone deacetylation at chromosomal CIITA-PIV.

The 19S ATPase S6a subunit plays an important role in regulating IFNγ-inducible CIITA expression. S6a associates with CIITA-PIV in response to IFNγ stimulation as demonstrated in ChIP assays (Truax, Koues, Mentel, & Greer, 2010), while S6a knockdown significantly inhibits IFNγ-induced CIITA and MHC class II expression in HeLa cells (Truax et al., 2010). ChIP assays also revealed that S6a knockdown significantly diminished IFNγ-inducible acetylation of histones H3 and H4 at CIITA-PIV with a preferential loss of acetylation at K18 and K8 in histones H3 and H4, respectively, which was correlated with decreased recruitment of STAT1 and IRF1 to the promoter (Truax et al., 2010). These data indicate the importance of 19S ATPase S6a in the assembly of transcription machinery at CIITA-PIV.

The ATP-dependent chromatin remodeling protein BRG1 is also required for IFNγ-inducible CIITA expression. Despite the presence of intact functional STAT1, IRF1, and USF1, IFNγ could not induce CIITA expression in BRG1-deficient SW13 cells (Pattenden, Klose, Karaskov, & Bremner, 2002). Meanwhile, exogenously expressed wild-type BRG1, but not an ATPase-deficient mutant, restored IFNγ-inducible endogenous CIITA expression in SW13 cells (Pattenden et al., 2002). Intriguingly, BRG1 inhibited rather than induced plasmid CIITA-PIV promoter activity (Pattenden et al., 2002), suggesting that BRG1 regulation of CIITA expression requires an appropriate chromosomal environment. ChIP assays revealed that BRG1 is recruited to CIITA-PIV in response to IFNγ, and in vivo DNase I footprinting and restriction enzyme accessibility assays showed that an IFNγ-inducible open chromatin structure at CIITA-PIV requires ATPase-intact BRG1 (Pattenden et al., 2002). Moreover, many interdependent distal enhancers are essential for BRG1-dependent IFNγ-inducible CIITA expression (Ni, Abou El Hassan, Xu, Yu, & Bremner, 2008), which likely involves remote elements as suggested by the finding that deletion of distal elements at −50, −16, −8, +40 or +59 kb (as relative to the start site at CIITA-IV) significantly impaired IFNγ-induced CIITA expression, even in the presence of BRG1 (Ni et al., 2008).

BRG1-dependent distal events, e.g., recruitment of BRG1, STAT1, and IRF1 to chromosomal DNA, as well as acetylation of histones H3 and H4, have been observed at the −50 and −36, −16, −8, +40 and +59 kb regions. DNase I accessibility assays also revealed that the open chromatin structure at distal sites −50 and −8 kb as well as the proximal CIITA-PIV region is BGR1 dependent (Ni et al., 2008). These observations suggest a complex control mechanism of *MHC2TA* transcription and that BGR1 may control *MHC2TA* transcription through long-range chromatin remodeling.

5.2 CIITA-PIII

CIITA-PIII primarily controls developmental and constitutive *MHC2TA* expression in B cells, $CD4^+$ T cells, monocytes, and plasmacytoid dendritic cells. CIITA-PIII expression is regulated by several elements, including site A, site B, activation response elements (ARE)1, ARE2, an Ets/ISRE-consensus element (site C), and an E-box. CREB1 binding to ARE2 and sometimes to the 5′-UTR of CIITA-PIII is critical for PIII transcription. In B cells, the Ets/ISRE-consensus element is bound by transcription factors PU.1 and IRF4, while the basic helix–loop–helix factor E47 binds the E-box. These transcription factors also play an important role in B cell differentiation and activation (van der Stoep, Quinten, & van den Elsen, 2002). MHC class II expression is developmentally regulated in B cells, is expressed early during the pre-B cell stage, increases upon B cell maturation, and is constitutively expressed in mature B cells (Scholl & Geha, 1994). However, when B cells differentiate into plasma cells, both MHC class II and CIITA expression are rapidly lost (Green, Yoon, & Boss, 2006). In contrast to B cells, wherein the CIITA-PIII promoter is occupied by multiple transcription factors, the PIII promoter in plasma cells is not occupied, although all requisite factors are intact (Green et al., 2006).

B lymphocyte-induced maturation protein (Blimp)1, also known as positive regulation domain I-binding factor (PRDI-BF)1, mediates CIITA silencing in plasma cells. Blimp1 was originally identified as a zinc finger-containing protein that can drive the maturation of B cells into immunoglobulin-secreting plasma cells, and Blimp1 expression is also induced during B cell differentiation into plasma cells (Turner, Mack, & Davis, 1994). Exogenously expressed Blimp1 suppressed CIITA mRNA expression and cell surface MHC class II expression in B cells (Piskurich et al., 2000). Blimp1 also inhibited CIITA-PIII and MHC class II transcription in plasmid-based reporter assays (Piskurich et al., 2000).

A Blimp1-binding site was identified in the CIITA-PIII promoter within the Ets/ISRE-consensus element (Piskurich et al., 2000). These data suggest that Blimp1 binding to the Ets/ISRE-consensus element of CIITA-PIII interferes with the binding between PU.1 and IRF4 in plasma cells. PRDI-BF1 also binds the ISRE site at CIITA-PIV to repress IFNγ-inducible CIITA expression (Chen et al., 2007). Moreover, Blimp1 is known to function by binding to DNA and serving as a scaffold to recruit chromatin remodeling factors, including HDACs (Yu, Angelin-Duclos, Greenwood, Liao, & Calame, 2000), histone H3 K9 dimethyltransferase G9a (Gyory, Wu, Fejer, Seto, & Wright, 2004), and histone H3 K4 demethylase LSD1 (Su et al., 2009), all of which are generally correlated with transcription repression. Thus, histone modification may also be involved in Blimp1-mediated CIITA silencing. Consistent with this hypothesis, specific histone modifications associated with accessible chromatin structure were observed at CIITA-PIII in wild-type B cells but not in plasma cells (Green et al., 2006). When primary B cells were differentiated ex vivo, histone H3 and H4 acetylation at CIITA-PIII was lost within 24 h. A reciprocal exchange of histone H3 K9 acetylation to methylation was observed during B cell differentiation to plasma cells (Green et al., 2006). The loss of histone acetylation correlated with an absence of transcription factor binding (Green et al., 2006). Intriguingly, loss of histone H3 and H4 acetylation was also observed at CIITA-PI, PII, and PIII during plasma cell differentiation, indicating that this silencing mechanism has a global effect (Green et al., 2006).

CIITA-III is also important during T cell activation (Holling, van der Stoep, Quinten, & van den Elsen, 2002). In unstimulated T cells, CIITA-PIII has histone modifications that are associated with repressive transcription, e.g., low levels of histone H3 and H4 acetylation, and repressive histone methylation markers (trimethyl (3Me)-K27 of histone H3 and 3Me-K20 of histone H4). These repressive histone markers are replaced by histone modifications that are associated with active transcription, e.g., high levels of histone H3 and H4 acetylation, and active histone methylation markers (3Me-K4-H3) in activated T cells, which are correlated with recruitment of RNA polymerase II to CIITA-PIII. Active histone modifications and recruitment of RNA polymerase II have also been observed at CIITA-PI, PII, and PIV, indicating that global chromatin remodeling across the *MHC2TA* locus occurs during T cell activation (van Eggermond et al., 2011).

In T cell leukemia, MHC class II silencing is often accompanied by CIITA silencing (Holling, Schooten, Langerak, & van den Elsen, 2004).

However, activation of a CIITA-PIII reporter plasmid was not affected in CIITA-deficient T cell leukemia cell lines (e.g., Jurkat, Molt4, CEM, and HSB2) (Holling et al., 2004), indicating that the CIITA silencing may be induced by epigenetic modifications. ChIP assays revealed low levels of histone H3 and H4 acetylation and repressive histone methylation markers at CIITA-PIII in these CIITA-deficient cell lines. In contrast, CIITA-expressing T cell leukemia cell lines (e.g., HH, HUT-78, and Karpas299) display a high level of histone H3 acetylation and 3Me-K4-H3, and a low level of 3Me-K27-H3 and 3Me-K20-H4 (van Eggermond et al., 2011). These data indicate that active chromatin remodeling is critical for CIITA expression in activated T cells and CIITA-expressing leukemia cells.

5.3 CIITA-PI

MHC2TA expression in immature dendritic cells is dependent on CIITA-PI and to a lesser extent PIII (Muhlethaler-Mottet et al., 1997). Transcription mediated by CIITA-PI in immature dendritic cells requires binding of the transcription factors PU.1, Sp1, nuclear factor k B (NF-kB), and IRF8 to corresponding binding sites in the promoter (Smith et al., 2011). Mutational analysis of the PU.1, NF-kB, and IRF8 binding sites demonstrated that these sites were critical for transcriptional activation of CIITA-PI. Mice lacking IRF8 displayed a bare CIITA-PI promoter, which was reoccupied by transcription factors after expression of IRF8 in vitro (Smith et al., 2011), indicating that IRF8 binding is a prerequisite for assembly of transcription factors at the CIITA-PI promoter and for CIITA-PI transcription. Production of CIITA-PI transcripts is induced during monocyte differentiation to immature dendritic cells. In vivo footprint analysis of CIITA-PI in immature dendritic cells revealed promoter occupation by multiple transcription factors. However, these DNA–protein interactions were lost upon dendritic cell maturation in correlation with CIITA and MHC class II silencing (Smith et al., 2011). Changes in the intracellular localization and cell surface expression of MHC class II genes are intimately associated with the function and maturation of dendritic cells (Cella, Engering, Pinet, Pieters, & Lanzavecchia, 1997; Pierre et al., 1997). In immature dendritic cells, MHC class II proteins are retained in intracellular compartments. Upon maturation, MHC class II-peptide complexes are assembled and transported to the cell surface. In contrast to increased surface expression of MHC class II proteins, de novo biosynthesis of MHC class II mRNA is silenced during dendritic cell maturation. This MHC class II silencing is due to a remarkably rapid reduction in CIITA-PI mRNA and protein synthesis, and to a lesser extent that of CIITA-PIII (Landmann et al., 2001). A variety of maturation stimuli (e.g., lipopolysaccharide, tumor necrosis factor α, CD40 ligand, IFNα, or infection with *Salmonella typhimurium* or Sendai virus) can induce CIITA silencing in human monocyte-derived dendritic cells and mouse bone marrow-derived dendritic cells (Landmann et al., 2001). Intriguingly, high amounts of histone H3 and H4 acetylation were found across the entire *MHC2TA* locus in immature dendritic cells, which was lost upon maturation (Landmann et al., 2001).

During differentiation of mouse bone marrow cells into dendritic cells, histone H3 and H4 acetylation at CIITA-PI increased over time. However, culturing of dendritic cells with tumor-conditioned medium (TCM) or interleukin (IL)-10 significantly inhibited histone acetylation at CIITA-PI, and this inhibition correlated with decreased mRNA expression of CIITA and MHC class II genes (Choi, Yu, Yoon, & Bae, 2009). These results provide insight into the observation that in patients with advanced cancer, dendritic cells express a significantly lower level of MHC class II genes than do healthy individuals (Choi et al., 2009). The downregulation of MHC class II gene expression in tumor microenvironments might be attributable to certain inhibitory factors, e.g., IL-10 in the TCM, which can inhibit CIITA expression (Choi et al., 2009). Moreover, Blimp1/PRDI-BF1 has also been shown to be involved in CIITA silencing during dendritic cell maturation. Moreover, Blimp1/PRDI-BF1 mRNA and protein expression is significantly induced during dendritic cell maturation (Smith et al., 2011). Blimp1/PRDI-BF1 abrogated the activity of IRF8-mediated activation of CIITA-PI in plasmid-based reporter gene assays (Smith et al., 2011). During dendritic cell maturation, recruitment of Blimp1/PRDI-BF1 to CIITA-PI, and, to a lesser extent, CIITA-PIII, was paralleled by time-dependent disassociation of transcription factors from the respective promoters (Smith et al., 2011). Blimp1/PRDI-BF1-associated chromatin remodeling factors HDAC2 and G9a also accumulate at CIITA-PI and CIITA-PIII, and this accumulation correlated with time-dependent decreases in histone H3 and H4 acetylation and acquisition of histone H3 K9 dimethylation and heterochromatin protein 1γ (HP1γ) (Smith et al., 2011). Thus, the ability of Blimp1/PRDI-BF1 to silence three type-specific CIITA promoters places it as a central regulator of the MHC class II-antigen presentation pathway.

6. EPIGENETIC IMMUNE ESCAPE BY MYCOBACTERIAL INFECTION

Macrophages are pivotal players in preventing mycobacterial infection, as they harbor the mycobacteria in cellular compartments and presentation of bacterial antigens to $CD4^+$ T cells via MHC class II molecules. Activated $CD4^+$ T cells secrete IFNγ to activate infected macrophages and induce microbicidal function to eliminate intracellular bacteria. However, a percentage of mycobacteria can survive and multiply within macrophages for many years. Indeed, one strategy by which mycobacteria evade immune attack is the blunting of macrophage responses to IFNγ signals. Mycobacteria-infected macrophages appear to be reluctant to upregulate MHC class II gene expression upon IFNγ stimulation, despite the presence of an intact Janus kinase (JAK)/STAT1 signaling pathway (Kincaid & Ernst, 2003; Pai, Convery, Hamilton, Boom, & Harding, 2003; Pennini, Pai, Schultz, Boom, & Harding, 2006; Wang, Curry, Zwilling, & Lafuse, 2005). In phorbol 12-myristate 13-acetate (PMA)-differentiated human THP-1 monocytes, *Mycobacterium tuberculosis* (*M. tuberculosis*) or *Mycobacterium avium* (*M. avium*) infection abolished IFNγ-induced *HLA-DRA* mRNA expression despite normal levels of IRF1 and STAT1 expression (Wang et al., 2005). Intriguingly, both *M. tuberculosis* and *M. avium* significantly inhibited IFNγ-induced histone acetylation in the *HLA-DRA* proximal promoter as shown by ChIP assays (Wang et al., 2005), while HDAC inhibitors restored the ability of mycobacteria-infected cells to express endogenous *HLA-DRA* in response to IFNγ (Wang et al., 2005). Although HDAC1/2 expression was not increased by mycobacterial infection, expression of Sin3A, a corepressor that associates with HDAC1/2, was upregulated in infected cells, and was recruited to the *HLA-DRA* promoter in mycobacteria-infected THP-1 cells upon IFNγ stimulation (Wang et al., 2005). In contrast, IFNγ-induced recruitment of the HAT coactivator CBP to the *HLA-DRA* proximal promoter was abolished by mycobacterial infection (Wang et al., 2005). These observations suggest that mycobacterial infection specifically inhibits IFNγ-inducible *HLA-DRA* expression by a mechanism that involves HDAC compressor complex formation at the *HLA-DRA* promoter, which leads to histone deacetylation and MHC class II silencing.

Moreover, *M. tuberculosis* or its 19 kDa lipoprotein inhibited IFNγ-induced CIITA mRNA expression in RAW264.7 macrophages, and this inhibition was reversible by treatment with the HDAC inhibitor sodium butyrate (Pennini et al., 2006). ChIP assays also revealed that IFNγ-inducible histone H3 and H4 acetylation at CIITA-PIV was abolished by *M. tuberculosis* or its 19 kDa lipoprotein (Pennini et al., 2006). However, the *M. tuberculosis* 19 kDa lipoprotein inhibited neither histone acetylation at CIITA-PIV nor CIITA mRNA expression in macrophages with defects in the Toll-like receptor 2 (TLR2) or mitogen-activated protein kinases (MAPK) pathways (Pennini et al., 2006). These results indicate that *M. tuberculosis* induces histone deacetylation at CIITA-PIV and represses IFNγ-inducible CIITA expression via a TLR2-mediated MAPK signaling pathway.

7. TUMOR IMMUNOSURVEILLANCE ESCAPE AND ANTITUMOR THERAPY WITH HDAC INHIBITORS

As described earlier, constitutive MHC class II expression is often silenced in T-cell malignancies (Holling et al., 2004; van Eggermond et al., 2011), and significantly downregulated in dendritic cells in cancer patients relative to healthy individuals (Choi et al., 2009). Moreover, several nonlymphocytic tumor cell lines do not express MHC class II genes in response to IFNγ stimulation (Kanaseki et al., 2003; Magner et al., 2000; Satoh et al., 2004). This finding suggests that the absence of MHC class II expression in lymphocyte malignancies or in nonlymphocyte tumor cells contributes to their escape from tumor immunosurveillance systems. $CD8^+$ T cells play a major role in targeting MHC class I-restricted tumor antigens. Meanwhile, an increasing number of studies suggest that tumor-specific reactive $CD4^+$ T cells are also important for enhancing the efficiency of tumor immunotherapy (Bennett et al., 1998; Schoenberger, Toes, van der Voort, Offringa, & Melief, 1998). MHC class II gene expression was correlated with decreased invasiveness of human tumors, and generally with a better prognosis (Cabrera, Ruiz-Cabello, & Garrido, 1995). The notion that MHC class II genes also play a role in tumor immunity is strongly supported by a study demonstrating that transfection of MHC class II genes and the costimulatory factor CD80 in MHC class II-negative tumor cells inhibited their tumorigenicity, and elicited immunity to subsequent challenge with wild-type tumor cells, which suggests that this approach may constitute an effective tumor cell vaccine (Pulaski & Ostrand-Rosenberg, 1998).

The loss of constitutive MHC class II expression in T-cell malignancies coincides with CIITA silencing that is associated with transcription-repressive histone markers such as low levels of histone acetylation across the CIITA locus (van Eggermond et al., 2011). In IFNγ-noninducible tumor cell lines, apart from defects in IFNγ receptors and the JAK/STAT signaling pathway, MHC class II silencing is most often attributable to a defect in the expression of IFNγ-inducible CIITA (Chou et al., 2005; Kanaseki et al., 2003; Satoh et al., 2004). Both MHC class II and CIITA silencing are closely related to a "closed" chromatin structure that is characterized by high levels of histone deacetylation. Thus, HDAC inhibitors have attracted much attention as a potential antitumor therapy to restore MHC class II-mediated tumor immunity in MHC class II-negative tumor cells, even though the rationale for current clinical trials involving HDAC inhibitors to treat cancer primarily focuses on their ability to inhibit growth and induce cell differentiation (Marks et al., 2001).

Treatment with TSA clearly restored the expression of IFNγ-inducible CIITA and HLA-DR expression in HLA-DR-negative squamous cell carcinoma (SCC) cell lines, and the acetylation status of histone H3 at the CIITA-PIV and *HLA-DR* promoters was closely associated with their gene expression (Kanaseki et al., 2003). Notably, TSA-restored MHC class II expression is not always accompanied by the induction of CIITA expression. For example, TSA restored *HLA-DR* expression in human neuroblastoma SK-N-MC cells, colon adenocarcinoma Colon 26 cells, B lymphoma RJ2.2.5 cells, and trophoblast SM9-1 cells in the absence of CIITA expression (Chou et al., 2005; Magner et al., 2000), indicating a direct effect of TSA to "open" the chromatin structure of MHC class II genes through a mechanism that does not require CIITA activity. Moreover, reconstitution of MHC class II gene expression in tumor cells does not necessarily enhance tumor immunity, and, in the absence of costimulatory factors, may even promote tumor progression by inducing anergy (Martin, Frelinger, & Ting, 1999). The ability of TSA to enhance the constitutive expression of both MHC class I and class II genes, as well as the costimulatory molecules CD40, CD80, and CD86 in tumor cells (Khan, Magner, & Tomasi, 2004; Magner et al., 2000), makes this molecule an attractive tumor immunotherapy.

TSA-treated tumor cells manifest as an effective tumor cell vaccine in vivo. TSA induces cell surface expression of MHC class II, CD40, CD80, and CD86 in J558 myeloma cells in vitro (Chou et al., 2005; Khan et al., 2004). The characteristics of TSA-treated J558 cells were assessed in antigen presentation assays using whole ovalbumin (ova) protein, an ova-peptide 322–338 specific to MHC class II, and ova-peptide E333A, which is defective in TCR binding, as a negative control. TSA-treated J558 cells presented both whole ova and ova-peptide 322–338 to stimulate significant proliferation of antigen-specific T cells compared to nontreated J558 cells. The presentation of whole ova protein and ova-peptide 322–338 to stimulate $CD4^+$ T cell proliferation by TSA-treated J558 cells was epitope specific, as no T cell activation was seen with E333A (Chou et al., 2005). These results demonstrated that TSA treatment alters tumor cells in a way that makes them effective in presenting MHC class II-restricted peptides and in processing and presenting whole protein antigens. Furthermore, when tumor cells (J558 cells and melanoma B16 cells) treated with TSA in vitro were inoculated into immunocompromised mice, the onset of tumor formation was delayed and tumor growth was reduced. Long-term tumor-specific immunity was also elicited when immunized animals were rechallenged with wild-type tumor cells. Splenic T cells derived from mice immunized with TSA-treated tumor cells lysed untreated tumor cells, whereas those from severe combined immunodeficiency (SCID) mice did not (Khan et al., 2004), suggesting that TSA-treated tumor cells were able to activate T cells in vivo. TSA treatment can convert tumor cells into effective APCs, and recent studies in MHC class II knockout animals suggest that direct antigen presentation by TSA-treated tumor cells was a component in the tumor immunity induced in vivo (Tomasi, Magner, & Khan, 2006). Although further studies are needed to dissect the mechanisms involved in epigenetic tumor vaccination, these studies showed that HDAC inhibitors that can remodel MHC class II and CIITA chromatin could be used to produce effective tumor cell vaccines. Epigenetic tumor vaccination is perhaps the most effective in preventing metastasis and recurrence of cancers following surgery.

8. CONCLUSION

MHC class II molecules play a central role in activating $CD4^+$ T cells and initiating the adaptive immune response. Therefore, MHC class II gene expression is tightly regulated by mechanisms that involve numerous upstream signaling pathways, DNA-specific transcription factors, and a variety of chromatin remodeling machineries. In particular, the discovery of various distal enhancer and insulator elements, as well as broad-ranging higher order chromatin interactions that occur across the MHC class II supergene locus, highlights the complex coordinated control of this multigene family. With this greater understanding of chromatin remodeling-mediated control of MHC class II gene expression, novel therapeutic targets and modalities are also emerging. For example, HDAC inhibitors alone or in combination with other agents for treating various tumors are currently in preclinical or clinical phase I/II trials (Tomasi et al., 2006). Future studies are required to direct the activity of epigenetic agents to specific gene targets while minimizing their global effect.

REFERENCES

Adamski, J., Ma, Z., Nozell, S., & Benveniste, E. N. (2004). 17beta-Estradiol inhibits class II major histocompatibility complex (MHC) expression: Influence on histone modifications and cbp recruitment to the class II MHC promoter. *Molecular Endocrinology*, *18*(8), 1963–1974. http://dx.doi.org/10.1210/me.2004-0098.

Ayer, D. E. (1999). Histone deacetylases: Transcriptional repression with SINers and NuRDs. *Trends in Cell Biology*, *9*(5), 193–198.

Bennett, S. R., Carbone, F. R., Karamalis, F., Flavell, R. A., Miller, J. F., & Heath, W. R. (1998). Help for cytotoxic-T-cell responses is mediated by CD40 signalling. *Nature*, *393*(6684), 478–480. http://dx.doi.org/10.1038/30996.

Beresford, G. W., & Boss, J. M. (2001). CIITA coordinates multiple histone acetylation modifications at the HLA-DRA promoter. *Nature Immunology*, *2*(7), 652–657. http://dx.doi.org/10.1038/89810.

Bhat, K. P., Truax, A. D., Brooks, J. K., & Greer, S. F. (2010). Association of the 19S proteasomal ATPases with the ATPase-binding domain of CIITA is essential for CIITA stability and MHC class II expression. *Immunology and Cell Biology*, *88*(8), 807–816. http://dx.doi.org/10.1038/icb.2010.45.

Bhat, K. P., Turner, J. D., Myers, S. E., Cape, A. D., Ting, J. P., & Greer, S. F. (2008). The 19S proteasome ATPase Sug1 plays a critical role in regulating MHC class II transcription. *Molecular Immunology*, *45*(8), 2214–2224. http://dx.doi.org/10.1016/j.molimm.2007.12.001.

Bradbury, E. M. (2002). Chromatin structure and dynamics: State-of-the-art. *Molecular Cell*, *10*(1), 13–19.

Cabrera, T., Ruiz-Cabello, F., & Garrido, F. (1995). Biological implications of HLA-DR expression in tumours. *Scandinavian Journal of Immunology*, *41*(4), 398–406.

Cella, M., Engering, A., Pinet, V., Pieters, J., & Lanzavecchia, A. (1997). Inflammatory stimuli induce accumulation of MHC class II complexes on dendritic cells. *Nature*, *388*(6644), 782–787. http://dx.doi.org/10.1038/42030.

Chen, H., Gilbert, C. A., Hudson, J. A., Bolick, S. C., Wright, K. L., & Piskurich, J. F. (2007). Positive regulatory domain I-binding factor 1 mediates repression of the MHC class II transactivator (CIITA) type IV promoter. *Molecular Immunology*, *44*(6), 1461–1470. http://dx.doi.org/10.1016/j.molimm.2006.04.026.

Choi, Y. E., Yu, H. N., Yoon, C. H., & Bae, Y. S. (2009). Tumor-mediated down-regulation of MHC class II in DC development is attributable to the epigenetic control of the CIITA type I promoter. *European Journal of Immunology*, *39*(3), 858–868. http://dx.doi.org/10.1002/eji.200838674.

Chou, S. D., Khan, A. N., Magner, W. J., & Tomasi, T. B. (2005). Histone acetylation regulates the cell type specific CIITA promoters, MHC class II expression and antigen presentation in tumor cells. *International Immunology*, *17*(11), 1483–1494. http://dx.doi.org/10.1093/intimm/dxh326.

Chou, S. D., & Tomasi, T. B. (2008). Spatial distribution of histone methylation during MHC class II expression. *Molecular Immunology*, *45*(4), 971–980. http://dx.doi.org/10.1016/j.molimm.2007.07.039.

Cressman, D. E., O'Connor, W. J., Greer, S. F., Zhu, X. S., & Ting, J. P. (2001). Mechanisms of nuclear import and export that control the subcellular localization of class II transactivator. *The Journal of Immunology*, *167*(7), 3626–3634.

Cuddapah, S., Jothi, R., Schones, D. E., Roh, T. Y., Cui, K., & Zhao, K. (2009). Global analysis of the insulator binding protein CTCF in chromatin barrier regions reveals demarcation of active and repressive domains. *Genome Research*, *19*(1), 24–32. http://dx.doi.org/10.1101/gr.082800.108.

Decristofaro, M. F., Betz, B. L., Rorie, C. J., Reisman, D. N., Wang, W., & Weissman, B. E. (2001). Characterization of SWI/SNF protein expression in human breast cancer cell lines and other malignancies. *Journal of Cellular Physiology*, *186*(1), 136–145. http://dx.doi.org/10.1002/1097-4652(200101)186:1<136::AID-JCP1010>3.0.CO;2-4.

Dorn, A., Fehling, H. J., Koch, W., Le Meur, M., Gerlinger, P., Benoist, C., et al. (1988). B-cell control region at the 5′ end of a major histocompatibility complex class II gene: Sequences and factors. *Molecular and Cellular Biology*, *8*(10), 3975–3987.

Dunn, K. L., Zhao, H., & Davie, J. R. (2003). The insulator binding protein CTCF associates with the nuclear matrix. *Experimental Cell Research*, *288*(1), 218–223.

Durand, B., Sperisen, P., Emery, P., Barras, E., Zufferey, M., Mach, B., et al. (1997). RFXAP, a novel subunit of the RFX DNA binding complex is mutated in MHC class II deficiency. *The EMBO Journal*, *16*(5), 1045–1055. http://dx.doi.org/10.1093/emboj/16.5.1045.

Dziembowska, M., Fondaneche, M. C., Vedrenne, J., Barbieri, G., Wiszniewski, W., Picard, C., et al. (2002). Three novel mutations of the CIITA gene in MHC class II-deficient patients with a severe immunodeficiency. *Immunogenetics*, *53*(10–11), 821–829. http://dx.doi.org/10.1007/s00251-001-0395-7.

Fontes, J. D., Kanazawa, S., Jean, D., & Peterlin, B. M. (1999). Interactions between the class II transactivator and CREB binding protein increase transcription of major histocompatibility complex class II genes. *Molecular and Cellular Biology*, *19*(1), 941–947.

Garvie, C. W., & Boss, J. M. (2008). Assembly of the RFX complex on the MHCII promoter: Role of RFXAP and RFXB in relieving autoinhibition of RFX5. *Biochimica et Biophysica Acta*, *1779*(12), 797–804. http://dx.doi.org/10.1016/j.bbagrm.2008.07.012.

Gialitakis, M., Kretsovali, A., Spilianakis, C., Kravariti, L., Mages, J., Hoffmann, R., et al. (2006). Coordinated changes of histone modifications and HDAC mobilization regulate the induction of MHC class II genes by Trichostatin A. *Nucleic Acids Research*, *34*(3), 765–772. http://dx.doi.org/10.1093/nar/gkj462.

Gomez, J. A., Majumder, P., Nagarajan, U. M., & Boss, J. M. (2005). X box-like sequences in the MHC class II region maintain regulatory function. *The Journal of Immunology*, *175*(2), 1030–1040.

Green, M. R., Yoon, H., & Boss, J. M. (2006). Epigenetic regulation during B cell differentiation controls CIITA promoter accessibility. *The Journal of Immunology*, *177*(6), 3865–3873.

Guerra, R. F., Imperadori, L., Mantovani, R., Dunlap, D. D., & Finzi, L. (2007). DNA compaction by the nuclear factor-Y. *Biophysical Journal*, *93*(1), 176–182. http://dx.doi.org/10.1529/biophysj.106.099929.

Gyory, I., Wu, J., Fejer, G., Seto, E., & Wright, K. L. (2004). PRDI-BF1 recruits the histone H3 methyltransferase G9a in transcriptional silencing. *Nature Immunology*, *5*(3), 299–308. http://dx.doi.org/10.1038/ni1046.

Hadnagy, A., Beaulieu, R., & Balicki, D. (2008). Histone tail modifications and noncanonical functions of histones: Perspectives in cancer epigenetics. *Molecular Cancer Therapeutics*, 7(4), 740–748. http://dx.doi.org/10.1158/1535-7163.MCT-07-2284.

Harton, J. A., Zika, E., & Ting, J. P. (2001). The histone acetyltransferase domains of CREB-binding protein (CBP) and p300/CBP-associated factor are not necessary for cooperativity with the class II transactivator. *The Journal of Biological Chemistry*, *276*(42), 38715–38720. http://dx.doi.org/10.1074/jbc.M106652200.

Hehlgans, T., & Strominger, J. L. (1995). Activation of transcription by binding of NF-E1 (YY1) to a newly identified element in the first exon of the human DR alpha gene. *The Journal of Immunology*, *154*(10), 5181–5187.

Holling, T. M., Schooten, E., Langerak, A. W., & van den Elsen, P. J. (2004). Regulation of MHC class II expression in human T-cell malignancies. *Blood*, *103*(4), 1438–1444. http://dx.doi.org/10.1182/blood-2003-05-1491.

Holling, T. M., van der Stoep, N., Quinten, E., & van den Elsen, P. J. (2002). Activated human T cells accomplish MHC class II expression through T cell-specific occupation of class II transactivator promoter III. *The Journal of Immunology*, *168*(2), 763–770.

Holtz, R., Choi, J. C., Petroff, M. G., Piskurich, J. F., & Murphy, S. P. (2003). Class II transactivator (CIITA) promoter methylation does not correlate with silencing of CIITA transcription in trophoblasts. *Biology of Reproduction*, *69*(3), 915–924. http://dx.doi.org/10.1095/biolreprod.103.017103.

Ikenoue, T., Inoki, K., Zhao, B., & Guan, K. L. (2008). PTEN acetylation modulates its interaction with PDZ domain. *Cancer Research*, *68*(17), 6908–6912. http://dx.doi.org/10.1158/0008-5472.CAN-08-1107.

Imai, S., Armstrong, C. M., Kaeberlein, M., & Guarente, L. (2000). Transcriptional silencing and longevity protein Sir2 is an NAD-dependent histone deacetylase. *Nature*, *403*(6771), 795–800. http://dx.doi.org/10.1038/35001622.

Jabrane-Ferrat, N., Fontes, J. D., Boss, J. M., & Peterlin, B. M. (1996). Complex architecture of major histocompatibility complex class II promoters: Reiterated motifs and conserved protein–protein interactions. *Molecular and Cellular Biology*, *16*(9), 4683–4690.

Jabrane-Ferrat, N., Nekrep, N., Tosi, G., Esserman, L. J., & Peterlin, B. M. (2002). Major histocompatibility complex class II transcriptional platform: Assembly of nuclear factor Y and regulatory factor X (RFX) on DNA requires RFX5 dimers. *Molecular and Cellular Biology*, *22*(15), 5616–5625.

Jabrane-Ferrat, N., Nekrep, N., Tosi, G., Esserman, L., & Peterlin, B. M. (2003). MHC class II enhanceosome: How is the class II transactivator recruited to DNA-bound activators? *International Immunology*, *15*(4), 467–475.

Janeway, C. A., Jr., Travers, P., Walport, M., & Shlomchik, M. J. (2001). The major histocompatibility complex and its functions. In *Immunobiology: The immune system in health and disease* (5th ed.). New York: Garland Science.

Jenuwein, T., & Allis, C. D. (2001). Translating the histone code. *Science*, *293*(5532), 1074–1080. http://dx.doi.org/10.1126/science.1063127.

Juan, L. J., Shia, W. J., Chen, M. H., Yang, W. M., Seto, E., Lin, Y. S., et al. (2000). Histone deacetylases specifically down-regulate p53-dependent gene activation. *The Journal of Biological Chemistry*, *275*(27), 20436–20443. http://dx.doi.org/10.1074/jbc.M000202200.

Kanaseki, T., Ikeda, H., Takamura, Y., Toyota, M., Hirohashi, Y., Tokino, T., et al. (2003). Histone deacetylation, but not hypermethylation, modifies class II transactivator and MHC class II gene expression in squamous cell carcinomas. *The Journal of Immunology*, *170*(10), 4980–4985.

Kara, C. J., & Glimcher, L. H. (1991). In vivo footprinting of MHC class II genes: Bare promoters in the bare lymphocyte syndrome. *Science*, *252*(5006), 709–712.

Kara, C. J., & Glimcher, L. H. (1993). Three in vivo promoter phenotypes in MHC class II deficient combined immunodeficiency. *Immunogenetics*, *37*(3), 227–230.

Khan, A. N., Magner, W. J., & Tomasi, T. B. (2004). An epigenetically altered tumor cell vaccine. *Cancer Immunology, Immunotherapy*, *53*(8), 748–754. http://dx.doi.org/10.1007/s00262-004-0513-0.

Kincaid, E. Z., & Ernst, J. D. (2003). Mycobacterium tuberculosis exerts gene-selective inhibition of transcriptional responses to IFN-gamma without inhibiting STAT1 function. *The Journal of Immunology*, *171*(4), 2042–2049.

Kong, X., Fang, M., Li, P., Fang, F., & Xu, Y. (2009). HDAC2 deacetylates class II transactivator and suppresses its activity in macrophages and smooth muscle cells. *Journal of Molecular and Cellular Cardiology*, *46*(3), 292–299. http://dx.doi.org/10.1016/j.yjmcc.2008.10.023.

Koues, O. I., Dudley, R. K., Truax, A. D., Gerhardt, D., Bhat, K. P., McNeal, S., et al. (2008). Regulation of acetylation at the major histocompatibility complex class II proximal promoter by the 19S proteasomal ATPase Sug1. *Molecular and Cellular Biology*, *28*(19), 5837–5850. http://dx.doi.org/10.1128/MCB.00535-08.

Krawczyk, M., Peyraud, N., Rybtsova, N., Masternak, K., Bucher, P., Barras, E., et al. (2004). Long distance control of MHC class II expression by multiple distal enhancers regulated by regulatory factor X complex and CIITA. *The Journal of Immunology*, *173*(10), 6200–6210.

Kretsovali, A., Agalioti, T., Spilianakis, C., Tzortzakaki, E., Merika, M., & Papamatheakis, J. (1998). Involvement of CREB binding protein in expression of major histocompatibility complex class II genes via interaction with the class II transactivator. *Molecular and Cellular Biology*, *18*(11), 6777–6783.

Kuzmichev, A., Zhang, Y., Erdjument-Bromage, H., Tempst, P., & Reinberg, D. (2002). Role of the Sin3-histone deacetylase complex in growth regulation by the candidate tumor suppressor p33(ING1). *Molecular and Cellular Biology*, *22*(3), 835–848.

Landmann, S., Muhlethaler-Mottet, A., Bernasconi, L., Suter, T., Waldburger, J. M., Masternak, K., et al. (2001). Maturation of dendritic cells is accompanied by rapid transcriptional silencing of class II transactivator (CIITA) expression. *The Journal of Experimental Medicine*, *194*(4), 379–391.

LeibundGut-Landmann, S., Waldburger, J. M., Krawczyk, M., Otten, L. A., Suter, T., Fontana, A., et al. (2004). Mini-review: Specificity and expression of CIITA, the master regulator of MHC class II genes. *European Journal of Immunology*, *34*(6), 1513–1525. http://dx.doi.org/10.1002/eji.200424964.

Linhoff, M. W., Harton, J. A., Cressman, D. E., Martin, B. K., & Ting, J. P. (2001). Two distinct domains within CIITA mediate self-association: Involvement of the GTP-binding and leucine-rich repeat domains. *Molecular and Cellular Biology*, *21*(9), 3001–3011. http://dx.doi.org/10.1128/MCB.21.9.3001-3011.2001.

Lochamy, J., Rogers, E. M., & Boss, J. M. (2007). CREB and phospho-CREB interact with RFX5 and CIITA to regulate MHC class II genes. *Molecular Immunology*, *44*(5), 837–847. http://dx.doi.org/10.1016/j.molimm.2006.04.004.

Louis-Plence, P., Moreno, C. S., & Boss, J. M. (1997). Formation of a regulatory factor X/X2 box-binding protein/nuclear factor-Y multiprotein complex on the conserved regulatory regions of HLA class II genes. *The Journal of Immunology*, *159*(8), 3899–3909.

Magner, W. J., Kazim, A. L., Stewart, C., Romano, M. A., Catalano, G., Grande, C., et al. (2000). Activation of MHC class I, II, and CD40 gene expression by histone deacetylase inhibitors. *The Journal of Immunology*, *165*(12), 7017–7024.

Majumder, P., & Boss, J. M. (2010). CTCF controls expression and chromatin architecture of the human major histocompatibility complex class II locus. *Molecular and Cellular Biology*, *30*(17), 4211–4223. http://dx.doi.org/10.1128/MCB.00327-10.

Majumder, P., Gomez, J. A., & Boss, J. M. (2006). The human major histocompatibility complex class II HLA-DRB1 and HLA-DQA1 genes are separated by a CTCF-binding enhancer-blocking element. *The Journal of Biological Chemistry*, *281*(27), 18435–18443. http://dx.doi.org/10.1074/jbc.M601298200.

Majumder, P., Gomez, J. A., Chadwick, B. P., & Boss, J. M. (2008). The insulator factor CTCF controls MHC class II gene expression and is required for the formation of long-distance chromatin interactions. *The Journal of Experimental Medicine*, *205*(4), 785–798. http://dx.doi.org/10.1084/jem.20071843.

Marks, P., Rifkind, R. A., Richon, V. M., Breslow, R., Miller, T., & Kelly, W. K. (2001). Histone deacetylases and cancer: Causes and therapies. *Nature Reviews. Cancer, 1*(3), 194–202. http://dx.doi.org/10.1038/35106079.

Martin, B. K., Frelinger, J. G., & Ting, J. P. (1999). Combination gene therapy with CD86 and the MHC class II transactivator in the control of lung tumor growth. *The Journal of Immunology, 162*(11), 6663–6670.

Masternak, K., Barras, E., Zufferey, M., Conrad, B., Corthals, G., Aebersold, R., et al. (1998). A gene encoding a novel RFX-associated transactivator is mutated in the majority of MHC class II deficiency patients. *Nature Genetics, 20*(3), 273–277. http://dx.doi.org/10.1038/3081.

Masternak, K., Muhlethaler-Mottet, A., Villard, J., Zufferey, M., Steimle, V., & Reith, W. (2000). CIITA is a transcriptional coactivator that is recruited to MHC class II promoters by multiple synergistic interactions with an enhanceosome complex. *Genes & Development, 14*(9), 1156–1166.

Masternak, K., Peyraud, N., Krawczyk, M., Barras, E., & Reith, W. (2003). Chromatin remodeling and extragenic transcription at the MHC class II locus control region. *Nature Immunology, 4*(2), 132–137. http://dx.doi.org/10.1038/ni883.

McKinsey, T. A., Kuwahara, K., Bezprozvannaya, S., & Olson, E. N. (2006). Class II histone deacetylases confer signal responsiveness to the ankyrin-repeat proteins ANKRA2 and RFXANK. *Molecular Biology of the Cell, 17*(1), 438–447. http://dx.doi.org/10.1091/mbc.E05-07-0612.

Moreno, C. S., Beresford, G. W., Louis-Plence, P., Morris, A. C., & Boss, J. M. (1999). CREB regulates MHC class II expression in a CIITA-dependent manner. *Immunity, 10*(2), 143–151.

Mori-Aoki, A., Pietrarelli, M., Nakazato, M., Caturegli, P., Kohn, L. D., & Suzuki, K. (2000). Class II transactivator suppresses transcription of thyroid-specific genes. *Biochemical and Biophysical Research Communications, 278*(1), 58–62. http://dx.doi.org/10.1006/bbrc.2000.3769.

Morris, A. C., Beresford, G. W., Mooney, M. R., & Boss, J. M. (2002). Kinetics of a gamma interferon response: Expression and assembly of CIITA promoter IV and inhibition by methylation. *Molecular and Cellular Biology, 22*(13), 4781–4791.

Mudhasani, R., & Fontes, J. D. (2002). The class II transactivator requires brahma-related gene 1 to activate transcription of major histocompatibility complex class II genes. *Molecular and Cellular Biology, 22*(14), 5019–5026.

Mudhasani, R., & Fontes, J. D. (2005). Multiple interactions between BRG1 and MHC class II promoter binding proteins. *Molecular Immunology, 42*(6), 673–682. http://dx.doi.org/10.1016/j.molimm.2004.09.021.

Muhlethaler-Mottet, A., Krawczyk, M., Masternak, K., Spilianakis, C., Kretsovali, A., Papamatheakis, J., et al. (2004). The S box of major histocompatibility complex class II promoters is a key determinant for recruitment of the transcriptional co-activator CIITA. *The Journal of Biological Chemistry, 279*(39), 40529–40535. http://dx.doi.org/10.1074/jbc.M406585200.

Muhlethaler-Mottet, A., Otten, L. A., Steimle, V., & Mach, B. (1997). Expression of MHC class II molecules in different cellular and functional compartments is controlled by differential usage of multiple promoters of the transactivator CIITA. *The EMBO Journal, 16*(10), 2851–2860. http://dx.doi.org/10.1093/emboj/16.10.2851.

Nagarajan, U. M., Louis-Plence, P., DeSandro, A., Nilsen, R., Bushey, A., & Boss, J. M. (1999). RFX-B is the gene responsible for the most common cause of the bare lymphocyte syndrome, an MHC class II immunodeficiency. *Immunity, 10*(2), 153–162.

Nakagawa, T., Bulger, M., Muramatsu, M., & Ito, T. (2001). Multistep chromatin assembly on supercoiled plasmid DNA by nucleosome assembly protein-1 and ATP-utilizing chromatin assembly and remodeling factor. *The Journal of Biological Chemistry, 276*(29), 27384–27391. http://dx.doi.org/10.1074/jbc.M101331200.

Ni, Z., Abou El Hassan, M., Xu, Z., Yu, T., & Bremner, R. (2008). The chromatin-remodeling enzyme BRG1 coordinates CIITA induction through many interdependent distal enhancers. *Nature Immunology*, *9*(7), 785–793. http://dx.doi.org/10.1038/ni.1619.

Ogryzko, V. V., Schiltz, R. L., Russanova, V., Howard, B. H., & Nakatani, Y. (1996). The transcriptional coactivators p300 and CBP are histone acetyltransferases. *Cell*, *87*(5), 953–959.

Osborne, A., Zhang, H., Yang, W. M., Seto, E., & Blanck, G. (2001). Histone deacetylase activity represses gamma interferon-inducible HLA-DR gene expression following the establishment of a DNase I-hypersensitive chromatin conformation. *Molecular and Cellular Biology*, *21*(19), 6495–6506.

Pai, R. K., Convery, M., Hamilton, T. A., Boom, W. H., & Harding, C. V. (2003). Inhibition of IFN-gamma-induced class II transactivator expression by a 19-kDa lipoprotein from Mycobacterium tuberculosis: A potential mechanism for immune evasion. *The Journal of Immunology*, *171*(1), 175–184.

Pattenden, S. G., Klose, R., Karaskov, E., & Bremner, R. (2002). Interferon-gamma-induced chromatin remodeling at the CIITA locus is BRG1 dependent. *The EMBO Journal*, *21*(8), 1978–1986. http://dx.doi.org/10.1093/emboj/21.8.1978.

Pennini, M. E., Pai, R. K., Schultz, D. C., Boom, W. H., & Harding, C. V. (2006). Mycobacterium tuberculosis 19-kDa lipoprotein inhibits IFN-gamma-induced chromatin remodeling of MHC2TA by TLR2 and MAPK signaling. *The Journal of Immunology*, *176*(7), 4323–4330.

Peterson, C. L., & Laniel, M. A. (2004). Histones and histone modifications. *Current Biology*, *14*(14), R546–R551. http://dx.doi.org/10.1016/j.cub.2004.07.007.

Phillips, J. E., & Corces, V. G. (2009). CTCF: Master weaver of the genome. *Cell*, *137*(7), 1194–1211. http://dx.doi.org/10.1016/j.cell.2009.06.001.

Pierre, P., Turley, S. J., Gatti, E., Hull, M., Meltzer, J., Mirza, A., et al. (1997). Developmental regulation of MHC class II transport in mouse dendritic cells. *Nature*, *388*(6644), 787–792. http://dx.doi.org/10.1038/42039.

Piskurich, J. F., Lin, K. I., Lin, Y., Wang, Y., Ting, J. P., & Calame, K. (2000). BLIMP-I mediates extinction of major histocompatibility class II transactivator expression in plasma cells. *Nature Immunology*, *1*(6), 526–532. http://dx.doi.org/10.1038/82788.

Pulaski, B. A., & Ostrand-Rosenberg, S. (1998). Reduction of established spontaneous mammary carcinoma metastases following immunotherapy with major histocompatibility complex class II and B7.1 cell-based tumor vaccines. *Cancer Research*, *58*(7), 1486–1493.

Raval, A., Howcroft, T. K., Weissman, J. D., Kirshner, S., Zhu, X. S., Yokoyama, K., et al. (2001). Transcriptional coactivator, CIITA, is an acetyltransferase that bypasses a promoter requirement for TAF(II)250. *Molecular Cell*, 7(1), 105–115.

Reith, W., LeibundGut-Landmann, S., & Waldburger, J. M. (2005). Regulation of MHC class II gene expression by the class II transactivator. *Nature Reviews. Immunology*, *5*(10), 793–806. http://dx.doi.org/10.1038/nri1708.

Roberts, C. W., & Orkin, S. H. (2004). The SWI/SNF complex—Chromatin and cancer. *Nature Reviews. Cancer*, *4*(2), 133–142. http://dx.doi.org/10.1038/nrc1273.

Ronchi, A., Bellorini, M., Mongelli, N., & Mantovani, R. (1995). CCAAT-box binding protein NF-Y (CBF, CP1) recognizes the minor groove and distorts DNA. *Nucleic Acids Research*, *23*(22), 4565–4572.

Satoh, A., Toyota, M., Ikeda, H., Morimoto, Y., Akino, K., Mita, H., et al. (2004). Epigenetic inactivation of class II transactivator (CIITA) is associated with the absence of interferon-gamma-induced HLA-DR expression in colorectal and gastric cancer cells. *Oncogene*, *23*(55), 8876–8886. http://dx.doi.org/10.1038/sj.onc.1208144.

Schoenberger, S. P., Toes, R. E., van der Voort, E. I., Offringa, R., & Melief, C. J. (1998). T-cell help for cytotoxic T lymphocytes is mediated by CD40-CD40L interactions. *Nature*, *393*(6684), 480–483. http://dx.doi.org/10.1038/31002.

Scholl, P. R., & Geha, R. S. (1994). MHC class II signaling in B-cell activation. *Immunology Today*, *15*(9), 418–422. http://dx.doi.org/10.1016/0167-5699(94)90271-2.
Sheppard, H. M., Harries, J. C., Hussain, S., Bevan, C., & Heery, D. M. (2001). Analysis of the steroid receptor coactivator 1 (SRC1)-CREB binding protein interaction interface and its importance for the function of SRC1. *Molecular and Cellular Biology*, *21*(1), 39–50. http://dx.doi.org/10.1128/MCB.21.1.39-50.2001.
Shiina, T., Hosomichi, K., Inoko, H., & Kulski, J. K. (2009). The HLA genomic loci map: Expression, interaction, diversity and disease. *Journal of Human Genetics*, *54*(1), 15–39. http://dx.doi.org/10.1038/jhg.2008.5.
Sisk, T. J., Nickerson, K., Kwok, R. P., & Chang, C. H. (2003). Phosphorylation of class II transactivator regulates its interaction ability and transactivation function. *International Immunology*, *15*(10), 1195–1205.
Sisk, T. J., Roys, S., & Chang, C. H. (2001). Self-association of CIITA and its transactivation potential. *Molecular and Cellular Biology*, *21*(15), 4919–4928. http://dx.doi.org/10.1128/MCB.21.15.4919-4928.2001.
Smith, M. A., Wright, G., Wu, J., Tailor, P., Ozato, K., Chen, X., et al. (2011). Positive regulatory domain I (PRDM1) and IRF8/PU.1 counter-regulate MHC class II transactivator (CIITA) expression during dendritic cell maturation. *The Journal of Biological Chemistry*, *286*(10), 7893–7904. http://dx.doi.org/10.1074/jbc.M110.165431.
Spilianakis, C., Papamatheakis, J., & Kretsovali, A. (2000). Acetylation by PCAF enhances CIITA nuclear accumulation and transactivation of major histocompatibility complex class II genes. *Molecular and Cellular Biology*, *20*(22), 8489–8498.
Steimle, V., Durand, B., Barras, E., Zufferey, M., Hadam, M. R., Mach, B., et al. (1995). A novel DNA-binding regulatory factor is mutated in primary MHC class II deficiency (bare lymphocyte syndrome). *Genes & Development*, *9*(9), 1021–1032.
Sterner, D. E., & Berger, S. L. (2000). Acetylation of histones and transcription-related factors. *Microbiology and Molecular Biology Reviews*, *64*(2), 435–459.
Su, S. T., Ying, H. Y., Chiu, Y. K., Lin, F. R., Chen, M. Y., & Lin, K. I. (2009). Involvement of histone demethylase LSD1 in Blimp-1-mediated gene repression during plasma cell differentiation. *Molecular and Cellular Biology*, *29*(6), 1421–1431. http://dx.doi.org/10.1128/MCB.01158-08.
Ting, J. P., & Trowsdale, J. (2002). Genetic control of MHC class II expression. *Cell*, *109*(Suppl.), S21–S33.
Tomasi, T. B., Magner, W. J., & Khan, A. N. (2006). Epigenetic regulation of immune escape genes in cancer. *Cancer Immunology, Immunotherapy*, *55*(10), 1159–1184. http://dx.doi.org/10.1007/s00262-006-0164-4.
Torok, M. S., & Grant, P. A. (2004). Histone acetyltransferase proteins contribute to transcriptional processes at multiple levels. *Advances in Protein Chemistry*, *67*, 181–199. http://dx.doi.org/10.1016/S0065-3233(04)67007-0.
Tosi, G., Jabrane-Ferrat, N., & Peterlin, B. M. (2002). Phosphorylation of CIITA directs its oligomerization, accumulation and increased activity on MHCII promoters. *The EMBO Journal*, *21*(20), 5467–5476.
Truax, A. D., Koues, O. I., Mentel, M. K., & Greer, S. F. (2010). The 19S ATPase S6a (S6′/TBP1) regulates the transcription initiation of class II transactivator. *Journal of Molecular Biology*, *395*(2), 254–269. http://dx.doi.org/10.1016/j.jmb.2009.10.035.
Turner, C. A., Jr., Mack, D. H., & Davis, M. M. (1994). Blimp-1, a novel zinc finger-containing protein that can drive the maturation of B lymphocytes into immunoglobulin-secreting cells. *Cell*, *77*(2), 297–306.
Tzortzakaki, E., Spilianakis, C., Zika, E., Kretsovali, A., & Papamatheakis, J. (2003). Steroid receptor coactivator 1 links the steroid and interferon gamma response pathways. *Molecular Endocrinology*, *17*(12), 2509–2518. http://dx.doi.org/10.1210/me.2002-0439.

van der Stoep, N., Quinten, E., & van den Elsen, P. J. (2002). Transcriptional regulation of the MHC class II trans-activator (CIITA) promoter III: Identification of a novel regulatory region in the 5′-untranslated region and an important role for cAMP-responsive element binding protein 1 and activating transcription factor-1 in CIITA-promoter III transcriptional activation in B lymphocytes. *The Journal of Immunology*, *169*(9), 5061–5071.

van Eggermond, M. C., Boom, D. R., Klous, P., Schooten, E., Marquez, V. E., Wierda, R. J., et al. (2011). Epigenetic regulation of CIITA expression in human T-cells. *Biochemical Pharmacology*, *82*(10), 1430–1437. http://dx.doi.org/10.1016/j.bcp.2011.05.026.

van Holde, K., & Yager, T. (2003). Models for chromatin remodeling: A critical comparison. *Biochemistry and Cell Biology*, *81*(3), 169–172. http://dx.doi.org/10.1139/o03-038.

Wang, Y., Curry, H. M., Zwilling, B. S., & Lafuse, W. P. (2005). Mycobacteria inhibition of IFN-gamma induced HLA-DR gene expression by up-regulating histone deacetylation at the promoter region in human THP-1 monocytic cells. *The Journal of Immunology*, *174*(9), 5687–5694.

Wright, K. L., Chin, K. C., Linhoff, M., Skinner, C., Brown, J. A., Boss, J. M., et al. (1998). CIITA stimulation of transcription factor binding to major histocompatibility complex class II and associated promoters in vivo. *Proceedings of the National Academy of Sciences of the United States of America*, *95*(11), 6267–6272.

Yang, X. J., & Seto, E. (2008). The Rpd3/Hda1 family of lysine deacetylases: From bacteria and yeast to mice and men. *Nature Reviews. Molecular Cell Biology*, *9*(3), 206–218. http://dx.doi.org/10.1038/nrm2346.

Yang, W. M., Yao, Y. L., Sun, J. M., Davie, J. R., & Seto, E. (1997). Isolation and characterization of cDNAs corresponding to an additional member of the human histone deacetylase gene family. *The Journal of Biological Chemistry*, *272*(44), 28001–28007.

Yu, J., Angelin-Duclos, C., Greenwood, J., Liao, J., & Calame, K. (2000). Transcriptional repression by blimp-1 (PRDI-BF1) involves recruitment of histone deacetylase. *Molecular and Cellular Biology*, *20*(7), 2592–2603.

Zhang, Y., & Dufau, M. L. (2003). Dual mechanisms of regulation of transcription of luteinizing hormone receptor gene by nuclear orphan receptors and histone deacetylase complexes. *The Journal of Steroid Biochemistry and Molecular Biology*, *85*(2–5), 401–414.

Zhu, X. S., Linhoff, M. W., Li, G., Chin, K. C., Maity, S. N., & Ting, J. P. (2000). Transcriptional scaffold: CIITA interacts with NF-Y, RFX, and CREB to cause stereospecific regulation of the class II major histocompatibility complex promoter. *Molecular and Cellular Biology*, *20*(16), 6051–6061.

Zika, E., Greer, S. F., Zhu, X. S., & Ting, J. P. (2003). Histone deacetylase 1/mSin3A disrupts gamma interferon-induced CIITA function and major histocompatibility complex class II enhanceosome formation. *Molecular and Cellular Biology*, *23*(9), 3091–3102.

CHAPTER FIVE

Mechanisms of Chromatin Remodeling and Repurposing During Extracellular Translocation

D.S. Pisetsky[1]
Duke University Medical Center, VA Medical Center, Durham, NC, United States
[1]Corresponding author: e-mail address: david.pisetsky@duke.edu

Contents

Abstract

Chromatin is a highly conserved molecular structure that provides genetic information to regulate cell function. Comprised of DNA, histones and interacting proteins, chromatin is inherently dynamic and subject to remodeling. While usually conceptualized as an intranuclear event, remodeling can also involve extracellular movement. Indeed, chromatin can translocate entirely from the inside to the outside of the cell during cell death processes that include apoptosis, necrosis, and NETosis. During these processes, DNA and proteins can undergo other changes impacting on their activity. Thus, during apoptosis, DNA can be cleaved, histones can be posttranslationally modified and a nuclear protein called HMGB1 (high mobility group box 1) can undergo redox changes. Outside the cell, chromatin components can display powerful immunological activities. These activities result from the ability of DNA and RNA, once taken up by immune cells, to activate internal nucleic acid sensors; the likely function of these sensors is to recognize nucleic acids from intracellular infection. Depending on redox state, the prototype alarmin HMGB1 can interact with a variety of immune receptors including Toll-like receptors. As such, extracellular chromatin can stimulate inflammation and drive the pathogenesis of immune-mediated diseases; in experimental models in animals, agents

Advances in Protein Chemistry and Structural Biology, Volume 106
ISSN 1876-1623
http://dx.doi.org/10.1016/bs.apcsb.2016.08.003

that bind chromatin components can block disease. Thus, extracellular chromatin can have far-reaching biological effects involving a form of molecular repurposing.

1. INTRODUCTION

Chromatin is a highly conserved molecular structure that provides essential genetic information for heredity as well as regulation of differentiated cell function. Comprised of DNA, histones as well as a wide variety of interacting proteins, chromatin is inherently dynamic, subject to modification, rearrangement, and remodeling in the confines of the cell nucleus; during cell division, the rearrangement is extreme as chromosomes condense and migrate into daughter cells. The plasticity of chromatin structure is impressive, with epigenetic modification just one of many changes that can occur as the organism develops and responds to its environment (Becker & Workman, 2013).

While chromatin remodeling is usually conceptualized as an intranuclear (and certainly intracellular) event, remodeling can have a broader context and involve extensive and far-reaching molecular movement. Indeed, depending on the cell process, chromatin can translocate entirely from the inside to the outside of the cell. Outside the cell, the functional activities of chromatin can undergo a dramatic transformation since components of chromatin can, when present in the extracellular space, display powerful immunological activities (Pisetsky, 2014).

The immune activities of chromatin have been categorized as a response to "danger" and are a key element of innate immunity. Danger signifies the challenges to homeostasis that result from cell injury or cell death whether caused by infection, physical or chemical trauma or metabolic stress (Kono & Rock, 2008; Venereau, Ceriotti, & Bianchi, 2015). With the breakdown of cells, many intracellular components—both large and small—can enter the extracellular space to serve as signaling elements to drive inflammation and to counteract events challenging the host. Chromatin as well as its components shows such mobility and immune activity.

As these translocation events of chromatin can be regulated and mediated by enzymes, they can be considered a facet of chromatin remodeling and repurposing, albeit in the setting of cell death. This chapter will consider the remodeling processes that affect chromatin during cell death and their immunological consequences.

2. CHROMATIN TRANSLOCATION IN VITRO

The nucleosome is the key structural element of chromatin and is comprised of a length of DNA (approximately 147 base pairs) wrapped around a core octamer of two molecules each of H2A, H2B, H3, and H4 (Cutter & Hayes, 2015). Between the nucleosomes is a stretch of inter-nucleosomal DNA which can vary in length depending on cell type. While chromatin is the basic nuclear structure in living cells, the translocation of chromatin to the extracellular space is a prominent occurrence during cell death and a feature of both apoptosis and necrosis. Dichotomizing death into two forms is oversimplified and, in some extent, outdated, although it has commonly been used to understand the immune response to cell death (Galluzzi et al., 2015).

Despite burgeoning information on the existence of a wide variety of death forms, the dichotomy between apoptosis and necrosis is a useful starting point for considering chromatin translocation as a form of remodeling; the recognition of other death forms is more recent and fewer studies have investigated in detail the fate of chromatin during these forms of cellular demise (Wallach, Kang, Dillon, & Green, 2016).

As currently viewed, apoptosis is a form of regulated or programmed cell death in which the cell undergoes a systematic disassembly; during this process, macromolecular components, via the action of enzymes called caspases, are cleaved, degraded, and rearranged (Kitazumi & Tsukahara, 2011; Prokhorova, Zamaraev, Kopeina, Zhivotovsky, & Lavrik, 2015; Wallach et al., 2016). Apoptotic cells shrink; membrane lipids "flip"; the nucleus collapses and fragments; and chromatin components migrate into structures called blebs. Blebs are bubble-like structures that form on the cell surface both early and late in the apoptotic process; blebbing renders the cell the appearance of a blackberry (Charras, 2008; Charras, Coughlin, Mitchison, & Mahadevan, 2008). While the function of blebbing is not known, the formation of these structures may provide a mechanism to adjust the surface–volume relationship of a cell during apoptotic shrinkage. Blebbing is not simply a passive process secondary to volume loss but rather is controlled by enzymes including the ROCK kinases (Coleman et al., 2001; Sebbagh et al., 2001).

Some of the most striking molecular changes in apoptosis involve chromatin itself. Through the actions of a series of nucleases, DNA undergoes cleavage, with at least some of DNA fragmented in pieces that correspond to lengths of the DNA in the nucleosomes (Kitazumi & Tsukahara, 2011; Prokhorova et al., 2015; Wallach et al., 2016). The resulting DNA in apoptotic cells shows “laddering” of lower molecular-weight species by gel electrophoresis. Only some DNA is laddered since gel electrophoresis reveals other DNA species that are high molecular-weight species despite cleavage from intact chromosomal DNA. The DNA in the ladder occurs in multiples of the length of DNA in the nucleosome (approximately 180 base pairs). DNA molecules with multiples of this size produce the ladder appearance.

In addition to cleavage of DNA during apoptosis, histones may undergo posttranslational modification (Fullgrabe, Hajji, & Joseph, 2010). Table 1 lists these modifications. An additional change in chromatin during apoptosis relates to increased interaction with the HMGB1 (high mobility group box 1) protein. HMGB1 is a nonhistone nuclear protein that can regulate transcription and nuclear architecture. HMGB1 is 215 amino acids long and contains two DNA-binding regions (A box and B box) along with a C-terminal tail that is rich in anionic residues. This structure likely facilitates interaction of HMGB1 with both DNA and histones. HMGB1, however, is subject to extensive posttranslational modification include acetylation and redox changes of key sulfydryl groups. Redox changes are a feature of apoptosis and have immunological consequences as will be discussed (Tang et al., 2016; Yang, Antoine, Andersson, & Tracey, 2013).

Like histone H1, HMGB1 can bind to DNA although the interactions are much weaker than that of the core histones that form the nucleosome. The interactions of HMGB1 with DNA are transient and, in the nucleus in the living cell, HMGB1 is highly mobile as shown by fluorescence loss in photobleaching (FLIP) (Scaffidi, Misteli, & Bianchi, 2002). With apoptosis, however, HMGB1 becomes more tightly adherent to chromatin and shows a dramatic decrease in intranuclear mobility. As such, HMGB1 may remain attached to chromatin during apoptosis whether chromatin remains in the

Table 1 Histone Modifications Associated with Apoptosis

Phosphorylation
Acetylation
Methylation
Ubiquitination

cell or leaves. The changes in the interaction of HMGB1 with chromatin likely involve posttranslational modification of proteins.

Another striking change in chromatin involves the migration of nuclear molecules into the blebs (Casciola-Rosen, Anhalt, & Rosen, 1994; Halicka, Bedner, & Darzynkiewicz, 2000; Wickman et al., 2013; Zirngibl et al., 2015). This process was originally discovered as part of experiments to explain the immunogenicity of nuclear molecules during the course of systemic lupus erythematosus (SLE). SLE is a prototypic autoimmune disease characterized by the production of antibodies to the cell nucleus or antinuclear antibodies (ANA). While ANAs can occur in many autoimmune or rheumatic diseases, the expression of antibodies to DNA is unique to SLE; antibodies to histones and nucleosomal structures are highly associated with antibodies to DNA as well although only anti-DNA antibodies are considered as criteria for the classification of patients with this disease (Tsokos, 2011). A possible explanation for the immunogenicity of nucleosomal molecules is their location in blebs at or near the cell surface in a form that can induce immune responses.

The changes in chromatin during apoptosis, rather than inflicting a death blow to cell, may have evolved to make the cell immunologically less active and more "safe." The attenuation of immune activity of a dead and dying cell may occur because key immune components are destroyed or modified (Galluzzi, Lopez-Soto, Kumar, & Kroemer, 2016; Martin, Henry, & Cullen, 2012). Oxidation of HMGB1 is an example of this situation (Antoine, Williams, Kipar, Laverty, & Park, 2010; Yang et al., 2012). Alternatively, with the cellular changes during apoptosis, a dead and dying cell and its remnants may be presented in a more "appetizing" form to facilitate phagocytosis. Since DNA and other nuclear components have important immune activities (see below), these remodeling events may have an evolutionary origin in immunity.

In contrast to apoptosis, necrosis, as classically defined, is a form of random cell death mediated by physical or chemical trauma. Unlike apoptosis, necrosis leads to neither shrinkage nor blebbing. Rather, a cell dying by necrosis swells because of osmotic disequilibrium and most likely bursts in the fashion of an overinflated balloon. The end result is a spillage of internal contents, including chromatin, which can then enter the extracellular space. In this circumstance, the chromatin contents have not undergone any of the changes that occur during apoptosis. Molecular changes arising during different death forms can be used to distinguish chromatin arising from different cell death processes. Thus, DNA arising during necrosis may have higher

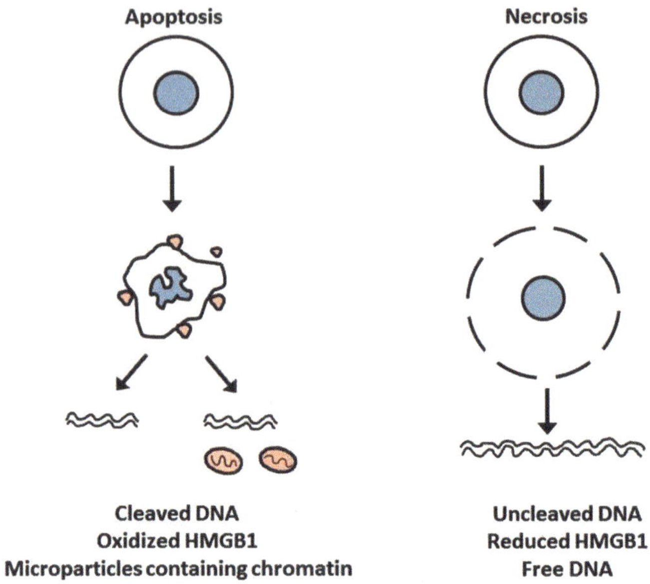

Fig. 1 Chromatin translocation in cell death.

molecular weight than that during apoptosis while apoptotic material may be characterized by histone acetylation and HMGB1 oxidation (Antoine et al., 2010; Jahr et al., 2001; van Bavel, Dieker, Tamboer, van der Vlag, & Berden, 2010; Yang et al., 2012). Fig. 1 illustrates chromatin translocation in apoptosis and necrosis.

Three other forms of death have been recently defined, adding to the mechanisms for cellular demise that can lead to chromatin modification and release. Necroptosis is a form of regulated cell death that involves a set of enzymes known as the RIP kinases. These enzymes orchestrate the death process and can determine the outcome of certain forms of cell stimulation that can variously lead to activation, apoptosis, or necroptosis (Kaczmarek, Vandenabeele, & Krysko, 2013; Linkermann & Green, 2014). The differentiation between these pathways and establishment of necroptosis can be made on the basis of the effects of chemical agents known as the necrostatins which inhibit RIP1 kinase (Degterev et al., 2005). It is likely that many forms of in vivo cell death considered to be necrosis are in fact necroptosis. Thus, the death induced by TNF (tumor necrosis factor) is likely necroptosis and not necrosis.

Among forms of cell death involving profound chromatin remodeling, NETosis is primarily a process of neutrophils although other myeloid cell populations can undergo this cellular program (Brinkmann et al., 2004; Brinkmann & Zychlinsky, 2012; Gupta & Kaplan, 2016). When stimulated by certain agents, neutrophils show a dramatic cellular response in which the nuclear membrane breaks down and the nuclear contents mix with those of granules to produce an elaborate structure called a NET (neutrophil extracellular trap). In the NET, which can include both nuclear and mitochondrial DNA, the DNA is decorated with antibacterial proteins (e.g., myeloperoxidase) from the granules, providing a structure to physically trap and kill bacteria and other microorganisms.

With NETosis, the high molecular-weight DNA matrix is released into the extracellular space with a process requiring reactive oxygen species and protein citrullination. The release of NETs can be demonstrated by confocal microscopy showing high molecular DNA in close association with granule proteins. The occurrence of NETosis can also be assessed from analysis of nucleosomes in the blood especially in settings of inflammation where numbers of neutrophils are high. While other death forms can lead to extracellular nucleosomes, assay of cell-free DNA and nucleosomes provides a convenient way to assess the occurrence of NETosis even if this assay is not specific (Sur Chowdhury et al., 2014).

Pyroptosis is a form of cell death that can occur during inflammasome activation (Bergsbaken, Fink, & Cookson, 2009; Sharma & Kanneganti, 2016). This activation usually requires two steps: priming by an agent such as LPS and then stimulation by ATP. While cells undergoing pyroptosis can release ctyokines such as IL-1 and IL-18, they will die. Since cell death can terminate replication of an infecting organism as well as reducing the number of cells to infect, pyroptosis can represent a strategy of host defense. Importantly, in this context, pyroptosis can lead to the release of HMGB1 and likely other chromatin components (Nystrom et al., 2013). In this case, the HMGB1 can display a partially reduced structure and therefore can drive inflammation via TLR 4.

3. RELEASE OF MICROPARTICLES

During the processes of cell death, changes in chromatin structure can involve modification of protein and nucleic acid structures; translocation into blebs; and extracellular translocation (Table 2). The last two processes are in fact related since at least some of the chromatin released during cell death is in the form of microparticles; at least some microparticles can derive from blebs and, during apoptosis, chromatin components relocate to blebs.

Table 2 Chromatin Modification and Remodeling During Apoptosis

DNA cleavage
Protein cleavage
Posttranslational modification of histones
Redox changes of HMGB1
Migration into blebs
Extracellular translocation

As currently defined, microparticles are small membrane-bound vesicles that are released from cells undergoing apoptosis as well as some cell types undergoing activation (Gyorgy et al., 2011; Pisetsky, Ullal, Gauley, & Ning, 2012). Microparticles can also be called microvesicles and are part of a spectrum of extracellular vesicles that are released from cells. Microparticles are usually considered to be 0.1–1.0 μm in diameter as defined by techniques such as flow cytometry. Since MPs emerge from the cell surface, they bear markers from their cell of origin, allowing their identification, and quantification. Among cellular constituents, MPs are an important source of extracellular nuclear molecules, including DNA, histones, and HMGB1 (Reich & Pisetsky, 2009; Schiller et al., 2008; Ullal et al., 2011). These molecules are accessible to antibody binding either because of their presence on the particle surface or the permeability of the particle that allows penetration of a protein the size of an antibody.

The presence of nuclear molecules in particles can be identified by flow cytometry using antibodies to specific molecules such as DNA, histones, or HMGB1. Since particles are small (approximately 1/100th to 1/10th the diameter of a cell), the concentration of constituent molecules can be low, making quantitation by flow cytometry technically challenging. If flow cytometry is not sufficiently sensitive, Western blotting can be used for protein identification. To assess the presence of DNA in particles, binding of dyes such as propidium iodide or SYTO13 can be used since the fluorescence yield of these dyes is high (Ullal, Pisetsky, & Reich, 2010).

Since particles contain a variety of chromatin components, it appears that at least certain features of this structure are retained during the translocation process. These components, however, could be altered by posttranslational modification or cleavage.

4. IN VITRO SYSTEMS FOR CHROMATIN TRANSLOCATION

The translocation of chromatin components occurs readily in in vitro systems in which primary cells or cell lines are induced to undergo cell death. Thus far, this translocation process has been studied most intensively for apoptosis, necrosis, and NETosis. Several approaches are available to assess the presence of chromatin components in the extracellular space. These techniques include biochemical assay of DNA by DNA-binding dyes; immunochemical assays of chromatin components by ELISA or Western blotting; flow cytometry of microparticles using either biochemical or immunochemical assays for chromatin components; or microscopic assay of cell culture supernatants using biochemical and immunochemical approaches to demonstrate the presence of NETs by staining for DNA and an attached protein such as myeloperoxidase.

In my laboratory, we have used biochemical approaches to demonstrate the generation of cell-free DNA and HMGB1 during apoptosis. Using Jurkat and U937 cell lines as models, these studies demonstrated a time-dependent process leading to an increase in the quantity of DNA in the tissue culture fluids as demonstrated by the dye PicoGreen; this release occurred in concert with the increase in the number of apoptotic cells as demonstrated by flow cytometry using annexin V and propidium iodide staining as measures of apoptosis (Choi, Reich, & Pisetsky, 2004). This release process could be blocked by caspase inhibitors, indicating a close relationship to the death process. As shown by gel electrophoresis of DNA isolated from tissue culture fluid, the DNA showed cleavage and laddering. While these studies did not assay the presence of histones, studies by other investigators indicate that these molecules also translocate into the extracellular space as death proceeds. With sufficient time of culture, cells can transition to a state called secondary necrosis marked by permeability changes that may allow the degree of nuclear and cytoplasmic constituents.

Given the extensive movement of chromatin during apoptosis and the change in the adherence of HMGB1 to this structure, a similar translocation of HMGB1 would be expected. The original studies on HMGB1 movement during cell death, however, indicated that, during apoptosis, HMGB1 remains in the cell nucleus; the retention in the nucleus presumably reflected a more firm attachment of HMGB1 to chromatin by posttranslational modification (Scaffidi et al., 2002). In contrast, with necrosis, a large amount of

HMGB1 emanated from cells as shown by Western blotting. In comparison to the amount of HMGB1 released during necrosis, the amount of HMGB1 released during apoptosis during these experiments thus seemed very limited at best.

In contrast to studies showing scant HMGB1 release during apoptosis, the studies by Bell et al. showed that this protein appeared in the extracellular space in significant quantities during apoptosis, albeit at a stage that could be called late apoptosis (Bell, Jiang, Reich, & Pisetsky, 2006). This result would be expected if HMGB1 became more tightly bound to chromatin during apoptosis, with chromatin translocation pulling HMGB1 along. The differences between various studies on HMGB1 translocation may reflect the different systems used to study this process in terms of cell lines and inducing agents. Another source of difference relates to the comparison in the amount of HMGB1 released during apoptosis and necrosis. If the amount of HMGB1 released during certain forms of necrosis is high, the amount released during apoptosis may appear low. Other studies have demonstrated the release of HMGB1 during apoptosis and the change in redox status (Kazama et al., 2008; Urbonaviciute et al., 2009).

Beyer et al. therefore performed studies to investigate HMGB1 release during necrosis, assessing a number of interventions that have been used in the literature as a model for this death form (Beyer et al., 2012). These interventions included freeze–thaw, heat, ethanol, and H_2O_2. In these experiments, HMGB1 release was measured by Western blotting; for comparison with the extent of HMGB1 release, the amount of DNA released was assessed by PicoGreen. As these studies demonstrated, the amount of HMGB1 released during necrosis varied markedly, with release greatest with freeze–thaw. In comparison, the amount of release with ethanol was very limited. The experiments with DNA showed a similar pattern although, in the case of freeze–thaw, extensive nuclease digestion reduced the amount of DNA present.

Together, these studies demonstrate that translocation of chromatin is a feature of both apoptotic and necrotic death although the amounts of this material released are highly dependent on the systems studied and the manner in which necrosis is modeled. While necrosis may be an unregulated process consequent to physical or chemical trauma, nevertheless, the kinetics of this type of death can vary as well as the type and extent of chromatin translocation. The use of extreme measures such as freeze–thaw, while leading to rapid and extensive cell death, may cause cellular disruption and destruction that exceeds those ordinarily observed even in the pathophysiologic settings. Nevertheless, since treatments such as freezing and hyperthermia have been used to treat cancer, such release of chromatin components may occur in patients and lead to systemic toxicity (Seifert et al., 1999).

A related setting for release of HMGB1 involves immune activation by agents such as the Toll-like receptors LPS and poly (I:C). As shown in seminal experiments on the role of HMGB1 as a late mediator of endotoxin, stimulation of TLR4 leads to HMGB1 translocation (Bonaldi et al., 2003). During this process, HMGB1 undergoes acetylation of nuclear localization signals, leading to migration of the protein into the cytoplasm with eventual release by a nonconventional secretory mechanism. These findings are important since they demonstrate a unique form of chromatin remodeling that may affect only one component of chromatin. They also indicate that posttranslational modification can alter the intracellular location of a chromatin component.

While TLR stimulation can lead to immune activation, it can also lead to cell death by processes that include pyroptosis. Thus, Jiang et al. showed that in cultures of the RAW264.7 macrophage cell line, LPS as well as poly (I:C) lead to extensive cell death along with HMGB1 translocation (Jiang, Bell, & Pisetsky, 2007; Jiang & Pisetsky, 2006). Importantly, the use of caspase inhibitors increased not decreased the amount of HMGB1 released. Since the actual pattern of death induced by agents such as LPS or cytokines such as TNF-α can vary depending on the state of the cells, these findings suggest that, in macrophages stimulated by TLR ligands, caspase inhibitors can induce an alternative death form (likely necroptosis) which displays even higher amounts of HMGB1 release than LPS or poly (I:C) treatment alone. Thus, in this setting of immune stimulation, HMGB1 translocation could result from either activation or death (Bonaldi et al., 2003; Jiang et al., 2007; Jiang & Pisetsky, 2006; Nystrom et al., 2013).

While cell death can be modeled with a cell line, in vivo cell death occurs in the setting of a tissue in which other cell types can be present. Among these cells are macrophages which can influence the extent of translocation via phagocytosis of dead and dying cells. To elucidate the potential effects of macrophages on the dynamics of extracellular chromatin express, Choi et al. studied the effects of culture of the RAW264.7 and J774 macrophage cell lines on the release of DNA from apoptotic and necrotic Jurkat cells (Choi, Reich, & Pisetsky, 2005). These results demonstrated that the extent of release depends on the ratio of phagocytic cells and dead and dying cells. Furthermore, the extent of release varied depending on whether the cells were apoptotic or necrotic. These findings suggest caution in the interpretation of levels of chromatin components in the blood or an extracellular fluid to infer the number of cells that have died.

In these model systems, the kinetics of chromatin release and microparticle release were very similar. Furthermore, in the stimulation of RAW264.7 cells with TLR ligands, the effects of caspase inhibitors were also similar, with caspase inhibitors leading to a marked increase in the number of particles produced. Further studies demonstrated that the particles released in culture contained both DNA as well as HMGB1, perhaps indicating the translocation of these molecules goes through blebs with subsequent release as particles (Gauley & Pisetsky, 2010; Spencer, Gauley, & Pisetsky, 2014; Spencer, Mobarrez, Wallen, & Pisetsky, 2014). A similar situation likely pertains in vivo as data suggest that extracellular chromatin component exists in both particulate and nonparticulate form.

The process of chromatin translocation is a distinguishing feature of NETosis although the end result is vastly different from that in apoptosis. Whereas chromatin is cleaved and rearranged during apoptosis, in NETosis, the chromatin remains intact. Thus, the presence of high molecular-weight DNA is essential to the NET structure, creating a web or mesh to entrap bacteria and other microorganisms (Brinkmann et al., 2004; Brinkmann & Zychlinsky, 2012; Gupta & Kaplan, 2016). In in vitro studies, NETs can be identified as extended strands which can be visualized by a DNA dye, with attached granule proteins demonstrated by immunochemical staining. As DNA and histones both have antibacterial activity, this structure provides a dramatic example of the repurposing of chromatin via remodeling in the extracellular space. In this setting, the chromatin acquires new components from cytoplasmic granules indicating the plasticity of this structure and its ability to bind proteins from outside the nucleus as a form of remodeling.

5. IN VIVO SYSTEMS FOR CHROMATIN TRANSLOCATION

In vivo systems to model chromatin movement during cell death are of two kinds. The first involves administration of an agent that can cause a significant amount of cell death, usually involving a specific tissue or organ. The liver provides an apt model for this purpose in view of the availability of a number of well-defined chemical agents that are hepatotoxic (Tran, Groben, & Pisetsky, 2008). Because the toxicity of these compounds is relatively specific for the liver, it is possible to assess more precisely the exit of chromatin release from the cell without potential complications that occur, for example, because of circulatory collapse or ischemia.

Drugs can be particularly effective in inducing hepatotoxicity, a critical feature underlying drug-induced liver injury. As shown most clearly with acetaminophen, liver cell death is marked by extensive release into the blood of DNA, histones and HMGB1. This release occurs coincident with the release of other specific markers of liver cell death such as ALT/AST. Furthermore, while acetaminophen death has been considered primarily necrosis, blood of animals with hepatotoxicity from high doses of acetaminophen also shows increased caspase levels, indicating the presence of apoptosis (Tran et al., 2008). Carbon tetrachloride can also induce hepatic injury along with the release of DNA into the blood.

In acetaminophen toxicity in both experimental models in animals as well as patients, analysis of HMGB1 isoforms can provide insights into the mechanisms underlying the chromatin translocation and remodeling of this model. As these studies demonstrate, HMGB1 release occurs in two phases as reflected in the pattern of posttranslational modifications. Thus, the initial phase of HMGB1 expression is characterized by the reduced isoform which is the form released from necrotic cells. In contrast, the second phase of HMGB1 release shows acetylation indicative of production by inflammatory cells. This pattern suggests that, following cell death from acetaminophen, an inflammatory response occurs, perhaps stimulated by products released from dead and dying cells. The products can drive inflammation and induce the production of an acetylated isoform (Antoine et al., 2013, 2009).

A related model for hepatotoxicity involves the use of a monoclonal antibody to the Fas molecule (Tran et al., 2008). Engagement of Fas induces profound and rapid apoptosis and striking amounts of liver cell death. With Fas engagement, liver cells undergo apoptosis and release into the blood large quantities of DNA and nucleosomes which can be assessed by biochemical assay of DNA or immunochemical assay of nucleosomes. Coincident with the release of DNA, large amounts of caspase 3 appear in the blood. These levels are much higher than those during acetaminophen toxicity and are indicative of an apoptotic mode of death. This release of DNA induced by anti-Fas is dose dependent with respect to the amounts of antibody administered. With the high doses, animals are sick and die and, while liver cell death occurs, involvement of other cell types is possible (Tran et al., 2008).

In contrasts to models of in situ cell death, administration of dead and dying cells allow the opportunity to track chromatin release in an animal that is otherwise intact and does not have the complicating factors that can result from systemic drug toxicity or the unknown effects of a drug on the clearance of dead and dying cells. A model developed by Jiang et al. involved administration of Jurkat T cells to female mice (Jiang, Reich, & Pisetsky, 2003). Since Jurkat cells have been derived from a human male with leukemia, the presence of a Y chromosome allows unambiguous identification and tracking of chromatin from dead and dying cells.

Studies with administration of Jurkat cells indicated that chromatin from both apoptotic and necrotic cells can appear in the blood in a time- and dose-dependent manner with polymerase chain reaction (PCR) analysis indicating that the DNA in the blood derives from the Jurkat cell. Interestingly, these studies indicated that, in both instances, the DNA in the blood showed laddering even though laddering is a feature of apoptosis and necrosis. These findings indicate that, even if a cell dies by necrosis, its DNA is nevertheless subject to nuclease degradation. This degradation can occur in the blood or possibly following uptake by phagocytic cells. Because of the cellular stress induced by uptake of a large burden of dead and dying material, attempts at phagocytosis of this material may lead to apoptosis of the phagocytic cell. Evidence suggesting this possibility came from experiments demonstrating the presence of murine as well as human DNA in the blood of mice receiving the Jurkat cells; the presence of murine DNA is indicative of some cell death consequent to the administration of a large number of dead cells (Jiang et al., 2003).

Further studies with this model showed that macrophages are required for a blood DNA response since elimination of macrophages by clodronate treatment prevented the expression of blood DNA following administration of dead cells. In the absence of macrophages, the blood levels did not rise suggesting that these cells mediated a process that rapidly translocates chromatin from the dead and dying cells into the blood (Jiang et al., 2003). Other studies were consistent with an active process since glucocorticoid treatment could also blunt the blood DNA response (Jiang & Pisetsky, 2004). These studies did not address, however, the translocation of histones or HMGB1 nor any changes in the HMGB1 isoform.

An interesting set of observations concerned the impact of sex on the process of chromatin translocation. Thus, male and female animals showed important differences in the clearance of dead and dying cells as reflected in the levels of blood DNA; levels of DNA in the blood of female mice receiving Jurkat cells were much higher than those of male mice, with castration of female mice leading to a pattern of blood DNA expression similar to that of males (Pisetsky & Jiang, 2006). While the basis of these differences is not known, a sex influence on chromatin translocation could be relevant to the setting of autoimmunity where high levels of extracellular nuclear molecules may be important in driving autoantibody production and immune complex formation. A male–female difference could also be relevant in the use of blood DNA as a biomarker in a wide variety of conditions.

Other models of this kind are directed to understanding the generation of extracellular DNA in the setting of malignancy. For this purpose, tumor cells can be transferred into nude mice for growth in a subcutaneous location (Gorges et al., 2012; Thierry et al., 2010). Since the DNA from the tumor can have mutations, the DNA in the blood of the recipient mice can be identified by PCR of cell-free DNA. DNA integrity can also be assessed by PCR using primers predicted to produce DNA of varying lengths. These models involve much more gradual generation of extracellular DNA than the bolus administration of Jurkat cells. Furthermore, since nude mice have an incomplete immune system, the process of DNA release may be affected.

6. IMMUNE ACTIVITIES OF CHROMATIN COMPONENTS

Chromatin remodeling as reflected in extracellular translocation also underlies chromatin repurposing since, as shown in in vitro and in vivo model systems, chromatin components can have intrinsic immunological activity once they have been released into the extracellular space in a cell-free form. These activities are part of the response to danger and can be classified using a number of different terms, including DAMP (damage-associated molecular pattern) and alarmin. A DAMP is a molecule (or structure) which can drive inflammation when released from an injured, dead or dying cell. DAMP release can occur with direct cell injury, whatever the origin, providing signaling elements to stimulate inflammation and recruit responding immune cells. DAMPs can be large or small molecules (Venereau et al., 2015).

A related term is an alarmin. As the name implies, an alarmin can alarm the immune system to meet danger arising from infection or injury. To an extent, an alarmin is a DAMP although the term is sometimes applied to those DAMPs that can induce chemotaxis or serve as adjuvants to promote specific immune responses. The nuclear protein HMGB1 is the classic alarmin. While the immune activity of HMGB1 was originally defined in the setting of LPS stimulation of immune cells, as described earlier, HMGB1 can also be released by dead and dying cells (Magna & Pisetsky, 2016).

Each of the components of chromatin has immune activity that allows its designation as a DAMP or alarmin. To define the immune activity of chromatin components, studies in general have used isolated purified molecules in model systems in vitro or in vivo. While these components each demonstrate activity alone, they may function as multicomponent complexes; inside the cell, these molecules are bound together and, with cell death or cell injury, these interactions may persist leading to the release of chromatin components as an ensemble in a higher order structure (Pisetsky, 2011). Furthermore, studies on the expression of different chromatin components in blood have indicated elevation of DNA, histones and HMGB1 in the same settings such as malignancy, infection or inflammatory disease (Abrams et al., 2013; Alhamdi et al., 2015; Chen, Kang, Fan, & Tang, 2014). The concurrent expression of these chromatin components suggests physical association although this possibility has not been explored.

Three chromatin components have important immune activity that can impact on events following cell death and extracellular release. For DNA, the situation is complicated because of the role of CpG (cytosine guanosine) motifs since unmethylated CpG motifs occur much more commonly in prokaryotic than eukaryotic DNA. Furthermore, the receptors involved in immune activation by DNA in cells such as macrophages, dendritic cells and B cells are on the inside of cells; inside the cell, the function of these receptors is likely to recognize either DNA from infecting organisms or damaged DNA that may have emerged from dysfunctional mitochondria or a failure of the nuclease systems that keep high concentrations of DNA out of the cytoplasm (Kawasaki, Kawai, & Akira, 2011; Miyake, Shibata, Ohto, & Shimizu, 2016; Roers, Hiller, & Hornung, 2016).

For DNA to stimulate immune cells, entry inside the cell is necessary. This entry usually requires a binding protein that can essentially transfect DNA into the intracellular space. This binding protein could be a chromatin component (e.g., HMGB1) although other molecules such as LL37, a defensin, can serve this function. Furthermore, in autoimmune disease, ANAs can bind to nuclear molecules to serve as transfection agents (Chamilos et al., 2012; Di Domizio, Dorta-Estremera, & Cao, 2013; Di Domizio et al., 2012; Means et al., 2005; Vallin, Perers, Alm, & Ronnblom, 1999). Once inside the cell, DNA can interact with Toll-like receptor 9 to activate immune cells; stimulation by other internal receptors is also possible. Like DNA, histones can induce immune activation although its interactions involve TLR2, 4 as well as the inflammasome.

Table 3 Immune Activities of HMGB1

Depend on redox changes involving cysteines 23, 45 and 106
Can interact with TLR2, 4 and RAGE
Can form complexes with cytokines and chemokines
Can form complexes with LPS
Can interact with chromatin components in NETs and microparticles

Immune stimulation by HMGB1 is complicated reflecting its origin by both activated and dying cells. Depending on posttranslational modifications during activation or cell death, the isoforms emerging from the cell differ structurally and functionally (Andersson, Antoine, & Tracey, 2014; Antoine, Harris, Andersson, Tracey, & Bianchi, 2014). As discussed earlier, these modifications concern a series of three cysteine residues. When released from dead cells, the three cysteines are reduced and HMGB1 can bind the chemokine CXCL12 and stimulate chemotaxis through the receptors. With oxidation, a disulfide bond involving cysteines 23 and 45 can form, allowing HMGB1 to stimulate TLR4. With complete oxidation, HMGB1 is inactive. Studies have also indicated that HMGB1 can activate immune cells via TLR2 and RAGE (receptor for advanced glycation end products); HMGB1 can also bind molecules like LPS and IL-1 to potentiate their activity (Table 3).

Thus, three major components of chromatin are immune active. The relevance of this activity is demonstrated by the association of increased levels of DNA, histones, nucleosomes and HMGB1 during diseases or in vivo models of a variety of diseases; the in vitro and in vivo induction of immune responses by administration of purified molecules to experimental models; and the efficacy of antibodies and other binding molecules to inhibit immune responses or disease manifestation in experimental models. Together, these observations suggest that translocation of chromatin can be a central event in the pathogenesis of infectious and inflammatory diseases.

7. REMODELED CHROMATIN AS A BIOMARKER

The translocation of chromatin to the extracellular space occurs widely in human disease since any disease ultimately leads to death of cells. This death can occur locally in the tissue and involve somatic cells or can occur systemically in the blood and involve lymphoid and myeloid cells. Since dead and dying cells can release their intracellular contents, the assay of these molecules has yielded some of the most valuable and venerable biomarkers in medicine. These markers include troponin for myocardial infarction and ALT/AST for liver disease. As now evident, chromatin is released in the same clinical settings as these other biomarkers. The significance of chromatin release may be greater than that of cytoplasmic proteins in terms of pathogenesis, however, since the chromatin components are immunologically active. Assay for the presence of chromatin components in the blood can therefore provide unique information on immune system activation that can accompany or follow tissue injury and death.

Another aspect of chromatin release in the clinical setting relates to the use of measurement of chromatin components in the blood to provide a very sensitive measure of cell damage (Akirav et al., 2011; Dwivedi et al., 2012; Lebastchi et al., 2013; Snyder, Kircher, Hill, Daza, & Shendure, 2016; Sun et al., 2015). In particular, a pool of extracellular DNA known as circulating cell-free DNA provides an important platform for monitoring events in different diseases, in some instances, using molecular probes to identify the cell of origin on the basis of patterns of DNA methylation. This type of analysis is very informative in a wide range of settings. In cancer, for example, mining circulating cell-free DNA can be used to assess diagnosis and prognosis as well as the response to cytotoxic therapy. In an entirely different setting, measurement of cell-free DNA can be used for prenatal screening since fetal DNA can be present in the blood of a mother. Finally, assay of circulating DNA can be used to assess the death of a particular cell type or tissue such as destruction of β cells in type 1 diabetes. Indeed, with the increasing power and sensitivity of DNA analysis from "liquid biopsies," it will soon be possible to survey events in individual tissues of the entire organism as a measure of health and disease.

8. REMODELED CHROMATIN AS A TARGET OF THERAPY

Once chromatin leaves the cell, it can acquire immunological activity to impact on the pathogenesis of infectious, inflammatory and malignant diseases among others. As such, chromatin components can be a target of therapy. Such therapy can be directed at the process of cell death which leads to this release; use of inhibitors of apoptosis and necroptosis are examples of this approach. Alternatively, strategies are available to modify more specifically the modeling processes (including translocation) as well as to inhibit the release of chromatin components. This approach has been explored most extensively with HMGB1 using antibodies to HMGB1 as well as a protein construct that contains one of the DNA-binding domains and can inhibit HMGB1 actions. Development of antibodies to this protein is complicated, however, because of the multiple isoforms that can be present and active. Antibodies to histones also have in vivo immunomodulatory action and can inhibit shock in mice induced by infection (Andersson & Tracey, 2011; Pisetsky, 2012; Xu et al., 2009).

Our laboratory has explored another platform to inhibit the extracellular immune activities of DNA. This approach involves a series of cationic molecules that can bind DNA as well as RNA on the basis of charge. Collectively called nucleic acid-binding polymers (NABPs) or nucleic acid scavengers, these molecules differ in chemistry and include such compounds as poly-L-lysine, protamine sulfate, hexadimethrine bromide (HDMBr) PAMAM dendrimers (polyamidoamine dendrimers, 1,4-diaminobutane core) (Holl, Shumansky, Pitoc, Ramsburg, & Sullenger, 2013; Lee et al., 2011). As shown in in vitro and in vivo experiments, the NABPs can block the activity of immunostimualtory forms of DNA known as CpG oligonucleotides; like bacterial DNA, these synthetic molecules contain unmethylated CpG dinucleotides which are important for the induction of immune response. These compounds show activity in a number of experimental models, presumably by binding the extracellular DNA and preventing its access to the internal receptors. These compounds can also inhibit the binding of anti-DNA antibodies to DNA in the setting of systemic lupus, thereby preventing the formation of pathogenic immune complexes (Stearns, Lee, Leong, Sullenger, & Pisetsky, 2012).

9. CONCLUSION

While ordinarily considered an intranuclear structure, chromatin can undergo translocation from the inside to the outside of the cell during cell death, allowing repurposing of its component molecules, including DNA, histones and HMGB1. Once in the extracellular space, the repurposed components can drive inflammation in a wide variety of clinical conditions. In addition to representing unique biomarkers, chromatin components can be targets of therapy using various strategies to block their access to immune receptors on both the inside and outside of cells. Future studies will refine the analysis of extracellular chromatins on the basis of epigenetic modification including DNA methylation and histone acetylation and define those clinical situations where therapy directed at chromatin is effective.

REFERENCES

Abrams, S. T., Zhang, N., Manson, J., Liu, T., Dart, C., Baluwa, F., ... Toh, C. H. (2013). Circulating histones are mediators of trauma-associated lung injury. *American Journal of Respiratory and Critical Care Medicine*, *187*(2), 160–169. http://dx.doi.org/10.1164/rccm.201206-1037OC.

Akirav, E. M., Lebastchi, J., Galvan, E. M., Henegariu, O., Akirav, M., Ablamunits, V., ... Herold, K. C. (2011). Detection of beta cell death in diabetes using differentially methylated circulating DNA. *Proceedings of the National Academy of Sciences of the United States of America*, *108*(47), 19018–19023. http://dx.doi.org/10.1073/pnas.1111008108.

Alhamdi, Y., Abrams, S. T., Cheng, Z., Jing, S., Su, D., Liu, Z., ... Toh, C. H. (2015). Circulating histones are major mediators of cardiac injury in patients with sepsis. *Critical Care Medicine*, *43*(10), 2094–2103. http://dx.doi.org/10.1097/ccm.0000000000001162.

Andersson, U., Antoine, D. J., & Tracey, K. J. (2014). The functions of HMGB1 depend on molecular localization and post-translational modifications. *Journal of Internal Medicine*, *276*(5), 420–424. http://dx.doi.org/10.1111/joim.12309.

Andersson, U., & Tracey, K. J. (2011). HMGB1 is a therapeutic target for sterile inflammation and infection. *Annual Review of Immunology*, *29*, 139–162. http://dx.doi.org/10.1146/annurev-immunol-030409-101323.

Antoine, D. J., Dear, J. W., Lewis, P. S., Platt, V., Coyle, J., Masson, M., ... Park, B. K. (2013). Mechanistic biomarkers provide early and sensitive detection of acetaminophen-induced acute liver injury at first presentation to hospital. *Hepatology*, *58*(2), 777–787. http://dx.doi.org/10.1002/hep.26294.

Antoine, D. J., Harris, H. E., Andersson, U., Tracey, K. J., & Bianchi, M. E. (2014). A systematic nomenclature for the redox states of high mobility group box (HMGB) proteins. *Molecular Medicine*, *20*, 135–137. http://dx.doi.org/10.2119/molmed.2014.00022.

Antoine, D. J., Williams, D. P., Kipar, A., Jenkins, R. E., Regan, S. L., Sathish, J. G., ... Park, B. K. (2009). High-mobility group box-1 protein and keratin-18, circulating serum proteins informative of acetaminophen-induced necrosis and apoptosis in vivo. *Toxicological Sciences*, *112*(2), 521–531. http://dx.doi.org/10.1093/toxsci/kfp235.

Antoine, D. J., Williams, D. P., Kipar, A., Laverty, H., & Park, B. K. (2010). Diet restriction inhibits apoptosis and HMGB1 oxidation and promotes inflammatory cell recruitment during acetaminophen hepatotoxicity. *Molecular Medicine*, *16*(11–12), 479–490. http://dx.doi.org/10.2119/molmed.2010.00126.

Becker, P. B., & Workman, J. L. (2013). Nucleosome remodeling and epigenetics. *Cold Spring Harbor Perspectives in Biology*, *5*(9), 1–19, http://dx.doi.org/10.1101/cshperspect.a017905.

Bell, C. W., Jiang, W., Reich, C. F., 3rd, & Pisetsky, D. S. (2006). The extracellular release of HMGB1 during apoptotic cell death. *American Journal of Physiology. Cell Physiology*, *291*(6), C1318–C1325. http://dx.doi.org/10.1152/ajpcell.00616.2005.

Bergsbaken, T., Fink, S. L., & Cookson, B. T. (2009). Pyroptosis: Host cell death and inflammation. *Nature Reviews. Microbiology*, 7(2), 99–109. http://dx.doi.org/10.1038/nrmicro2070.

Beyer, C., Stearns, N. A., Giessl, A., Distler, J. H., Schett, G., & Pisetsky, D. S. (2012). The extracellular release of DNA and HMGB1 from Jurkat T cells during in vitro necrotic cell death. *Innate Immunity*, *18*(5), 727–737. http://dx.doi.org/10.1177/1753425912437981.

Bonaldi, T., Talamo, F., Scaffidi, P., Ferrera, D., Porto, A., Bachi, A., ... Bianchi, M. E. (2003). Monocytic cells hyperacetylate chromatin protein HMGB1 to redirect it towards secretion. *EMBO Journal*, *22*(20), 5551–5560. http://dx.doi.org/10.1093/emboj/cdg516.

Brinkmann, V., Reichard, U., Goosmann, C., Fauler, B., Uhlemann, Y., Weiss, D. S., ... Zychlinsky, A. (2004). Neutrophil extracellular traps kill bacteria. *Science*, *303*(5663), 1532–1535. http://dx.doi.org/10.1126/science.1092385.

Brinkmann, V., & Zychlinsky, A. (2012). Neutrophil extracellular traps: Is immunity the second function of chromatin? *Journal of Cell Biology*, *198*(5), 773–783. http://dx.doi.org/10.1083/jcb.201203170.

Casciola-Rosen, L. A., Anhalt, G., & Rosen, A. (1994). Autoantigens targeted in systemic lupus erythematosus are clustered in two populations of surface structures on apoptotic keratinocytes. *Journal of Experimental Medicine*, *179*(4), 1317–1330.

Chamilos, G., Gregorio, J., Meller, S., Lande, R., Kontoyiannis, D. P., Modlin, R. L., & Gilliet, M. (2012). Cytosolic sensing of extracellular self-DNA transported into monocytes by the antimicrobial peptide LL37. *Blood*, *120*(18), 3699–3707. http://dx.doi.org/10.1182/blood-2012-01-401364.

Charras, G. T. (2008). A short history of blebbing. *Journal of Microscopy*, *231*(3), 466–478. http://dx.doi.org/10.1111/j.1365-2818.2008.02059.x.

Charras, G. T., Coughlin, M., Mitchison, T. J., & Mahadevan, L. (2008). Life and times of a cellular bleb. *Biophysical Journal*, *94*(5), 1836–1853. http://dx.doi.org/10.1529/biophysj.107.113605.

Chen, R., Kang, R., Fan, X. G., & Tang, D. (2014). Release and activity of histone in diseases. *Cell Death & Disease*, *5*, e1370. http://dx.doi.org/10.1038/cddis.2014.337.

Choi, J. J., Reich, C. F., 3rd, & Pisetsky, D. S. (2004). Release of DNA from dead and dying lymphocyte and monocyte cell lines in vitro. *Scandinavian Journal of Immunology*, *60*(1–2), 159–166. http://dx.doi.org/10.1111/j.0300-9475.2004.01470.x.

Choi, J. J., Reich, C. F., 3rd, & Pisetsky, D. S. (2005). The role of macrophages in the in vitro generation of extracellular DNA from apoptotic and necrotic cells. *Immunology*, *115*(1), 55–62. http://dx.doi.org/10.1111/j.1365-2567.2005.02130.x.

Coleman, M. L., Sahai, E. A., Yeo, M., Bosch, M., Dewar, A., & Olson, M. F. (2001). Membrane blebbing during apoptosis results from caspase-mediated activation of ROCK I. *Nature Cell Biology*, *3*(4), 339–345. http://dx.doi.org/10.1038/35070009.

Cutter, A. R., & Hayes, J. J. (2015). A brief review of nucleosome structure. *FEBS Letters*, *589*(20 Pt. A), 2914–2922. http://dx.doi.org/10.1016/j.febslet.2015.05.016.

Degterev, A., Huang, Z., Boyce, M., Li, Y., Jagtap, P., Mizushima, N., … Yuan, J. (2005). Chemical inhibitor of nonapoptotic cell death with therapeutic potential for ischemic brain injury. *Nature Chemical Biology*, *1*(2), 112–119. http://dx.doi.org/10.1038/nchembio711.

Di Domizio, J., Dorta-Estremera, S., & Cao, W. (2013). Methylated BSA mimics amyloid-related proteins and triggers inflammation. *PloS One*, *8*(5), e63214. http://dx.doi.org/10.1371/journal.pone.0063214.

Di Domizio, J., Dorta-Estremera, S., Gagea, M., Ganguly, D., Meller, S., Li, P., … Cao, W. (2012). Nucleic acid-containing amyloid fibrils potently induce type I interferon and stimulate systemic autoimmunity. *Proceedings of the National Academy of Sciences of the United States of America*, *109*(36), 14550–14555. http://dx.doi.org/10.1073/pnas.1206923109.

Dwivedi, D. J., Toltl, L. J., Swystun, L. L., Pogue, J., Liaw, K. L., Weitz, J. I., … Liaw, P. C. (2012). Prognostic utility and characterization of cell-free DNA in patients with severe sepsis. *Critical Care*, *16*(4), R151. http://dx.doi.org/10.1186/cc11466.

Fullgrabe, J., Hajji, N., & Joseph, B. (2010). Cracking the death code: Apoptosis-related histone modifications. *Cell Death and Differentiation*, *17*(8), 1238–1243. http://dx.doi.org/10.1038/cdd.2010.58.

Galluzzi, L., Bravo-San Pedro, J. M., Vitale, I., Aaronson, S. A., Abrams, J. M., Adam, D., … Kroemer, G. (2015). Essential versus accessory aspects of cell death: Recommendations of the NCCD 2015. *Cell Death and Differentiation*, *22*(1), 58–73. http://dx.doi.org/10.1038/cdd.2014.137.

Galluzzi, L., Lopez-Soto, A., Kumar, S., & Kroemer, G. (2016). Caspases connect cell-death signaling to organismal homeostasis. *Immunity*, *44*(2), 221–231. http://dx.doi.org/10.1016/j.immuni.2016.01.020.

Gauley, J., & Pisetsky, D. S. (2010). The release of microparticles by RAW 264.7 macrophage cells stimulated with TLR ligands. *Journal of Leukocyte Biology*, *87*(6), 1115–1123. http://dx.doi.org/10.1189/jlb.0709465.

Gorges, T. M., Schiller, J., Schmitz, A., Schuetzmann, D., Schatz, C., Zollner, T. M., … von Ahsen, O. (2012). Cancer therapy monitoring in xenografts by quantitative analysis of circulating tumor DNA. *Biomarkers*, *17*(6), 498–506. http://dx.doi.org/10.3109/1354750x.2012.689133.

Gupta, S., & Kaplan, M. J. (2016). The role of neutrophils and NETosis in autoimmune and renal diseases. *Nature Reviews. Nephrology*, *12*(7), 402–413. http://dx.doi.org/10.1038/nrneph.2016.71.

Gyorgy, B., Szabo, T. G., Pasztoi, M., Pal, Z., Misjak, P., Aradi, B., … Buzas, E. I. (2011). Membrane vesicles, current state-of-the-art: Emerging role of extracellular vesicles. *Cellular and Molecular Life Sciences*, *68*(16), 2667–2688. http://dx.doi.org/10.1007/s00018-011-0689-3.

Halicka, H. D., Bedner, E., & Darzynkiewicz, Z. (2000). Segregation of RNA and separate packaging of DNA and RNA in apoptotic bodies during apoptosis. *Experimental Cell Research*, *260*(2), 248–256. http://dx.doi.org/10.1006/excr.2000.5027.

Holl, E. K., Shumansky, K. L., Pitoc, G., Ramsburg, E., & Sullenger, B. A. (2013). Nucleic acid scavenging polymers inhibit extracellular DNA-mediated innate immune activation without inhibiting anti-viral responses. *PloS One*, *8*(7), e69413. http://dx.doi.org/10.1371/journal.pone.0069413.

Jahr, S., Hentze, H., Englisch, S., Hardt, D., Fackelmayer, F. O., Hesch, R. D., & Knippers, R. (2001). DNA fragments in the blood plasma of cancer patients: Quantitations and evidence for their origin from apoptotic and necrotic cells. *Cancer Research*, *61*(4), 1659–1665.

Jiang, W., Bell, C. W., & Pisetsky, D. S. (2007). The relationship between apoptosis and high-mobility group protein 1 release from murine macrophages stimulated with lipopolysaccharide or polyinosinic-polycytidylic acid. *Journal of Immunology*, *178*(10), 6495–6503.

Jiang, N., & Pisetsky, D. S. (2004). The effect of dexamethasone on the generation of plasma DNA from dead and dying cells. *American Journal of Pathology*, *164*(5), 1751–1759. http://dx.doi.org/10.1016/s0002-9440(10)63733-9.

Jiang, W., & Pisetsky, D. S. (2006). The role of IFN-alpha and nitric oxide in the release of HMGB1 by RAW 264.7 cells stimulated with polyinosinic-polycytidylic acid or lipopolysaccharide. *Journal of Immunology*, *177*(5), 3337–3343.

Jiang, N., Reich, C. F., 3rd, & Pisetsky, D. S. (2003). Role of macrophages in the generation of circulating blood nucleosomes from dead and dying cells. *Blood*, *102*(6), 2243–2250. http://dx.doi.org/10.1182/blood-2002-10-3312.

Kaczmarek, A., Vandenabeele, P., & Krysko, D. V. (2013). Necroptosis: The release of damage-associated molecular patterns and its physiological relevance. *Immunity*, *38*(2), 209–223. http://dx.doi.org/10.1016/j.immuni.2013.02.003.

Kawasaki, T., Kawai, T., & Akira, S. (2011). Recognition of nucleic acids by pattern-recognition receptors and its relevance in autoimmunity. *Immunological Reviews*, *243*(1), 61–73. http://dx.doi.org/10.1111/j.1600-065X.2011.01048.x.

Kazama, H., Ricci, J. E., Herndon, J. M., Hoppe, G., Green, D. R., & Ferguson, T. A. (2008). Induction of immunological tolerance by apoptotic cells requires caspase-dependent oxidation of high-mobility group box-1 protein. *Immunity*, *29*(1), 21–32. http://dx.doi.org/10.1016/j.immuni.2008.05.013.

Kitazumi, I., & Tsukahara, M. (2011). Regulation of DNA fragmentation: The role of caspases and phosphorylation. *The FEBS Journal*, *278*(3), 427–441. http://dx.doi.org/10.1111/j.1742-4658.2010.07975.x.

Kono, H., & Rock, K. L. (2008). How dying cells alert the immune system to danger. *Nature Reviews. Immunology*, *8*(4), 279–289. http://dx.doi.org/10.1038/nri2215.

Lebastchi, J., Deng, S., Lebastchi, A. H., Beshar, I., Gitelman, S., Willi, S., … Herold, K. C. (2013). Immune therapy and beta-cell death in type 1 diabetes. *Diabetes*, *62*(5), 1676–1680. http://dx.doi.org/10.2337/db12-1207.

Lee, J., Sohn, J. W., Zhang, Y., Leong, K. W., Pisetsky, D., & Sullenger, B. A. (2011). Nucleic acid-binding polymers as anti-inflammatory agents. *Proceedings of the National Academy of Sciences of the United States of America*, *108*(34), 14055–14060. http://dx.doi.org/10.1073/pnas.1105777108.

Linkermann, A., & Green, D. R. (2014). Necroptosis. *New England Journal of Medicine*, *370*(5), 455–465. http://dx.doi.org/10.1056/NEJMra1310050.

Magna, M., & Pisetsky, D. S. (2016). The alarmin properties of DNA and DNA-associated nuclear proteins. *Clinical Therapeutics*, *38*(5), 1029–1041. http://dx.doi.org/10.1016/j.clinthera.2016.02.029.

Martin, S. J., Henry, C. M., & Cullen, S. P. (2012). A perspective on mammalian caspases as positive and negative regulators of inflammation. *Molecular Cell*, *46*(4), 387–397. http://dx.doi.org/10.1016/j.molcel.2012.04.026.

Means, T. K., Latz, E., Hayashi, F., Murali, M. R., Golenbock, D. T., & Luster, A. D. (2005). Human lupus autoantibody-DNA complexes activate DCs through cooperation of CD32 and TLR9. *Journal of Clinical Investigation*, *115*(2), 407–417. http://dx.doi.org/10.1172/jci23025.

Miyake, K., Shibata, T., Ohto, U., & Shimizu, T. (2016). Emerging roles of the processing of nucleic acids and Toll-like receptors in innate immune responses to nucleic acids. *Journal of Leukocyte Biology*, *100*, 1–9. http://dx.doi.org/10.1189/jlb.4MR0316-108R.

Nystrom, S., Antoine, D. J., Lundback, P., Lock, J. G., Nita, A. F., Hogstrand, K., … Applequist, S. E. (2013). TLR activation regulates damage-associated molecular pattern isoforms released during pyroptosis. *The EMBO Journal*, *32*(1), 86–99. http://dx.doi.org/10.1038/emboj.2012.328.

Pisetsky, D. (2011). Cell death in the pathogenesis of immune-mediated diseases: The role of HMGB1 and DAMP–PAMP complexes. *Swiss Medical Weekly*, *141*, w13256. http://dx.doi.org/10.4414/smw.2011.13256.

Pisetsky, D. S. (2012). Antinuclear antibodies in rheumatic disease: A proposal for a function-based classification. *Scandinavian Journal of Immunology*, *76*(3), 223–228. http://dx.doi.org/10.1111/j.1365-3083.2012.02728.x.

Pisetsky, D. S. (2014). The translocation of nuclear molecules during inflammation and cell death. *Antioxidants & Redox Signaling*, *20*(7), 1117–1125. http://dx.doi.org/10.1089/ars.2012.5143.

Pisetsky, D. S., & Jiang, N. (2006). The generation of extracellular DNA in SLE: The role of death and sex. *Scandinavian Journal of Immunology*, *64*(3), 200–204. http://dx.doi.org/10.1111/j.1365-3083.2006.01822.x.

Pisetsky, D. S., Ullal, A. J., Gauley, J., & Ning, T. C. (2012). Microparticles as mediators and biomarkers of rheumatic disease. *Rheumatology*, *51*(10), 1737–1746. http://dx.doi.org/10.1093/rheumatology/kes028.

Prokhorova, E. A., Zamaraev, A. V., Kopeina, G. S., Zhivotovsky, B., & Lavrik, I. N. (2015). Role of the nucleus in apoptosis: Signaling and execution. *Cellular and Molecular Life Sciences*, *72*(23), 4593–4612. http://dx.doi.org/10.1007/s00018-015-2031-y.

Reich, C. F., 3rd, & Pisetsky, D. S. (2009). The content of DNA and RNA in microparticles released by Jurkat and HL-60 cells undergoing in vitro apoptosis. *Experimental Cell Research*, *315*(5), 760–768. http://dx.doi.org/10.1016/j.yexcr.2008.12.014.

Roers, A., Hiller, B., & Hornung, V. (2016). Recognition of endogenous nucleic acids by the innate immune system. *Immunity*, *44*(4), 739–754. http://dx.doi.org/10.1016/j.immuni.2016.04.002.

Scaffidi, P., Misteli, T., & Bianchi, M. E. (2002). Release of chromatin protein HMGB1 by necrotic cells triggers inflammation. *Nature*, *418*(6894), 191–195. http://dx.doi.org/10.1038/nature00858.

Schiller, M., Bekeredjian-Ding, I., Heyder, P., Blank, N., Ho, A. D., & Lorenz, H. M. (2008). Autoantigens are translocated into small apoptotic bodies during early stages of apoptosis. *Cell Death and Differentiation*, *15*(1), 183–191. http://dx.doi.org/10.1038/sj.cdd.4402239.

Sebbagh, M., Renvoize, C., Hamelin, J., Riche, N., Bertoglio, J., & Breard, J. (2001). Caspase-3-mediated cleavage of ROCK I induces MLC phosphorylation and apoptotic membrane blebbing. *Nature Cell Biology*, *3*(4), 346–352. http://dx.doi.org/10.1038/35070019.

Seifert, J. K., Stewart, G. J., Hewitt, P. M., Bolton, E. J., Junginger, T., & Morris, D. L. (1999). Interleukin-6 and tumor necrosis factor-alpha levels following hepatic cryotherapy: Association with volume and duration of freezing. *World Journal of Surgery*, *23*(10), 1019–1026.

Sharma, D., & Kanneganti, T. D. (2016). The cell biology of inflammasomes: Mechanisms of inflammasome activation and regulation. *Journal of Cell Biology*, *213*(6), 617–629. http://dx.doi.org/10.1083/jcb.201602089.

Snyder, M. W., Kircher, M., Hill, A. J., Daza, R. M., & Shendure, J. (2016). Cell-free DNA comprises an in vivo nucleosome footprint that informs its tissues-of-origin. *Cell*, *164*(1–2), 57–68. http://dx.doi.org/10.1016/j.cell.2015.11.050.

Spencer, D. M., Gauley, J., & Pisetsky, D. S. (2014). The properties of microparticles from RAW 264.7 macrophage cells undergoing in vitro activation or apoptosis. *Innate Immunity*, *20*(3), 239–248. http://dx.doi.org/10.1177/1753425913492552.

Spencer, D. M., Mobarrez, F., Wallen, H., & Pisetsky, D. S. (2014). The expression of HMGB1 on microparticles from Jurkat and HL-60 cells undergoing apoptosis in vitro. *Scandinavian Journal of Immunology*, *80*(2), 101–110. http://dx.doi.org/10.1111/sji.12191.

Stearns, N. A., Lee, J., Leong, K. W., Sullenger, B. A., & Pisetsky, D. S. (2012). The inhibition of anti-DNA binding to DNA by nucleic acid binding polymers. *PloS One*, 7(7), e40862. http://dx.doi.org/10.1371/journal.pone.0040862.

Sun, K., Jiang, P., Chan, K. C., Wong, J., Cheng, Y. K., Liang, R. H., … Lo, Y. M. (2015). Plasma DNA tissue mapping by genome-wide methylation sequencing for noninvasive prenatal, cancer, and transplantation assessments. *Proceedings of the National Academy of Sciences of the United States of America*, *112*(40), E5503–E5512. http://dx.doi.org/10.1073/pnas.1508736112.

Sur Chowdhury, C., Giaglis, S., Walker, U. A., Buser, A., Hahn, S., & Hasler, P. (2014). Enhanced neutrophil extracellular trap generation in rheumatoid arthritis: Analysis of underlying signal transduction pathways and potential diagnostic utility. *Arthritis Research & Therapy*, *16*(3), R122. http://dx.doi.org/10.1186/ar4579.

Tang, Y., Zhao, X., Antoine, D., Xiao, X., Wang, H., Andersson, U., … Lu, B. (2016). Regulation of posttranslational modifications of HMGB1 during immune responses. *Antioxidants & Redox Signaling*, *24*(12), 620–634. http://dx.doi.org/10.1089/ars.2015.6409.

Thierry, A. R., Mouliere, F., Gongora, C., Ollier, J., Robert, B., Ychou, M., … Molina, F. (2010). Origin and quantification of circulating DNA in mice with human colorectal cancer xenografts. *Nucleic Acids Research*, *38*(18), 6159–6175. http://dx.doi.org/10.1093/nar/gkq421.

Tran, T. T., Groben, P., & Pisetsky, D. S. (2008). The release of DNA into the plasma of mice following hepatic cell death by apoptosis and necrosis. *Biomarkers*, *13*(2), 184–200. http://dx.doi.org/10.1080/13547500701791719.

Tsokos, G. C. (2011). Systemic lupus erythematosus. *New England Journal of Medicine*, *365*(22), 2110–2121. http://dx.doi.org/10.1056/NEJMra1100359.

Ullal, A. J., Pisetsky, D. S., & Reich, C. F., 3rd. (2010). Use of SYTO 13, a fluorescent dye binding nucleic acids, for the detection of microparticles in in vitro systems. *Cytometry. Part A*, 77(3), 294–301. http://dx.doi.org/10.1002/cyto.a.20833.

Ullal, A. J., Reich, C. F., 3rd, Clowse, M., Criscione-Schreiber, L. G., Tochacek, M., Monestier, M., & Pisetsky, D. S. (2011). Microparticles as antigenic targets of antibodies to DNA and nucleosomes in systemic lupus erythematosus. *Journal of Autoimmunity*, *36*(3–4), 173–180. http://dx.doi.org/10.1016/j.jaut.2011.02.001.

Urbonaviciute, V., Meister, S., Furnrohr, B. G., Frey, B., Guckel, E., Schett, G., … Voll, R. E. (2009). Oxidation of the alarmin high-mobility group box 1 protein (HMGB1) during apoptosis. *Autoimmunity*, *42*(4), 305–307.

Vallin, H., Perers, A., Alm, G. V., & Ronnblom, L. (1999). Anti-double-stranded DNA antibodies and immunostimulatory plasmid DNA in combination mimic the endogenous IFN-alpha inducer in systemic lupus erythematosus. *Journal of Immunology*, *163*(11), 6306–6313.

van Bavel, C. C., Dieker, J. W., Tamboer, W. P., van der Vlag, J., & Berden, J. H. (2010). Lupus-derived monoclonal autoantibodies against apoptotic chromatin recognize acetylated conformational epitopes. *Molecular Immunology*, *48*(1–3), 248–256. http://dx.doi.org/10.1016/j.molimm.2010.08.003.

Venereau, E., Ceriotti, C., & Bianchi, M. E. (2015). DAMPs from cell death to new life. *Frontiers in Immunology*, *6*, 422. http://dx.doi.org/10.3389/fimmu.2015.00422.

Wallach, D., Kang, T. B., Dillon, C. P., & Green, D. R. (2016). Programmed necrosis in inflammation: Toward identification of the effector molecules. *Science*, *352*(6281), 52–59, http://dx.doi.org/10.1126/science.aaf2154. aaf2154.

Wickman, G. R., Julian, L., Mardilovich, K., Schumacher, S., Munro, J., Rath, N., … Olson, M. F. (2013). Blebs produced by actin-myosin contraction during apoptosis release damage-associated molecular pattern proteins before secondary necrosis occurs. *Cell Death and Differentiation*, *20*(10), 1293–1305. http://dx.doi.org/10.1038/cdd.2013.69.

Xu, J., Zhang, X., Pelayo, R., Monestier, M., Ammollo, C. T., Semeraro, F., … Esmon, C. T. (2009). Extracellular histones are major mediators of death in sepsis. *Nature Medicine*, *15*(11), 1318–1321. http://dx.doi.org/10.1038/nm.2053.

Yang, H., Antoine, D. J., Andersson, U., & Tracey, K. J. (2013). The many faces of HMGB1: Molecular structure–functional activity in inflammation, apoptosis, and chemotaxis. *Journal of Leukocyte Biology*, *93*(6), 865–873. http://dx.doi.org/10.1189/jlb.1212662.

Yang, H., Lundback, P., Ottosson, L., Erlandsson-Harris, H., Venereau, E., Bianchi, M. E., … Antoine, D. J. (2012). Redox modification of cysteine residues regulates the cytokine activity of high mobility group box-1 (HMGB1). *Molecular Medicine*, *18*, 250–259. http://dx.doi.org/10.2119/molmed.2011.00389.

Zirngibl, M., Furnrohr, B. G., Janko, C., Munoz, L. E., Voll, R. E., Gregory, C. D., … Herrmann, M. (2015). Loading of nuclear autoantigens prototypically recognized by systemic lupus erythematosus sera into late apoptotic vesicles requires intact microtubules and myosin light chain kinase activity. *Clinical and Experimental Immunology*, *179*(1), 39–49. http://dx.doi.org/10.1111/cei.12342.

CHAPTER SIX

Epigenetic Changes in Chronic Inflammatory Diseases

O. Fogel*,†, C. Richard-Miceli*,‡, J. Tost†,1
*Université Paris Descartes, Hôpital Cochin—Assistance Publique—Hôpitaux de Paris, EULAR Centre of Excellence, Paris, France
†Laboratory for Epigenetics and Environment, Centre National de Génotypage, CEA-Institut de Génomique, Evry, France
‡Immunoregulation Unit, Unité Mixte Pasteur/APHP, Institut Pasteur, Paris, France
[1]Corresponding author: e-mail address: tost@cng.fr

Contents

Abstract

The number of people diagnosed with chronic inflammatory diseases has increased noteworthy in the last 40 years. Spondyloarthritis (SpA), inflammatory bowel diseases (IBD), and psoriasis are the most frequent chronic inflammatory diseases, resulting from a combination of genetic predisposition and environmental factors. Epigenetic modifications include DNA methylation, histone modifications, and small and long noncoding RNAs. They are influenced by environmental exposure, life-style, and aging and have recently been shown to be altered in many complex diseases including inflammatory diseases. While epigenetic modifications have been well characterized in other diseases such as cancer and autoimmune diseases, knowledge on changes

Advances in Protein Chemistry and Structural Biology, Volume 106
ISSN 1876-1623
http://dx.doi.org/10.1016/bs.apcsb.2016.09.003

in inflammatory diseases is lagging behind with some disease-specific differences. While the DNA methylation profile of different cell types in patients with IBD has been relatively well described, less is known on changes implicated in psoriasis, and no systematic genome-wide studies have so far been performed in SpA. In this chapter, we review in detail the reported changes in patterns of DNA methylation and posttranslational histone modifications in chronic inflammatory diseases highlighting potential connections between disease-associated pathophysiological changes such as the dysbiosis of the microbiome or genetic variations associated with disease susceptibility and the epigenome. We also discuss important parameters of meaningful epigenetic studies such as the use of well defined, disease-relevant cell populations, and elude on the potential future of engineering of the epigenome in inflammatory diseases.

1. INTRODUCTION

The number of people diagnosed with inflammatory diseases has increased noteworthy in the last 40 years and this fast-growing number has been impossible to explain by Mendelian inheritance only. Among them, spondyloarthritis (SpA), inflammatory bowel diseases (IBD), and psoriasis are the most frequent chronic inflammatory diseases, resulting from a combination of genetic predisposition and environmental factors. Even if a familial segregation exists, large-scale genetic studies have failed to explain the observed heritability of these diseases. For example, in ankylosing spondylitis, only 25% of heritability can be explained by currently known risk loci, suggesting that 75% of the disease risk remains unknown (Reveille, 2012). This lack of knowledge is called "the missing heritability." Among other hypotheses, the epigenome might hold at least part to the answer of this missing heritability. Epigenetics refers to the mechanisms that modulate gene expression without altering the DNA sequence. Epigenetic processes involve DNA methylation, histone modifications that modulate chromatin structure or noncoding RNAs (ncRNAs) such as microRNAs or PIWI-interacting (pi)RNAs that are involved in the regulation of the posttranscriptional steps. All these mechanisms can be affected by exposure to environmental factors and are transmissible from cell generation to cell generation, but can also be reversed. They are important for controlling patterns of gene expression during the cell cycle, proper mammalian development, cell differentiation, function and homeostasis, and explain how a multicellular organism can derive from a unique genome. In females, epigenetic modifications contribute largely to the random inactivation of one of the two X chromosomes. Aberrant epigenetic patterns have been found in a large number of diseases including cancer (Baylin & Jones, 2011), but more recently also in other complex diseases, including autoimmune diseases, neurodegenerative, psychiatric, and metabolic disorders (Absher et al., 2013; Miceli-Richard et al., 2016; Nielsen & Tost, 2012; Ronn & Ling, 2015; Voelter-Mahlknecht, 2016; Yu et al., 2015). Epigenetic changes seem to be involved in the early steps of inflammatory dysregulation, facilitating the onset of the disease.

The influence of nongenetic factors has been well demonstrated in inflammatory diseases. Studies on monozygotic (MZ) twins show a variable degree of discordance in terms of phenotypic traits or susceptibility to the disease. Environmental factors must therefore play a crucial role. Because of its propensity to be modulated by external stimuli, epigenetics is at the crossroads between genetics and environment.

This domain of research has experienced a large increase in interest in the last years due to improved technical possibilities especially with advances in standardized DNA methylation arrays and next-generation sequencing (NGS) technologies. Further locus-specific DNA methylation assays have reached a level of technical maturity, which allows their implementation in clinical diagnostics and the FDA and other regulatory agencies have recently approved first assays using DNA methylation-based-biomarkers. Investigating the influence of aberrant epigenetic modifications has provided new insights into the pathogenesis of complex disorders. Also, epigenetic marks could be promising biomarkers for many applications. In this review, we summarize and discuss the current state of research and knowledge on epigenetic modification in chronic inflammatory diseases.

2. EPIGENETIC PROCESSES

2.1 DNA Methylation

DNA methylation occurs through the covalent addition of a methyl group to the 5′ carbon of the cytosine ring in the context of CpG dinucleotides (Tost, 2010). Cytosine methylation in other sequence contexts is rare in humans and other mammals and has only been observed in embryonic stem cells. The CpG dinucleotide is often clustered in particular regions of the genome called CpG islands located in the promoter or the first exon of approximately two-thirds of all genes. In the genome, most CpGs are methylated whereas CpG islands remain mainly unmethylated (Bird, 2002). Of note, an unmethylated promoter does not yield information about the actual transcriptional status of the gene, which can be either actively transcribed, transcribed at low levels or be silent, while a gene with a methylated promoter is in most cases transcriptionally silent. DNA methyltransferases (DNMT) catalyze the addition of methyl group on cytosine (Law & Jacobsen, 2010). DNMT1 methylates mainly preexisting methylated profiles contributing to the maintenance of DNA methylation patterns, whereas DNMT3A and DNMT3B methylate unmethylated substrates contributing to de novo methylation. DNA methylation is particularly important for the regulation of developmental and tissue-specific gene expression. As the methylome is specific for a given cellular state, the analysis of tissues such as blood, that are constituted of a heterogeneous mixture of different cell types each with its own methylome, represents a suboptimal approach for DNA methylation analysis as observed changes in DNA methylation patterns might simply reflect changes in the cell proportions in the analyzed samples. Although, great progress has been made on statistical algorithms aiming at the correction of cellular heterogeneity at the computational level using either reference epigenomes or reference-free methods (Houseman et al., 2012, 2016), the methylation profile, or methylome, should be studied in cell-sorted using genome-wide approaches. Major international efforts are currently ongoing to map the methylome of various human tissues and cell types at single base resolution (Adams et al., 2012; Roadmap Epigenomics et al., 2015; Schultz et al., 2015). The variety of epigenetic modifications in mammalian DNA has recently increased further with the discovery of 5-hydroxymethylcytosine (5hmC) and its oxidative derivatives, 5-formyl, and 5-carboxylcytosine in mammalian neurons and embryonic stem cells, which are formed from 5-methylcytosines by a catalytic oxidation mediated by the TET proteins (Kriaucionis & Heintz, 2009; Tahiliani et al., 2009). The 5hmC modification has been found in nearly all mouse embryonic tissues, but with the exception of brain tissue and bone marrow only very low levels (0.2–0.05% of all cytosines) are detectable in adult tissue (Globisch et al., 2010; Ruzov et al., 2011). While a role of the Tet2 enzyme has recently been demonstrated in the resolution of inflammation by directly contributing to the repression of interleukin (IL) 6 (Zhang, Zhao, et al., 2015), there is not yet much knowledge on the presence of altered hydroxymethylation in inflammatory diseases. We will therefore not further mention hydroxymethylation of cytosines within this review, but the reader should keep in mind that this does exist, might play a role in inflammatory diseases and—as pointed out in Section 2.4—most of the commonly used methods for DNA methylation analysis based on sodium bisulfite conversion will not be able to distinguish between DNA methylation and hydroxymethylation and results may confound the presence of the respective marks.

2.2 Histone Modifications

The nucleosome is the basic subunit of chromatin comprised of 146 base pairs of DNA wound around an octamer of histones H2A, H2B, H3, and H4 consisting of two copies of each histone and linked by H1 proteins. This complex is called euchromatin, when relatively loosely attached to the DNA, and heterochromatin when the assembly is highly compacted and corresponds to a silenced chromatin limiting the accessibility of transcription factors to the DNA. Thus, histone state-dependent DNA compaction is a prime mechanism for the regulation of gene expression. Different posttranslational histone modifications exist including acetylation, methylation, ubiquitination, phosphorylation, sumoylation, or citrulination. Among them, acetylation and methylation are the most studied posttranslational modifications and are tightly controlled (Zhang, Cooper, & Brockdorff, 2015).

The addition of an acetyl group to a lysine residue of histone is catalyzed by histone acetyl transferases (HATs) and the removal by histone deacetylases (HDACs). Histone acetylation opens chromatin and is associated with transcriptional activity while histone deacetylation correlates with chromatin compaction. HATs and HDACs play important roles in cellular functions such as proliferation or apoptosis leading to therapeutic approaches especially with HDAC inhibitors.

Addition of a methyl group on lysines (K) or arginines (R) by histone methyltransferases (HMTs) leads to activation or repression of transcriptional activity according to the position of the targeted amino acid. Particularly, methylation of lysine 9 and 27 of H3 (H3K9 and H3K27) are repressive, whereas methylation of H3K4, H3K36, and H3K79 are associated with initiation and elongation of transcription. There are currently a large number of therapeutic agents under clinical investigation or in development that are focused on modifying the activity of enzymes posing or removing these posttranslational modifications to control gene expression and potentially correct aberrant gene expression patterns.

2.3 Noncoding RNAs

ncRNAs are involved in cell proliferation and differentiation and function as RNA silencing and posttranscriptional regulator of gene expression.

MicroRNAs (miRNAs) are produced from single-stranded RNA that folds back on themselves to form short hairpins (van Rooij, 2011). Changes in miRNA expression have been shown to contribute to the regulation of inflammatory responses in humans and can regulate the release of proinflammatory chemokines (Marques-Rocha et al., 2015).

Messenger RNAs (mRNAs) are silenced by several processes after recognition by miRNA via base pairing of complementary sequences preferentially located in 3′UTR of the mRNA (van Rooij, 2011). To date, more than 1200 miRNA have been identified in human. A given miRNA may have hundreds of different mRNA targets, and a given target might be regulated by multiple miRNAs. This mechanism is powerful and it was estimated that 30–90% of the genome can be regulated by miRNAs. However, the function of only a few miRNAs has been clearly identified. Among them, several miRNAs are able to modulate DNA methylation or histone modifications. Conversely, these two processes are able to up- or down-regulate miRNA production (Samuel et al., 2016; Shen et al., 2016). As the description of altered miRNA expression in inflammatory diseases is an intensively investigated field of research on its own and beyond the scope of this review, the interested reader is referred to several excellent recent reviews summarizing the role of miRNAs in the different chronic inflammatory diseases (Chapman & Pekow, 2015; Hawkes et al., 2016; Kalla et al., 2015; Li, Wong, Shen, Chan, & Wu, 2016; Xia & Zhang, 2014).

2.4 Methods of Analysis

Methods for DNA methylation analysis are based on three main principles (Tost, 2008): (1) The use of methylation-sensitive or methylation-dependent restriction endonucleases, i.e., enzymes that depend on or are blocked by methylated cytosines in their recognition sequence, respectively. (2) The methylated fraction of a genome can be enriched by precipitation with a bead-immobilized antibody specific for 5-methylcytosine or by affinity purification of methylated DNA with methyl-binding domain (MBD) proteins. (3) The most widely used approach consists of the chemical modification of genomic DNA with sodium bisulfite. This chemical reaction induces hydrolytic deamination of nonmethylated cytosines to uracils, while methylated cytosines are resistant to conversion under the chosen reaction conditions (Frommer et al., 1992). This method allows the retention of the methylation status that would otherwise be lost during PCR amplification and translates the methylation signal into a sequence difference. After performing PCR, the methylation status at a given position is manifested in the ratio C (former methylated cytosine) to T (former nonmethylated cytosine) and can be analyzed as a virtual C/T polymorphism spanning the entire allele frequency spectrum from 0% to 100% in the bisulfite-treated DNA.

NGS-based approaches have revolutionized the analysis of epigenetic modifications in general and DNA methylation in particular (Table 1). Although potentially confounded by the presence of 5hmC, whole-genome bisulfite sequencing can be considered as the current gold standard for the genome-wide identification of differentially methylated regions (DMRs) at single-nucleotide resolution (Urich, Nery, Lister, Schmitz, & Ecker, 2015). However, the high quantitative and spatial resolution comes currently at a high cost and requires substantial sequencing to obtain a proper and even coverage. Reduced-representation bisulfite sequencing (RRBS) requires significantly less sequencing and CpG-rich regions are relatively well covered, however many CpG poor regions might be missed, so that its utility will depend on the application (Meissner et al., 2005). Methylated DNA immunoprecipitation (MeDIP) (Taiwo et al., 2012) and MBD protein-based approaches (Brinkman et al., 2010; Jung et al., 2015) do have a relatively low resolution and while MeDIP suffers from a high general background signal, MBD-seq focus on mainly CpG-rich regions such as CpG islands, which is less the case for MeDIP. Alternative technologies for the analysis of a large number of CpGs include a number of capture approaches or microfluidic amplification approaches (Table 1).

With whole-genome bisulfite sequencing being not yet affordable at a large scale, and the low resolution of antibody and methyl-binding protein enrichment of methylated regions, epigenotyping technologies have emerged as an alternative tool for the identification of differentially methylated regions (DMRs) and DNA methylation-based biomarkers. Epigenotyping technologies such as the Infinium Human Methylation 27K, 450K or the recent EPIC BeadChip (Illumina Inc., CA, USA) generate a methylation-state specific "pseudo-SNP" through bisulfite conversion translating thereby differences in the DNA methylation patterns into sequence differences that can be analyzed using quantitative genotyping methods (Bibikova et al., 2009; Moran et al., 2016; Sandoval et al., 2011). The 450K BeadChip expanded dramatically the genome coverage analyzing more than 480,000 CpG sites covering 99% of all RefSeq genes with an average of 17 probes per gene (Sandoval et al., 2011) and the recent EPIC BeadChip increases the coverage with another 400,000 probes mainly in intergenic gene-regulatory regions (Moran et al., 2016). Thus probes are

Table 1 Methods Commonly Used for the Analysis of Genome-Wide and Locus-Specific DNA Methylation Patterns

Technology	Methylation Detection	Coverage CpGs	Principle	Single-Nucleotide Resolution (Y/N)	Throughput	Read-Out Platform	Reference
WGBS/ MethylC-seq	Bisulfite conversion	29 M	Adaptor ligation before or after bisulfite treatment of genomic DNA prior to sequencing	Y	+	NGS	Miura, Enomoto, Dairiki, and Ito (2012) and Urich et al. (2015)
RRBS	Bisulfite conversion	1.6 M	Restriction digest allows for size selection of small fragments, used for library construction, bisulfite conversion, and sequencing	Y	++	NGS	Guo et al. (2015)
MeDIP-seq	Antibody-based enrichment	17 M	Library construction, enrichment of methylated DNA using an antibody against 5mC followed by sequencing	N	++	NGS	Taiwo et al. (2012)
MBD-seq	Affinity purification	17 M	Enrichment of methylated DNA using a methyl-binding domain protein before or after a library construction followed by sequencing	N	++	NGS	Brinkman et al. (2010) and Jung et al. (2015)
Agilent SureSelect^{XT} methylation capture	Bisulfite conversion	3.7 M	Libraries are prepared from genomic DNA fragments, hybridized to the capture probes, eluted, bisulfite converted, and amplified before being sequenced by a NGS	Y	++	NGS	Borno et al. (2012)
SeqCapEpi CpGiant enrichment	Bisulfite conversion	5.5 M	Libraries are prepared from genomic DNA fragments, bisulfite converted, preamplified hybridized to the capture probes, eluted, and amplified before being sequenced by a NGS	Y	++	NGS	Li et al. (2015)
Infinium BeadChips	Bisulfite conversion	480K/ 840K	Genotyping of bisulfite converted DNA	Y	+++	Microarray	Moran, Arribas, and Esteller (2016) and Sandoval et al. (2011)

Amplicon BS-Seq	Bisulfite conversion	100s–1000s	Regions of interest are PCR amplified from bisulfite converted DNA and libraries are constructed and sequenced by NGS (MiSeq/IonTorrent)	Y	+++	NGS	Korbie et al. (2015) and Masser, Berg, and Freeman (2013)
Pyrosequencing	Bisulfite conversion	10s	Bisulfite conversion, PCR amplification, and sequencing-by-synthesis method. Incorporation of a complementary nucleotide releases a phosphate group that will be involved in the ATP-driven oxidation of luciferin. The light emitted from this reaction is monitored	Y	++	Pyrosequencer	Tost and Gut (2007)
MassCLEAVE/EpiTYPER	Bisulfite conversion	10s	T7-promoter-tagged bisulfite-specific PCR followed by ssRNA transcription and base-specific cleavage by RNase A. The mixture is analyzed quantitatively by MALDI-TOF-MS	Y/N	+++	Mass spectrometry	Ehrich et al. (2005)
MS-HRM	Bisulfite conversion	<10	High-resolution melting curve analysis following quantitative real-time PCR RT-PCR yields quantitative methylation levels	N	++	Real-time thermocycler	Wojdacz, Dobrovic, and Hansen (2008)
MS-qPCR	Methylation-sensitive restriction enzymes	1	Digestion of DNA with methylation-sensitive restriction enzymes and subsequent amplification of undigested target DNA monitored quantitatively in real time	Y/N	++	Real-time thermocycler	Oakes, La Salle, Robaire, and Trasler (2006)
MSP-PCR	Bisulfite conversion	<10	At least one CpG site at or near the 3′ of the amplification primers ensures amplification of methylated DNA only. By including non-CpG cytosines in the primer sequence and having stringent annealing temperatures, amplification of incomplete converted- and nontarget DNA is limited	N	++	Electrophoresis gel	Herman, Graff, Myohanen, Nelkin, and Baylin (1996)

Continued

Table 1 Methods Commonly Used for the Analysis of Genome-Wide and Locus-Specific DNA Methylation Patterns—cont'd

Technology	Methylation Detection	Coverage CpGs	Principle	Single-Nucleotide Resolution (Y/N)	Throughput	Read-Out Platform	Reference
MethyLight	Bisulfite conversion	<10	Cleavage of a fluorescent hydrolysis probe designed to hybridize between MSP primers, principle based on the TaqMan chemistry. Detects one methylated molecule in background of 10,000 molecules	N	+++	Real-time thermocycler	Eads et al. (2000)
HeavyMethyl	Bisulfite conversion	480K/ 840K	Blockers, designed to bind unmethylated DNA only, ensure the amplification of methylated DNA with methylation independent amplification primers. A probe hybridizing to CpG sites containing a fluorophore and a quencher is designed to detect the amplification of methylated molecules	N	+++	Real-time thermocycler	Cottrell et al. (2004)

WGBS: Whole-genome bisulfite sequencing; RRBS: reduced-representation bisulfite sequencing; MBD: methyl-binding domain; MeDIP: methylated DNA immunoprecipitation; MS-HRM: methylation-sensitive high resolution melting; MSP: methylation-specific PCR > Y: yes, N: No, +: 10's of samples can be conveniently analyzed using the technology; ++: 100's of samples can be conveniently analyzed using the technology; +++: 1000's of samples can be conveniently analyzed using the technology.

no longer only focused on CpG islands, but distributed over various functional elements that are more prone to alter their DNA methylation status in function of environmental conditions or in complex diseases other than cancer. These arrays have the potential for large-scale high-throughput studies as they use highly standardized protocols that can be implemented with a large degree of automation into existing genotyping pipelines. These epigenotyping assays have been found to be accurate and display high correlation to RRBS (Bock et al., 2010) as well as locus-specific quantitative assays such as MethyLight (Campan et al., 2011). However, these arrays do analyze only a small number of the 28 million CpG sites of the human genome and there are also currently no commercial arrays for the analysis of nonhuman samples available. The 450K/EPIC arrays provide nonetheless a good compromise between coverage, throughput, cost, resolution, and accuracy permitting genome-wide epigenome analysis by epigenotyping and have been rapidly adapted by the community including for the analysis of DNA methylation changes in inflammatory diseases (Adams et al., 2014; Cooke et al., 2012; Gervin et al., 2012; Harris et al., 2014, 2012; Hasler et al., 2012; Hughes et al., 2014; Nimmo et al., 2012; Roberson et al., 2012).

Following the interest in DNA methylation for the various clinical applications, many technologies have been developed for the quantitative analysis of DNA methylation patterns or levels at specific loci including pyrosequencing (Tost & Gut, 2007), next/second generation sequencing (NGS)-based amplicon sequencing (Korbie et al., 2015; Masser et al., 2013), or real-time qPCR methods (Cottrell et al., 2004; Eads et al., 2000) (Table 1). Most of them rely on the conversion of genomic DNA with sodium bisulfite for the detection of DNA methylation patterns. In a recent worldwide technology comparison study investigating the reliability, reproducibility, and accuracy of technologies for locus-specific DNA methylation analysis, a number of technologies were found to yield accurate and reproducible results especially pyrosequencing and NGS-based amplicon bisulfite sequencing (Blueprint Consortium, 2016).

Global changes in the distribution of histone modifications can be assessed using fluorescent or mass spectrometry-based assays. However, in most cases, the genomic context of a specific histone modification is interrogated using chromatin immunoprecipitation (ChIP) (Gilmour & Lis, 1984). ChIP is a well-established method in cellular biology to study the specific interaction between a protein of interest and genomic DNA and has been extensively used to identify transcription factor-binding sites (Gerstein et al., 2012). Briefly, proteins are cross-linked to the DNA by chemicals (normally formaldehyde) in order to conserve the in vivo chromatin architecture. Chromatin is extracted and randomly fragmented by sonication into 200–600 base pair fragments. Then, DNA–protein complexes are immunoprecipitated using a specific antibody and protein A/G agarose resin. Finally, covalent cross-links are reversed by heating and DNA is purified. At this point, a small amount of purified DNA is available that can be subsequently analyzed by qPCR for the detection and quantification of the analyzed modification at a locus of interest using specific primers, microarrays (ChIP-on-chip), or NGS (ChIP-seq) for a genome-wide picture of the DNA–protein-binding events (i.e., identification of all binding sites of a transcription factor; mapping of a histone modification on the entire genome at very high resolution) (Barski et al., 2007; Gerstein et al., 2012; Mikkelsen et al., 2007). ChIP-seq has since been performed at production scale in the ENCODE, modENCODE, and Roadmap epigenomics mapping projects and has become the workhorse for the genome-wide mapping of the occupancy of DNA by transcription factors and posttranslationally modified histones.

3. EPIGENETIC MODIFICATION INFLUENCE MEDIATORS OF INFLAMMATION

3.1 Immune Cells

3.1.1 Monocytes–Macrophages

Monocytes–macrophages are the major cells of innate immunity. Macrophage subsets have been classified as classic M1 or alternative M2 macrophages. Activated M1 macrophages are considered as proinflammatory macrophages and are induced by IFNγ and bacterial products, such as LPS (lipopolysaccharides), and play a key role during host defense. However, uncontrolled M1 activity sustains chronic inflammatory diseases and causes collateral damages (Hoeksema, Stoger, & de Winther, 2012). Conversely, M2 macrophages, induced by IL4, IL10, or helminthes products, decrease inflammatory responses, and promote tissue repair (Gordon & Martinez, 2010). Because of this dual role, research on potential therapies for inflammatory disease has focused on ways to converse M1 macrophages into activated M2 macrophages. Helminthes have shown to provide protection in a murine IBD model by inducing M2 macrophages (Hoeksema et al., 2016). Several groups have shown that epigenetic processes modulate macrophage polarization through the balance between activating and repressive modifications on histones located in the enhancer area of key genes (Van den Bossche, Neele,

Hoeksema, & de Winther, 2014). Macrophages lacking HDAC3 show an M2-like phenotype and are hyperresponsive to IL4 (Mullican et al., 2011). HDAC inhibitors reverse the immunoregulatory effects of helminthes products in inflammatory diseases (Hoehe et al., 2000). Thus, the expression of histone-modifying enzymes is a key regulator in macrophage polarization and a manipulation of these epigenetic enzymes could be an interesting target for intervention.

Human monocytes are a heterogeneous cell population discriminated according to their CD14- and CD16-surface expression, and present distinct roles. Intermediate $CD14^{++}$ $CD16^{+}$ monocytes display a unique phenotype in terms of DNA methylation, which is distinct from the other monocyte subsets that might contribute to a potential role in inflammatory conditions (Zawada et al., 2016). Therefore, manipulation of epigenetic features could be of interest to modulate the inflammatory response in combination with other effective treatments.

3.1.2 T Cells

T helper $CD4^{+}$ cells have clearly been identified as key players in inflammatory diseases. The Th1 and Th17 pathways are involved in inflammatory conditions whereas the Th2 pathway plays a major role in allergic inflammation. The Th17 subtype is of special interest for inflammatory diseases as these cells may interact specifically with other adaptive immune cells in the environment of the disease manifestations such as the mucosa or epithelium to contribute to the defense against extracellular pathogens. Given their role at the interface with extracellular environment, there is strong evidence for an involvement of Th17 cells in psoriasis and IBD (Karczewski, Dobrowolska, Rychlewska-Hanczewska, & Adamski, 2016; Ueno, Ghosh, Hung, Li, & Jijon, 2015). The remarkable plasticity of Th17 cells to covert into T regulatory cells may explain their dual protective/inflammatory role in mucosal immunity especially with regards to microbiota tolerance (Ueno et al., 2015). The role of regulators such as cytokines in the differentiation of Th0 cells is well known, but changes in the epigenetic landscape are necessary in order to complete the differentiation process. There is increasing evidence for the role of DNA methylation in governing cytokine expression developmentally before and after Th0 differentiation, acting via gene silencing to organize signaling and transcription factor networks important for the maturation of cytokine expression in Th1/Th2/Th17 cells (Ansel, Lee, & Rao, 2003). Epigenetic changes at the level of DNA methylation and histone modifications within the *IFNγ* locus or the *IL4* locus have been widely studied during Th1 or Th2 differentiation, respectively (Ansel, Djuretic, Tanasa, & Rao, 2006; Djuretic et al., 2007; Nielsen & Tost, 2012) and functional $CD4^+$ T cells are required for the epigenetic remodeling of cytotoxic $CD8^+$ T cells upon antigen activation (Northrop, Thomas, Wells, & Shen, 2006). The development of the differentiated Th17 cells also relies on epigenetic modifications. TGFβ and IL6 regulate Th1 cell conversion into the Th17 subset via expression of Runx1 and enhanced H3K9 acetylation (Liu et al., 2015). Also, several miRs influence development of the Th17 lineage by targeting *BCL6*, *STAT1*, and *SOCS1*: some of them enhance Th17 differentiation whereas others promote Treg formation (Ueno et al., 2015). Finally, H3K4me3 modifications at the *IL17A* promoter and H3K27me3 at *IL4* or *IFNγ* promote the Th17 pathway. However, a bivalent activating/repressive chromatin environment characterized by the simultaneous presence of H3K4 and H3K27 trimethylation at gene-regulatory regions of some transcription factors was found in Th17 cells suggesting a reversible epigenetic state (Ueno et al., 2015).

3.1.3 B Cells

In contrast to autoimmune diseases, B cells have been little studied in inflammatory diseases. Although the pathophysiological knowledge does not support a major involvement of these cells in inflammatory diseases, B cells might still be relevant for some aspects. One of the adaptive immune effector mechanisms within the gut is the immunoglobulin A-producing B cell system that provides a first-line defense within the mucosa (Brandtzaeg, 2010). IBD lesions have been found to be markedly skewed toward IgG production (Brandtzaeg, Carlsen, & Halstensen, 2006). In a model of innate intestinal inflammation, B cells suppressed bowel inflammation (Mizoguchi, Mizoguchi, Takedatsu, Blumberg, & Bhan, 2002). In addition, circulating B cells from Crohn's disease (CD) and ulcerative colitis (UC) patients have recently been observed to demonstrate elevated basal levels of activation, expressing surface Toll-like receptor 2 and secreting high levels of IL8 (Noronha et al., 2009). Epigenetic mechanisms are involved at all stages of B cell development from immunoglobulin class switch DNA recombination through AID modulation and somatic hypermutation, to plasma cell or memory B cell differentiation (Bao & Cao, 2016; Zan & Casali, 2015). Butyrate, which is generated in the colon by commensal bacteria, has shown to modulate immunoglobulin class switch through its direct histone deacetylase inhibitor activity (White et al., 2014). Thus, therapeutic intervention targeting epigenetic modifications could potentially be relevant through their impact on B cells.

3.1.4 NK Cells

NK cells are crucial components of the innate immune system phenotypically characterized by $CD56^+$ $CD16^+$ and $CD3^-$ expression in humans. They represent 10% of the cells in the total PBMC population. Their role in viral infection and cancer as killers of abnormal cells has been well established. NK cells act through the release of cytotoxic granules as well as production of cytokines regulated by the balance between activating (NKG2D, NKp46, and FcγRIII) and inhibitory receptors (KIR, c-type lectine receptors) on the cell surface. Existing evidence suggests that NK cells are also involved in inflammatory disease. In IBD, NK cells are increased in the inflamed mucosa of patients and NK cell differentiation is accelerated in the lamina propria (Yadav, Chen, & Liu, 2011). Genetic association studies have implicated both *KIR* and *HLA-C* loci in IBD development, two genes involved in NK cell function (Wilson et al., 2010). NK cells can act on adaptive T- and B-lymphocytes and maintain immune homeostasis. Only few studies have investigated the influence of epigenetic modifications on the differentiation and maturation of NK cells. In these cells, *KIR* genes exhibited the major hallmarks of epigenetic activation, which were active histone status (absence of H3K9 dimethylation and presence of H4K8 acetylation) and were tightly regulated by DNA methylation of a CpG cluster in the promoter region (Santourlidis, Graffmann, Christ, & Uhrberg, 2008). Recently, adaptive NK cells have been described, which lead after CMV infection to an expansion of a different subset of NK cells that exhibit diverse functions. Bisulfite sequencing of genomic DNA of each cell subset showed different patterns of methylation especially on the *FCER1G* and *SH2D1B* (which encodes EAT2) promoters involved in NK activation and signal transduction (Schlums et al., 2015). Genome-wide DNA methylation analysis revealed that the methylation profile of adaptive $EAT2^-$ NK cells diverged from both early (2372 differentially methylated regions (DMRs)) and late mature NK cells with an enrichment of hypomethylated regions and strongly correlated with that of effector T cells (only 61 DMRs) (Schlums et al., 2015). To the best of our knowledge, no methylome analyses have been published on sorted NK cells in the context of inflammatory disease.

3.2 Resident and Target Cells

In contrast to autoimmune disease where the dysfunction is related to the immune cells, resident cells are directly causal in inflammatory conditions. The onset of the disease is linked to a deregulation or an altered homeostasis of these resident cells. The mechanism by which the disease is initiated and/or progresses has not been clearly identified, but current evidence supports a role of environmental factors including potentially infectious triggers, toxic exposition, or UV radiation. Furthermore, the intrinsic properties of the tissue that becomes inflamed play probably an important role in the preference of the disease site. Epigenetic mechanisms could provide responses to these unsolved interrogations.

In psoriasis or IBD, the resident cell is an epithelium (keratinocyte or colonic cell); a highly differentiated tissue derived from stem cells that underwent multiple differentiation steps under extrinsic condition orchestrated by epigenetic regulation in order to establish tissue-specific gene expression patterns (Abhishek & Palamadai Krishnan, 2016; Cavazza et al., 2016). The skin commensal flora and the intestinal microbiota can influence the function and the differentiation of epithelial cells through the modulation of the epigenome, transmitted from a generation to the next until mature cells are obtained (Cortese, Lu, Yu, Ruden, & Claud, 2016). The loss of this tightly controlled regulation could lead to inflammatory conditions and carcinogenesis. Several studies have shown that aging tissues show some degree of CpG island methylation in somatic gene different from their status at young age. This age-related methylation is likely the precursor for aberrant methylation patterns seen under pathological conditions (Ahuja, Li, Mohan, Baylin, & Issa, 1998). The fact that epithelium is a tissue with high turnover could accelerate age-related methylation and explain the preferential site for inflammatory diseases. Furthermore, age-related expression of genes is crypt specific (Cohn, Roth, Birkenmeier, & Gordon, 1991). This finding suggests a potential role of resident cells leading to an epigenetic mosaicism. Furthermore, mouse models with dextran sodium sulfate (DSS)-induced colitis showed DNA methylation changes in the colonic epithelium even in the absence of immune cells such as B or T cells providing convincing evidence that resident cells are solely responsible for the initial inflammatory response (Katsurano et al., 2012).

The DNA methylation landscape evolves throughout life (Kellermayer et al., 2010). Most cancers gain methylation at a number of gene-regulatory elements and these can already be observed in early abnormal lesions of colonic or breast epithelium (dysplasia) representing the continuum from a normal tissue to cancer (Fleischer et al., 2014; Issa, 2002; Subramaniam, Mizoguchi, & Mizoguchi, 2016). Thus, tissue-specific epigenetic variability, differential exposure to environmental factors, rates of proliferation, and intrinsic susceptibility to abnormal methylation are the key determinants for establishing the risk to develop inflammatory conditions or cancer (Allan, 2015). Finally, the combination of genetic variation leading to an increased susceptibility to disease as evidenced in genome-wide association study (GWAS) and the epigenome confers the total risk of developing inflammatory disease by integrating predisposing and acquired risk factors (Kellermayer, 2012).

4. EPIGENETICS IN CHRONIC INFLAMMATORY DISEASES

A number of potential clinical applications of epigenetics in diagnostics and therapeutics are currently receiving attention. Diagnostic applications of epigenetics include the use of biomarkers to confirm diagnosis and predict response to treatment. Epigenetic processes are involved in gene regulation of the different components of inflammation. In this part, we describe the current knowledge on epigenetic changes in chronic inflammatory diseases with special focus on DNA methylation and histone modifications. The epigenetic changes predisposing to cancer progression during the course of inflammatory diseases have been studied extensively (Fleisher et al., 2000; Gloria et al., 1996; Hsieh et al., 1998; Issa, Ahuja, Toyota, Bronner, & Brentnall, 2001; Sato, Harpaz, et al., 2002; Sato, Shibata, et al., 2002), but will not be discussed here.

4.1 Inflammatory Bowel Disease

IBD are relapsing conditions of the gastrointestinal tract characterized by chronic inflammation leading to abdominal pain, diarrhea, and alteration of the quality of life. Patients with uncontrolled IBD have an increased risk of developing colitis-associated colorectal cancer (CA-CRC) (Ahmadi, Polyak, & Draganov, 2009; Ekbom, Helmick, Zack, & Adami, 1990). Improved control of the disease is an important public health issue. The two major types of IBD are ulcerative colitis (UC) and Chron's disease (CD). Current knowledge supports the idea that IBD are caused by a complex interaction between the host that encompasses the intestinal microbiota, the immune system, the genetic background, and environmental factors such as diet, drugs, or smoking and the age of onset has considerably decreased during the last decades proving strong arguments for nongenetic factors (Hou, Abraham, & El-Serag, 2011; Jaenisch & Bird, 2003; Kellermayer, 2012; Kugathasan & Amre, 2006; Scarpa & Stylianou, 2012). Genetic studies have shown many loci associated with IBD, but the genetic factors account for only a modest proportion of disease heritability (Ellinghaus, Bethune, Petersen, & Franke, 2015). Furthermore, metabolic profiling provided strong evidence for an altered methylation metabolism in IBD model systems and folate supplementation did resolve at least partially colitis in a murine model, while treatment with a reversible *S*-adenosylhomocysteine hydrolase inhibitor led to disease exacerbation (Kominsky et al., 2011). Therefore, epigenetic factors that have been widely studied in IBD in the last decade may provide new insights into the pathogenesis of IBD.

Initially, most studies have focused on the development of neoplasia in the context of chronic UC or CD, based on the known role of altered DNA methylation patterns in sporadic colorectal cancer. Only recently, there has been a surge of interest to investigate DNA methylation patterns in the IBD pathogenesis itself.

Because of its role in carcinogenesis, DNA methylation has been the most studied epigenetic modification and some efforts have been focused on potential alterations in the enzymatic machinery that is responsible for establishment and maintenance of DNA methylation patterns, notably the DNMT. A GWAS metaanalysis revealed an association of IBD with a genetic polymorphism in the *DNMT3A* gene (Franke et al., 2010). Furthermore, higher expression levels of *DNMT1* and *DNMT3B* were reported in actively inflamed UC mucosa compared to paired quiescent samples (Saito et al., 2011). These reports highlight the potential role of DNA methylation in IBD pathogenesis.

From 1996 to 2010, most studies were limited to candidate gene approaches according to pathophysiological hypotheses and initially referred to carcinogenesis (Table 2). The very first study reported a global DNA hypomethylation in rectal mucosa from UC patients compared with healthy controls (HCs) (Gloria et al., 1996). DNA hypomethylation was also observed in patients with extensive UC compared with patients with an inactive UC. These results suggested a correlation between hypomethylated DNA profiles and proliferative activity of the colonic mucosa potentially explained by the in situ inflammation, a local background that may predispose patients to further develop colorectal neoplasm. Nevertheless, no complementary functional analyses were performed at this stage (Gloria et al., 1996). In the next years, five genes, all previously implicated in

Table 2 Candidate Gene Studies Analyzing Locus-Specific DNA Methylation Patterns in Inflammatory Bowel Diseases

Study (Year)	Disease	Aim	Studied Population	Studied Tissue	Comparative Group	Gene(s)	Methylation Status	Main Results	Functional Studies
Gloria et al. (1996)	UC	Mechanism of carcinogenesis	Extensive and long standing UC ($n=26$)	Rectal mucosa	Healthy controls ($n=11$)	NA	Hypomethylation	Global DNA hypomethylation and proliferative activity increased in UC vs HC	NA
Hsieh et al. (1998)	UC	Mechanism of carcinogenesis	Extensive and long standing colitis ($n=3$)	Serial biopsies ($n=89$) of the colon on colectomy specimen ($n=3$)	According to the level of dysplasia	*CDKN2A/p16*	Hypermethylation	Association between hypermethylation and the severity of the neoplastic progression (from low dysplasia to cancerous lesions)	NA
Fleisher et al. (2000)	UC +CD	Mechanism of carcinogenesis	UC and CD neoplasia	Dysplasia or carcinoma ($n=148$)	Matched normal control tissue from the same patients	*hMLH1*	Hypermethylation	Association between *hMLH1* hypermethylation and microsatellite instability (MSI)	Lack of hMLH1 protein expression in hypermethylated and MSI-high IBD neoplasms (immunohistochemical staining)
Wheeler et al. (2001)	UC	Mechanism of carcinogenesis	UC-associated colorectal cancer ($n=14$)	Colorectal cancer tissue	Sporadic colorectal cancer ($n=14$)	*CDH1*	Hypermethylation	No significant difference between UC-associated colorectal cancer and sporadic colorectal cancer	Reduced E-cadherin expression (immunohistochemical staining)
Issa et al. (2001)	UC	Mechanism of carcinogenesis	UC ($n=18$) with ($n=12$) or without ($n=6$) dysplasia and cancer	Colonic mucosa	Partly paired samples from the same patients and controls ($n=5$)	*ER*, *MYOD*, *CSPG2*, *CDKN2A/p16*, *hMLH1*, *MINT2*	Hypermethylation of *ER*, *MYOD*, *CDKN2A*, *CSPG2*	Hypermethylation in dysplastic/cancerous tissue compared to controls, large differences in pathologically normal tissue from UC cases compared to controls (field defect)	NA

Continued

Table 2 Candidate Gene Studies Analyzing Locus-Specific DNA Methylation Patterns in Inflammatory Bowel Diseases—cont'd

Study (Year)	Disease	Aim	Studied Population	Studied Tissue	Comparative Group	Gene(s)	Methylation Status	Main Results	Functional Studies
Azarschab et al. (2002)	UC	Mechanism of carcinogenesis	Patients with long standing UC (n=26)	Colonic mucosa	Patient with irritable bowel disease (n=15)	*CDH1*	Hypermethylation	Increased methylation with the different stages of UC-associated cancer; *CDH1* methylation: 0% HC, 6% nondysplastic UC, and 93% dysplastic UC	Decreased expression of E-cadherin in dysplastic UC sections (immunohistochemical staining)
Sato, Shibata, et al. (2002)	UC	Mechanism of carcinogenesis	UC-associated colorectal cancer (n=47)	Colonic mucosa with adenocarcinoma or dysplasia	Nonneoplastic mucosae from the same patients (n=5)	*HPP1*	Hypermethylation	Increased methylation status with the different stage of UC-associated cancer: *HPP1* methylation in 50% of adenocarcinomas and 40% of dysplasias. No *HPP1* methylation in normal mucosae	Restoration of *HPP1* expression in the colon cancer cell line HCT116 by 5-Aza-dC (demethylating agent)
Sato, Harpaz, et al. (2002)	UC	Mechanism of carcinogenesis	UC-associated colorectal cancer (n=40)	Colonic mucosa with adenocarcinoma or dysplasia	Ileal mucosa of normal control samples (n=30)	*CDKN2A/* $p14^{ARF}$	Hypermethylation	Increased methylation status with the different stage of UC-associated cancer: *CDKN2A/*$p14^{ARF}$ methylation in 50% of adenocarcinomas, 33% of dysplasias, and 60% of nonneoplastic UC mucosae. 3.7% of *CDKN2A/*$p14^{ARF}$ methylation in normal tissue	NA
Tahara et al. (2009a)	UC	Disease severity	UC patients (n=84)	Rectal inflammatory mucosa from UC patients	Paired normal terminal ileum from UC patients (n=23)	*PAR2*	Hypermethylation	Methylation level associated with the severity of the disease (total colitis vs rectal colitis)	NA

Gonsky et al. (2009)	IBD	Disease pathogenesis	IBD patients	T lymphocytes of the lamina propria vs T lymphocytes from the peripheral blood	Healthy controls	*IFNg*	Hypomethylation	Hypomethylation of *IFNg* in lamina propria T cells compared to peripheral blood. No difference according to diseased status of the subjects (IBD or HC)	Increased expression of *IFNg* associated with hypomethylated profiles of regions flanking the *IFNg* transcriptional start site
Tahara et al. (2009b)	UC	Disease pathogenesis	UC patients ($n=83$)	Rectal inflammatory mucosa from UC patients	Paired normal terminal ileum from UC patients ($n=17$)	*ABCB1*	Hypermethylation	Increased methylation level of *ABCB1* in rectal inflammatory mucosa compared to paired controls of the normal terminal ileum	NA
Lobaton et al. (2014)	UC +CD	Disease pathogenesis/ remission	CD ($n=9$) and UC ($n=29$), serial biopsies	Biopsy	NA	*TGFB2*, *SLIT2*	Hypermethylation	*SLIT2* methylation associated with persistence and acute inflammation	NA
Balasa et al. (2010)	UC +CD	Disease pathogenesis	CD ($n=18$) and UC ($n=20$)	PBMCs	Healthy controls ($n=49$)	*IRF5*	Hypomethylation	No significant difference between UC, CD, and HC	NA

IBD: Inflammatory bowel disease; UC: ulcerative colitis; CD: Crohn's disease; PBMCs: peripheral blood mononuclear cell; NA: not applicable; MSI: microsatellite instability.

carcinogenesis (*CDKN2A/p16*^{INK4A}, *CDH 1*, *hMLH1*, *HPP1*, and *CDKN2A/p14*^{ARF}), were found to be hypermethylated in the colonic epithelium of dysplastic or cancerous lesions of patients with UC or CD, with lower protein expression of hMLH1 and CDH1 in these lesions as a consequence of the hypermethylation (Azarschab, Porschen, Gregor, Blin, & Holzmann, 2002; Fleisher et al., 2000; Hsieh et al., 1998; Sato, Harpaz, et al., 2002; Sato, Shibata, et al., 2002; Wheeler et al., 2001). More recently, a number of candidate gene studies have focused on the IBD pathogenesis rather than on carcinogenesis and described new differentially methylated genes (*PAR2*, *ER*, *IFNγ*, *IRF5*, and *ABCB1*) that are mostly involved in innate immunity (Balasa et al., 2010; Gonsky, Deem, & Targan, 2009; Tahara et al., 2009a, 2009b; Wang et al., 2008).

Nevertheless, these candidate gene approaches were sometimes disappointing, providing negative results (Balasa et al., 2010) and together with the exciting expansion in technology available for genome-wide analyses. Ten epigenome-wide associations studies (EWAS) have been performed since 2011 in IBD with ever increasing coverage following the different generations of technological tools available (Adams et al., 2014; Cooke et al., 2012; Harris et al., 2014, 2012; Hasler et al., 2012; Koizumi et al., 2012; Lin et al., 2011; Lin et al., 2012; Nimmo et al., 2012). The most striking observation resulting from a comparative analysis of these EWAS is that differentially methylated sites or regions are very difficult to be replicated by independent groups. This can be explained by the great heterogeneity existing between these studies regarding the analyzed population (adult or pediatric, MZ twins discordant for the disease), the analyzed tissue (PBMCs or EBV immortalized B cells or colonic mucosa), and the reference population (HCs or macroscopically normal intestinal tissue from the same IBD patients) (Table 3). Also the location of the inflamed tissue within the colon might be a potential confounding factor, as a recent study showed highly divergent DNA methylation profiles between the sigmoid and the ascending colon, whereby the location in the colon showed more profound DNA methylation changes compared to the changes observed between IBD patients and controls at the same location (Kraiczy et al., 2016).

The analyzed tissue and the potential use of PBMCs as proxy for the inflamed colonic epithelial cells are still an issue of concern. Focusing on 22 genes involved in inflammatory response and autoimmunity, Karatzas et al. showed a moderate to strong correlation between methylation levels in colon biopsies and in peripheral blood both in CD and UC that could lead to a specific methylation signature (Karatzas et al., 2014). While some

Table 3 Overview on Genome-Wide Studies Analyzing DNA Methylation Patterns in Inflammatory Diseases

Study	Disease	Studied Population	Comparative Group	Method	Differentially Methylated Genes
Lin et al. (2012)	CD ($n=9$) and UC ($n=9$)	B cell lines from IBD patients (EBV transformed)	B cell lines (EBV transformed) from healthy siblings	GoldenGate BeadArray	IBD associated: *BCL3*, *IL12B*, *PPARG*, *TJP2*, *SMAD2*, *HCK*, *LMTK2*, *CASP2*, *SOX1*, *COL18A1*, *LMO1* CD specific: *SEPT5*, *GUCY2F*, *CLDN4*, *SERPINE1*, *LMTK2*, *MST1R*, *BGN*, *MXI1*, *IL16*, *GADD45A*, *EPHA2*, *RHOH*, *LIF*, *IL10* UC specific: *EYA4*, *DHCR24*, *EPS8*, *TBX1*, *TEK*, *PPAT*, *BAX*, *SEPT9*, *IL12B*, *SMARCB1*, *PPARG*, *CD81*, *ARHGDIB*, *MUC1*, *PLAT*, *HLA-DOB*, *ELK3*, *ERN1*, *MOS*, *LIG3*, *P2RX7*, *IL16*, *POMC*, *PECAM1*
Lin et al. (2011)	CD ($n=9$) and UC ($n=17$)	Diseased intestinal tissues from IBD patients	Macroscopically normal intestinal tissue from the same IBD patients	GoldenGate BeadArray	IBD associated: *BGN*, *SERPINAS*, *TNFRSF1A*, *AATK*, *GABRA5*, *MAPK10*, *STAT5* CD specific: *DLCI*, *LTB4R*, *APOC2*, *MPO*, *MFAP4*, *PECAM1*, *KRT13*, *MEG3*, *AFF3*, *RHO*, *MPL*, *SPARC*, *NEU1*, *LAT*, *FABP3*, *SPI1*, *GNAS*, *TJP2*, *IL18BP*, *S100A4*, *DLC1*, *FGF2*, *PDGFR*, *NOTCH4* UC specific: *NOTCH3*, *CDH17*, *PAD14*, *TNFSF8*, *MEST*, *HLA-DQA2*, *AATK*, *EPHX1*, *CLDN4*, *PIK3R1*, *FRK*, *HOXB2*, *FMR1*
Koizumi et al. (2012)	UC ($n=14$)	Nonneoplastic colon epithelial biopsies	Normal colonic mucosal tissues from HC	MS-AFLP	133 hypermethylated probes within 111 genes and 118 hypomethylated probes within 93 genes
Cooke et al. (2012)	CD ($n=32$) and UC ($n=32$)	Rectal biopsies from CD and UC patients with quiescent or active rectal inflammation	HC ($n=16$)	27K BeadChip	Inflamed CD, inflamed UD, or noninflamed UC vs controls > 30 loci including *THRAP2*, *FANCC*, *GBGT1*, *DOK2*, *TNFSF4*, *TNFSF12*, and *FUT7*

Continued

Table 3 Overview on Genome-Wide Studies Analyzing DNA Methylation Patterns in Inflammatory Diseases—cont'd

Study	Disease	Studied Population	Comparative Group	Method	Differentially Methylated Genes
Nimmo et al. (2012)	CD ($n=21$)	PBMCs	HC ($n=19$)	27K BeadChip	Hypomethylated: *CD22*, *CEBPE*, *ADPRH*, *SPI1*, *SSBP2*, *FLJ22746*, *RABGEF1*, *IL21R*, *RNASE3*, *CASP10*, *C10orf91*, *IL19*, *CTSG*, *SFTPD*, *ACOT11*, *TLR8*, *FYB*, *ICAM2*, *SLC5A8*, *CD300LF*, *GLRX2*, *RNASE2*, *SH3TC1*, *CD59*, *YIPF4*, *SH3TC2*, *PLEKHA6* Hypermethylated: *APBA2*, *MAPK13*, *CCDC47*, *ABI3*, *SLAMF7*, *PRF1*, *FASLG*, *CCL4*, *TNFSF12-TNFSF13*, *NDFIP1*, *ST6GALNAC1*, *S100A13*, *RIPK3*, *CD163*, *HRAS*, *RNF36*, *PROK1*, *ALOX12B*, *EVL*, *CFLAR*, *PPP6C*
Hasler et al. (2012)	Monozygotic twins discordant for UC ($n=10$) and 50 unrelated UC patients for pyrosequencing validation	Intestinal mucosa biopsies	Unaffected twin in the screening panel and HC for the validation panels	27K BeadChip and Nimblegen custom 385K array for validation	Loci with pyrosequencing validation: *CFI*, *FLNA*, *HKDC1*, *IGHG1*, *MT1H*, *PTN*, *SLC7A7*, *SPINK4*, *THY1*, *TK1*
Harris et al. (2012)	Monozygotic twins concordant for CD ($n=4$), UC ($n=3$) , discordant CD ($n=4$), UC ($n=7$) for MSAM and discordant monozygotic twins for CD ($n=3$), UC ($n=3$) , treatment naive pediatric CD ($n=14$), UC ($n=8$), HC ($n=14$) for 450K BeadChip	PBMCs	HC ($n=14$) for 450K BeadChip only	MSAM and 450K BeadChip, pyrosequencing for validation	*TEPP*

Harris et al. (2014)	Treatment naïve CD ($n=15$) and UC ($n=9$), remission UC ($n=2$)	Colonic biopsy	HC ($n=22$)	450K BeadChip	182 CD-associated probes (108 genes), 3365 UC-associated probe (2243 genes) 54% overlap between CD and UC 86 DMRs associated with gene expression changes in UC, e.g., *IFITM1*, *ITGB2*, *S100A9*, *SLP1*, *SAA1*, *STAT3*
Kang et al. (2016)	UC ($n=8$)	Colonic biopsy	HC ($n=3$)	450K BeadChip + MSP for validation ($n=79$)	48 Probes, validated *FAM217B*, *KIAA1614*, *RIBC2*
Kraiczy et al. (2016)	Treament naïve pediatric patients UC ($n=10$) and CDS ($n=10$)	Purified colonic epithelial cells	HC ($n=10$)	450K BeadChip + pyrosequencing for validation	233 differentially methylated probes; 94 in regions with varying methylation levels during colonic development Genes include: *MUC2*, *TL3*, *TLR4*, *FUT3*, *FUT6*, *PIGR*, *TFF3*, *HLA-C*, *HLA-DMA*, *HLA-DMB*, *HLA-DP1*, *HLA-DRA*, *TDC7L1*, *WIF1*, *OLFM4*
Adams et al. (2014)	Pediatric CD ($n=36$)	PBMCs	HC ($n=36$)	450K BeadChip	Loci with significant differential methylation profiles (probes or regions): *RPS6KA2*, *ARHGEF3*, *CFI*, *TNFSF10*, *SOCS3*, *CX3CR1*, *SBNO2*, *IL18RAP*, *VMP1/MIR21*, *ZBTB16*, *TNF*, *RUNX3*, *SLC15A4*, *TOLLIP*, *CD247*, *ITGB2*, *PRF1*, *GPR56*, *UBASH3A*, *DIABLO*, *ZFYVE28*, *PIEZO1*, *WRAP73*, *ZBTB12*, *PRF1*, *CETP*, *HGF*
Karatzas, Mantzaris, Safioleas, and Gazouli (2014)	CD ($n=12$) and UC ($n=12$)	PBMCs	HC ($n=12$)	EpiTect Methyl II DNA restriction kit	Hypermethylated (UC vs controls): *CXCL14*, *CXCL5*, *GATA3*, *IL17C*, *IL4R* No significant differences between CD and controls
Hughes et al. (2014)	BD ($n=16$)	PBMCs	HC ($n=16$)	450K BeadChip	Hypomethylated: *RAC1*, *RGS14*, *HLA-DRB6*, *HLA-DRB5*, *BCYRN1*, *GJB1*, *GDAP2* Hypermethylated: *MRI1*, *BAIAP2L1*, *JRK*, *LCE1C*, *SMYD3*, *C3ORF39*, *NBPF4*, *TUBB8*

Continued

Table 3 Overview on Genome-Wide Studies Analyzing DNA Methylation Patterns in Inflammatory Diseases—cont'd

Study	Disease	Studied Population	Comparative Group	Method	Differentially Methylated Genes
Gervin et al. (2012)	Monozygotic twins discordant for psoriasis ($n=27$)	$CD4^+$ ($n=17$) and $CD8^+$ ($n=13$) sorted cells (from PBMCs)	Unaffected twin	27K BeadChip	Genes with a correlated difference in DNA methylation and gene expression in $CD4^+$ T cells: Down regulated: *LDHC, TNFSF11, CSF2, GIMAPS, PTHLH, DRD1, SHKBP1, ARHGAP25, LILRB1, BCL9, NVL, LOC202459, CYP27B1,KIAA0652, MRPL49, CTSW, FLJ36116, CTNNA1,HLA-DPA1, NLN,DPM1, CCL3L1* Upregulated: *IL13,MGC3207, GIMAP1, NPL, EDARADD, TCP11L1, LIME1, TNFRSF9, GSTT1, PHEX, CCL1, FLI1, ZNF622, CTSH, ANP32E, CLSTN3, SCN4A, OSBPL7, DDX43, JAK3, SFRS5, SLAMF7, HIST1H1B, CD53, SAMD10, SPESP1, LPIN2, MEF2C, HRIHFB2122, ALOX5AP*
Park, Han, Park, Kim, and Kim (2014)	Psoriasis ($n=15$)	$CD4^+$ sorted cells (from PBMCs)	HC ($n=11$)	MeDIP-seq	Hypermethylated genes in $CD4^+$ T cells: *ALOX12, ANKRD30A,HOXA5, EXOC3L2, TPTE, AGAP2, SPESP1, TDRD9, MEST, C19orf77, LOC100130872,KCNQ1DN, C6orf201, TP73, C19orf34, DCAF4L1, NCRNA00164, KCNAB3,FANK1,SOHLH1, C4orf44, ZNF469,UCN, DIO3, CCDC154, OTUD6A, HIST1H1A, DYSFIP1, PPAPDC3, KRTCAP3, FTMT, CATSPER2, KCNG2, DUSP22, HDGFL1*
Roberson et al. (2012)	Psoriatic skin ($n=12$)	Skin biopsies	Normal skin of psoriasis patients ($n=8$) and HC ($n=10$)	27K BeadChip	1108 differentially methylated CpG in psoriasis patients vs HC The top 10 most significant CpG sites with differential methylation in psoriatic vs normal skin (HC): *FOLR1, MCF2L, S100A9, C10orf99, S100A12, KYNU, EFNA3, HSPB2, IFI27, CHMP7*

HumanMethylation450 DNA BeadChip (485,000 methylation sites covering all genes); 27K BeadChip: HumanMethylation27 BeadChip (27,578 CpG and >14,000 genes); GoldenGate BeadArray: Illumina GoldenGate BeadArray (1505 CpG sites—807 genes), MSAM: methylation-specific amplification microarray (Shen et al., 2007); MS-AFLP: methylation-sensitive amplified fragment length polymorphism (Yamamoto & Yamamoto, 2004); MeDIP-seq: methylated DNA immunoprecipitation and sequencing.

EWAS in PBMCs were able to identify differentially methylated genes (Nimmo et al., 2012), other EWAS using for example PBMCs from a number of MZ twins discordant for IBD, as well as unrelated pediatric IBD patients failed to identify differentially methylated genes in PBMCs with the exception of a single gene (*TEPP*) suggesting that PBMCs might not be well suitable to identify IBD-related methylation changes (Harris et al., 2012).

Nevertheless, an EWAS performed on 10 MZ twin pairs discordant for UC, identified 61 differentially methylated loci with corresponding differentially expressed transcripts (Hasler et al., 2012). Among these, *BRINP3* has been independently replicated showing a hypermethylated profile of the promoter region with a corresponding downregulation of the *BRINP3* transcript in the rectal mucosa of UC patients (Smith et al., 2014). No difference in *BRINP3* expression was observed in circulating leukocytes suggesting a tissue-specific alteration in DNA methylation patterns of this gene. Low levels of *BRINP3* expression were previously reported in a microarray study performed on biopsies taken from quiescent UC patients suggesting that *BRINP3* is not induced by inflammation, but rather specific for colonic mucosal tissue in UC (Planell et al., 2013). Interestingly, GWAS have never identified *BRINP3* as associated with IBD, confirming that epigenomic studies can successfully identify new loci in chronic inflammatory diseases.

Some EWAS performed in IBD reported a significant enrichment of methylation changes within IBD-associated loci identified through genetic studies suggesting the potential link between germline genetic variations and epigenetic risk factors in IBD (Adams et al., 2014; Cooke et al., 2012; Nimmo et al., 2012). An EWAS performed on whole blood samples from CD patients compared with controls identified 35 differentially methylated loci located within 50 kb of one of the 71 known CD-associated loci (Nimmo et al., 2012). Some of these sites were hypomethylated ($n=21$); others were hypermethylated ($n=14$). In another EWAS on rectal biopsies from IBD patients with active rectal inflammation compared with normal colon mucosa from HCs, 7 differentially methylated probes in CD-associated loci (including *TAP1*, *PTRF*, *PSMB8*, *IL8RB*, *PKLR*, *C6orf25*, and *KCNK4*) and 11 in UC-associated loci (including *KCNK4*, *ICAM3*, *C6orf25*, *CARD9*, *IL8RB*, *IL8RA*, *C6orf27*, *TNNI2*, *CDH1*, *ADA*, and *SMPD3*) were detected (Cooke et al., 2012). Interestingly, *CARD9*, which is an IBD-associated locus identified through GWAS approaches, has been found to be differentially methylated in both PBMCs

(Nimmo et al., 2012) and in inflamed mucosa from UC patients compared with normal mucosa from HCs (Cooke et al., 2012).

The study of DNA methylation profiles in IBD has delivered some key messages. Similar to genetic studies and GWAS, the identification of differentially methylated genes potentially involved in the pathogenesis of IBD is critical to improve our knowledge on the inflammatory deregulation leading to the disease (Nimmo et al., 2012). DNA methylation changes in colonic epithelial cells are accelerated in IBD because of higher cell turnover in the context of local inflammation (Issa et al., 2001). The increased DNA methylation observed in colonic cells of patients with colitis could lead to genetic instability and cancer. This result illustrates the continuum from a normal mucosa to inflamed tissue and dysplasia or IBD-related colorectal cancer. Many of the identified DMRs are found in regions exhibiting dynamic methylation changes throughout the development and differentiation of the intestinal epithelium (Kraiczy et al., 2016) and this increased plasticity might render them more susceptible to changes during inflammation. Ultimately, it will be of great interest to prevent the occurrence of aberrant DNA methylation patterns or restore normal DNA methylation levels at an early stage of epigenetic deregulation in colonic mucosa, which otherwise can ultimately lead to cancerous lesions. The different studies have suggested new therapeutic targets as well as provided support for potential new therapeutic options through the modulation of epigenetic changes.

Very recently it was shown, that aberrant DNA methylation patterns in a DSS-induced colitis mouse model can be transmitted through the paternal lineage. DNA methylation changes can be detected in the colonic cells and the sperm of the fathers, but also in the colonic epithelial cells of the offspring, and lead to metabolic changes and an increased susceptibility to DSS-induced colitis in the offspring (Tschurtschenthaler et al., 2016) providing new evidence for novel mechanisms of transmission and heritability of the inflammatory disease.

Moreover, differences in the methylation profile between noninflamed tissue and mucosa from HCs confer a possible tool for diagnosis. Although there seems to be little overlap between the DNA methylation profile in blood cells and colonic mucosa and some blood-based studies failed to detect any significant methylation changes, it will be important to develop in the future a specific methylation signature as noninvasive biomarker for diagnosis, disease prognosis, or prediction of treatment outcome. As shedding of intestinal epithelial cells and cell death is increased in IBD

(Blander, 2016), and dying cells release short DNA fragments displaying tissue-specific DNA methylation patterns, methylation analysis of cell-free circulating DNA might be an option to detect colonic DNA methylation changes without the need for biopsies (Tost, 2016b).

Posttranslational histones modifications have been less studied in IBD. Chromatin compaction within the nucleus is regulated by the degree of acetylation/deacetylation of histone residues and thereby regulates among other factors the access of transcription factors to the DNA. Acetylation is mediated by opposing activities of HDACs and HATs. HDAC inhibitors shift this balance toward an accumulation of hyperacetylated histones, thus modifying gene expression in a cell- and gene-specific manner, and have shown promising results by significantly reducing disease severity in murine models of systemic lupus erythematosus and arthritis (Lee et al., 2015; Li et al., 2013; Reilly et al., 2008). The potential protective effect of HDAC inhibitors has been reported in different experimental mice models of colitis (DSS- or 2,4,-trinitrobenzene sulfonic acid (TNBS)-induced colitis) through the induction of apoptosis and the suppression of proinflammatory cytokines (Glauben et al., 2006). Inhibition of HDACs is also able to increase *FoxP3* gene expression and to improve the suppressive function of regulatory T cells in DSS-induced colitis leading to a milder disease compared to untreated mice (de Zoeten, Wang, Sai, Dillmann, & Hancock, 2010). Conversely, in two different rat models of experimental colitis (TNBS- and DSS-induced colitis), a global increase in histone 4 (H4)-acetylation was reported in inflamed mucosa compared to noninflamed tissue especially on lysine K8 (H4K8) and K12 (H4K12) (Tsaprouni, Ito, Powell, Adcock, & Punchard, 2011). Similar results were found in inflamed mucosa and Peyer's patches of patients with CD (Tsaprouni et al., 2011). However, as discussed by the authors, it was not clear whether H4 acetylation was causal or downstream to activation of inflammation, but suggests that HAT inhibitors may be an interesting therapeutic approach in IBD (Tsaprouni et al., 2011). However, data on the association of increased histone acetylation with disease is controversial with conflicting reports and will need further investigation. A recent work found that lysine acetyltransferase 2B (KAT2B) was downregulated in inflamed mucosa of IBD patients leading to a decreased H4K5 acetylation and IL10 expression (Bai et al., 2016). Thus, the equilibrium between HAT and HDAC activities should be relevant in disease, but the regulation of this balance remains unclear. In addition, HDACs and HATs directly modulate the acetylation state of nonhistone proteins involved in the regulation of gene expression such as transcription factors like NFkB or tumor suppressor genes (Kiernan et al., 2003). In a mice model of colitis, it has been shown that Sirt2 deficiency was associated with a more severe DSS-induced colitis compared with the wild-type littermates. Sirt2, a member of the NAD^+-dependent deacetylases, has been shown to directly bind and deacetylate the p65 subunit of NFkB. Sirt2 deficiency induces increased levels of proinflammatory cytokines through hyperacetylation of NFkB that affects macrophage polarization and limits M2 antiinflammatory differentiation (Lo Sasso et al., 2014).

Because of its location in the gut, the inflamed cells in IBD are in close relation with the extracellular environment. Bacteria of the gut flora can produce epigenetically active metabolites such as folate, biotin, and short chain fatty acids such as butyrate, propionate, and acetate. Short chain fatty acid can increase HAT activity, which is regulated through the tricarboxylic acid cycle and inhibit HDAC activity, and have been shown to exert antiinflammatory effects and improve epithelial barrier function in a number of animal models and delay the development of cancerous lesions (Mischke & Plosch, 2016). Multiple reports support the hypothesis that dysbiosis, which denotes the unfavorable alteration of the gut microbiota composition, may have a key role in IBD (Couturier-Maillard et al., 2013). Among the bacterial components of the microbiome, *Roseburia* is able to produce butyrate, a HDAC inhibitor that can decrease intestinal inflammation and has demonstrated positive effects on the treatment of refractory distal UC (Berni Canani, Di Costanzo, & Leone, 2012; Fofanova, Petrosino, & Kellermayer, 2016; Machiels et al., 2014; Vernia et al., 2003). Moreover, components of the microbiome change during life and a specific combination could predispose to IBD. A recent work has demonstrated an association between microbiome composition and DMRs in colonic mucosal cells from IBD patients compared to age-matched HCs (Harris et al., 2016). For example, DNA methylation of *KHDC3L* was correlated with increased abundance of *Roseburia* or *Bacteroides* and decreased abundance of *Streptococcus*. Because of the modification of the bacterial component in the gut of UC patients, characterized notably by reduced representation of *Roseburia*, the methylation of the promoter of this gene tended to be decreased in UC. Furthermore, reduced levels of *Roseburia* and others butyrate-producing bacteria in IBD can lead to an increased inflammation through the lack of butyrate. Thus, it is very probable that the host epigenome and the commensal microbiota act in concert to modulate IBD.

4.2 Spondyloarthritis

In contrast to IBD, the epigenetic component of SpA has been poorly studied to date. To the best of our knowledge, no genome-wide analyses of DNA methylation profiles have so far been published in SpA and only a single candidate gene study on *SOCS1* has been reported (Lai et al., 2014). Forty-three serum samples from ankylosing spondylitis (AS) patients were compared to six serum samples from HCs representing thus a very limited cohort. The authors did not detect any methylation of *SOCS1* in HC whereas AS patients showed varying levels of *SOCS1* methylation as assessed by methylation-specific PCR. The significance of the difference was confirmed using a more resolutive and quantitative technology (bisulfite pyrosequencing). Significantly higher methylation levels of *SOCS1* were observed in patients with structural damage, and with systemic inflammation measured by acute phase reactant (erythrocyte sedimentation rate and C-reactive protein) or by serum levels of IL6 or TNFα. The authors proposed that an uncontrolled inflammation could result in the activation of the JAK/STAT signaling pathway leading to the epigenetic silencing of *SOCS1*, which is a negative regulator of JAK/STAT signaling and in turn intensifies the JAK/STAT signaling through a negative loop. In that case, epigenetic changes may be secondary to inflammation and not the etiologic mechanism of the disease (Lai et al., 2014).

Several studies have focused on histone modifications in AS. Both HAT and HDAC activities were found to be decreased in PBMCs from AS patients ($n=21$) compared to PBMCs from HCs ($n=38$). TNFα levels weakly correlated with HDAC activity (Toussirot et al., 2013). The same group further reported that SpA patients treated with NSAIDs had increased Sirtuin 1 activity (SIRT1, a class III HDAC) in both nuclear and cytoplasmic sites compared to controls that could favor increased serum levels of IL6 (Wendling et al., 2015). A complex mechanism involving HDAC3 and miRNA regulation has also been reported in AS (Jiang & Wang, 2016). PBMCs from 20 AS patients and 20 HCs were included in the study and HDAC3 expression (mRNA and protein) was increased in AS, whereas miR-130a was decreased. MiR-130a downregulation led to an increased expression of TNFα in PBMCs. HDAC3 was demonstrated to bind the promoter of *miR-130a* in PBMCs and the authors suggested that HDAC3-mediated histone deacetylation contributes to the regulation of miR-130a expression (Jiang & Wang, 2016).

4.3 Psoriasis

Several lines of evidence support the involvement of epigenetic factors in the pathogenesis of psoriasis. First, the concordance rate among MZ twins is only 35–72% (Ballestar, 2010; Bowcock, 2005; Grjibovski, Olsen, Magnus, & Harris, 2007). Second, environmental factors that have been reported to influence the course of the disease include traumatism, stress, infection, alcohol consumption, drugs, and smoking status. Most of them are recognized to induce epigenetic changes (Alegria-Torres, Baccarelli, & Bollati, 2011; Ladd-Acosta, 2015). Third, DNMT1 expression, required for maintaining DNA methylation patterns, was found to be higher in psoriatic PBMCs compared to PBMCs from HCs whereas the methyl-binding domain protein 2 and the methyl-CpG-binding protein 2 (MeCP2) were both significantly downregulated in psoriatic PBMCs (Zhang, Su, Chen, Zhao, & Lu, 2010).

Few gene candidate approaches have been performed in psoriasis. The promoter region of *SHP1* (*PTPN6*) was found to be differentially methylated in psoriasis (Ruchusatsawat, Wongpiyabovorn, Shuangshoti, Hirankarn, & Mutirangura, 2006). This study was performed on microdissected paraffin-embedded tissues from psoriatic patients and normal controls. The promoter 2 of *SHP1* was found demethylated (68%) in keratinocytes of psoriatic patients compared with normal skin (95%) with a corresponding increased expression of *SHP1* isoform II in case of a demethylated promoter. Methylation studies of genes involved in the regulation of cell cycle (*CDKN2B/p15*, *CDKN2A/p16*, and *CDKN1A/p21*) have shown lower promoter methylation of *CDKN2B* and *CDKN1A* and increased transcription levels in bone marrow mononuclear cells from psoriatic patients (Zhang, Zhang, Li, Yin, & Niu, 2009) and *CDKN2A* hypermethylation in psoriatic epidermis and decreased expression of *CDKN2A* mRNA (Chen et al., 2008). Unfortunately, the consequences of altered methylation profiles at the *SHP1*, *CDKN2B*, *CDKN2A*, and *CDKN1A* promoters in the pathogenesis of psoriasis are still unknown. Furthermore, an increase in global DNA methylation levels was reported in psoriatic PBMCs compared to HCs (Zhang et al., 2010).

Several genome-wide methylation studies have been further performed in disease-relevant cell types. The study of epigenetic variation in MZ twins discordant for psoriasis is a very interesting approach to investigate the role of epigenetic mechanisms involved in the disease (Ballestar, 2010). Gervin et al. performed a methylation study on $CD4^+$ and $CD8^+$ sorted cells from 27 MZ pairs discordant for psoriasis (Gervin et al., 2012). Unfortunately, the authors did not find any significant differentially methylated sites between unaffected and affected twins, possibly due to the low number of analyzed MZ pairs. However, in $CD4^+$ cells, a combined analysis of DNA methylation and gene expression identified a correlation between methylation profiles and gene expression for several loci (Table 3). Some of these genes have also been reported to be associated with psoriasis susceptibility in previous GWAS (*IL13*, *TNFSF11*, *PTHLH*, and *ALOX5AP*). In dermal mesenchymal stem cells (MSCs), aberrant methylation (96 hyper- and 234 hypomethylation) was described for genes involved in cell signaling, cell communication and migration, or keratinocytes differentiation and proliferation, underlying the potential role of an epigenetic deregulation of MSCs in the pathogenesis of psoriasis (Hou et al., 2013). In naïve $CD4^+$ T cells, 26 hypomethylated DNA regions compared to HCs were found that displayed an at least four-fold difference when enriching for methylated regions using an MDB-based sequencing approach in male patients and controls (Han et al., 2012). Most DMRs were associated with other epigenomic signals like histone modifications or transcription factor-binding sites. When focusing on gene-related sequence features 124 genes were identified to be hypermethylated within the promoter regions, most of them ($n=121$) being immune-related genes located on the X chromosome (Han et al., 2012). Using MeDIP-seq technology, these results were confirmed on sorted $CD4^+$ T cells (Park et al., 2014). Hypermethylated profiles in the promoter regions were found in $CD4^+$ T cells from 15 psoriasis patients compared to 11 HCs. Thirty-five genes were differentially methylated (Table 3), but only four of them had a significant association between DNA hypermethylation and decreased gene expression (*PPAPDC3*, *TP73*, *FANK1*, and *CATSPER2*).

Further genome-wide high-throughput analyses have been performed on psoriatic skin lesions compared to uninvolved or normal skin. Roberson et al. detected 1108 differentially methylated CpG sites when comparing psoriatic skin ($n=12$) and normal skin ($n=10$) (Roberson et al., 2012). Among those most differentially methylated sites, 12 CpGs mapped to the epidermal differentiation complex that is critical for epidermal development. These sites showed decreased methylation levels in psoriatic skin compared to normal skin and mapped close to genes upregulated in psoriasis (*S100A3*, *S100A5*, *S100A7*, *S100A12*, *SMCP*, *SPRR2A*, *SPRR2D*, *SPRR2E*, and *LCE3A*). A subset of these hypomethylated loci was associated with an increased expression of the nearby genes such as *KYNU*, *OAS2*, and *SERPINB3*, whose high expression was previously demonstrated to differentiate psoriasis from other skin lesions (Guttman-Yassky et al., 2009). These results emphasized the potential close relationship between methylation profiles of the diseased skin and the expression of a set a disease-associated genes. Furthermore, uninvolved skin from psoriatic patients showed intermediate levels of differential CpG methylation without any correlation with the expression of nearby genes suggesting that some differentially methylated loci could involve ncRNAs or distant regulatory elements such as enhancers rather than genes themselves (Roberson et al., 2012). MeDIP-Seq analysis revealed a higher number of hypermethylated genes rather than hypomethylated genes in psoriatic skin samples. Gene ontology analysis showed that these aberrantly methylated genes belonged to the immune system, cell cycle regulation, and apoptotic mechanisms (Zhang et al., 2013). In a three-stage EWAS (I, paired psoriatic (PP) and uninvolved skin (NP) sample; II, psoriatic skin (PP) vs normal skin (NN); and III, replication cohort PP vs NN), Zhou identified 262 DMRs and a robust association (sites with methylation differences across all three stages) for nine skin-specific DMRs and psoriasis: *S100A8*, *AHDC1*, *DLGAP4*, *ECE1*, *CYP2S1*, *SPIRE2*, *MAN1C1*, *EIF2C2*, and *ATG7* (Zhou et al., 2016). These associations were not found in PBMCs suggesting the existence of skin-specific methylation profiles in the pathogenesis of psoriasis. DMRs concerned genes mapping to the epidermal differentiation complex region (*S100A* family), *CYP2S1* involved in aberrant cell growth and differentiation of psoriatic cells or *EIF2C2*, which participates in microRNA-mediated mRNA silencing (Zhou et al., 2016).

The impact of treatment on epigenetic marks has been poorly studied to date. UVB phototherapy is a well-established first-line treatment for extensive psoriatic lesions and is effective in about 70% of patients. Gu et al. performed a genome-wide methylation analysis to assess changes in the DNA methylation profiles induced by phototherapy (Gu, Nylander, Coates, Fahraeus, & Nylander, 2015). In this study, 482,421 CpG loci were assessed on skin samples taken from psoriatic lesions of 12 patients before and after UVB treatment. After 2–3 months of treatment 2108 CpG sites were significantly modified by phototherapy. The corresponding genes were mainly involved in inflammatory response, cytoskeleton organization, response to hormone stimulus, regulation of cell motion, and regulation of programmed cell death. Most DNA methylation changes were located in distal gene-regulatory elements such as enhancers. Moreover, DNA methylation profiles of psoriatic samples tended to become more similar to normal skin after phototherapy, especially for patients demonstrating a good response to treatment. Unfortunately, early changes (after 1 month of treatment) in methylation profiles were unable to predict response to UVB treatment in this small cohort of patients. Methylation changes induced by anti-TNF were assessed on skin biopsies from a limited number of patients ($n=5$) before and after 1 month of adalimumab treatment (Roberson et al., 2012). The authors observed that the mean methylation levels of treated samples increased and tended to become similar to that of uninvolved skin, similar to what was described after phototherapy treatment. These very preliminary results suggest that the assessment of a limited number of CpG loci in psoriatic skin lesions before treatment might in the future help to predict the response to different treatments. Nevertheless, larger number of patients must be studied in order to provide more robust conclusions.

Zhang et al. assessed histone modifications among 30 psoriasis patients and 20 HCs (Zhang, Su, Zhao, Huang, & Lu, 2011). This study performed on PBMCs assessed the acetylation level of H3/H4 and the methylation profile of H3K4/H3K27. The authors reported a global histone H4 hypoacetylation in psoriatic patients, but no significant differences in H3 acetylation or H3K4/H3K27 methylation. The expression level of *P300* and *CBP*, two histone acetyltransferases, and SIRT1, a histone deacetylase, were found to be decreased at the mRNA level in PBMCs from psoriatic patients compared with HCs while *HDAC1*, and the HMTs *SUV39H1* and *EZH2* were found to be increased. However, the establishment of a more repressive chromatin environment was—as stated earlier—not supported by global changes in the histone methylation. The use of the heterogeneous cell populations of PBMCs in both studies raises nonetheless questions on the functional relevance of the observed expression changes. Increased *HDAC1* mRNA expression was also reported in psoriatic skin (Tovar-Castillo et al., 2007), but no difference in HDAC1, HDAC2, and HDAC3 activity was evidenced in psoriatic lesions in a more recent publication (Ekman & Enerback, 2016), suggesting that HDACi may not be a relevant treatment in psoriasis.

4.4 Epigenetic Studies in Other Chronic Inflammatory Diseases

Epigenetic studies in other chronic inflammatory disease (sarcoidosis, Behcet's disease (BD)) are still limited to date, at least regarding the analysis of DNA methylation profiles or histone modifications.

Sarcoidosis is a chronic and systemic granulomatous disease characterized by noncaseating epithelioid cell granuloma typically affecting the lung, eye, lymph nodes, liver, and other organ systems. Aging-related alterations of subtelomeric methylation patterns in sarcoidosis patients have been observed (Maeda, Guan, Higuchi, Oyama, & Makino, 2009). Telomeres are a structure composed of thousands of hexamer repeats (TTAGGG/AATCCC) at the ends of eukaryotic chromosomes. Telomere shortening is frequently used to indicate cellular senescence as telomeres shorten with every cell division due to the mechanism of DNA replication and to the inability of complete DNA duplication at the chromosome ends. Telomere length can therefore be considered as a biological clock. Telomere shortening is accelerated in several chronic conditions (e.g., type 2 diabetes mellitus, Alzheimer's disease, obesity) and by environmental factors (smoking, for example). Maeda et al. observed an age-related attrition of telomere length in peripheral leucocytes from both controls and sarcoidosis patients, but such telomere shortening was increased among sarcoidosis patients and associated with subtelomeric hypomethylation, an event occurring earlier among patients compared to HCs (Maeda et al., 2009).

BD is a multisystem and chronic inflammatory condition characterized by mucocutaneous lesions and vasculitis. As other chronic inflammatory diseases, BD is a complex and multifactorial disease with a genetic background (a strong association with HLA-B51, for example), but also involves environmental and epigenetic factors in its pathogenesis. Several works have reported miRNAs deregulation in BD, but are outside the scope of the present review (Choi et al., 2015; Deng et al., 2015; Qi et al., 2013, 2014). Yuksel et al. analyzed the methylation status of interspersed repetitive sequences such as *LINE1* and Alu elements and observed a hypomethylation of Alu sequences in neutrophils and PBMCs of inactive BD patients compared with HCs (Yuksel et al., 2016). The methylation patterns of *LINE1* sequences were identical between BD patients and controls in both neutrophils and PBMCs.

Hughes et al. were the first to describe modifications in genome-wide methylation patterns in leucocytes from patients with BD (Hughes et al., 2014). Sixteen untreated male BD patients were analyzed and methylation profiles compared to age, sex, and ethnicity-matched controls. The methylation study was performed using the HumanMethylation450 BeadChip on sorted cells (monocytes and $CD4^+$ T cells). Numerous sites were found differentially methylated between patients and controls (383 in monocytes and 125 in $CD4^+$ T cells), with an enrichment of genes associated with cytoskeletal remodeling. The authors reported a reversal of aberrant DNA methylation observed between patients and controls after treatment, especially in sorted monocytes. This study suggests that reversible epigenetic changes in cytoskeletal remodeling genes might play an important role in the pathogenesis of BD and could lead to the identification of new targets for disease monitoring and for BD treatment.

5. CONCLUSIONS

While still in its infancy, the study and understanding of epigenetic factors involved in chronic inflammatory diseases will help us to better understand disease pathogenesis and to identify new disease pathways and potentially novel targets for treatment. However, current approaches have some limitations. First, many studies have been performed on a limited number of patients and controls and results failed to be independently replicated. Second, differentially methylated sites or regions are not always assessed for correlation with changes in gene expression or altered transcription factor binding, thereby not allowing to establish a clear link between the observed methylation profiles and functional consequences for most identified loci. Third, results on PBMCs are difficult to extrapolate because they reflect a mixture of cells with diverse functions. It is important to perform new analysis on purified and well-defined disease-relevant cell populations. Last, many studies provide an epigenetic profile at a precise time point after disease onset rather than describing the temporal and spatial changes of the epigenome leading to disease development. It is therefore unclear whether the observed profiles are the cause or consequence of the disease, and whether they are influenced by environmental factors, by age or treatment. Future longitudinal studies involving repeated biological sampling whenever possible and larger number of patients and matched controls with detailed phenotypic characterization will undoubtedly answer these questions.

Several studies performed in chronic inflammatory diseases have demonstrated a close relationship between genetics and epigenetics. Some of the genes with altered DNA methylation profiles or changes in histone modifications have previously been identified in genetic studies as disease-associated genes. Farh et al. reported a fine-mapping algorithm (probabilistic identification of causal SNPs), built to identify candidate causal variants for 21 chronic autoimmune and inflammatory diseases, among which figured psoriasis, SpA, CD, and UC (Farh et al., 2015). In all of these chronic affections, the authors reported an enrichment of SNPs in gene-regulatory elements of specific cell types (among which monocytes, Th0, Th1, Th2, and Th17) and found that 90% of causal variants were noncoding, with 60% mapping to immune-cell enhancers, many of which gained histone acetylation and transcribed enhancer-associated RNAs upon immune stimulation. These data suggest that studying the epigenetic regulation in chronic inflammatory disease will allow a better understanding of their pathogenesis, but also establish a link between genetics and epigenetics that could explain the functional consequences of several SNPs located within noncoding regions. For example, two SNPs (respectively, rs11209032 in the *IL-23R/IL-12Rβ2* intergenic region and rs4648889 upstream of the *RUNX3* promoter) significantly associated with AS, were shown to alter the level of H3K4me1 methylation leading to a decreased activity of the corresponding gene. *RUNX3* mRNA was significantly decreased in patients who carried the AS-risk A allele (Roberts et al., 2016; Vecellio et al., 2016). Further precise and combined analysis of epigenetic changes and disease-associated variants in specific cell types will allow a better understanding of the functional consequences of genetic markers associated with chronic inflammatory diseases.

In the next years, the major objective of epigenetic research will be to identify epigenetic changes that are present before the clinically measurable onset of the disease. Reversing the phenotype induced by the epigenetic deregulation under environmental pressure could provide the "window of opportunity" for therapy to treat the disease before its onset. With the recent advances in engineering of the epigenome (Tost, 2016a), we do now have the tools in hand to interrogate the functionality and causality of epigenetic changes in inflammatory and other diseases to discern if the observed changes are causing or caused by the disease. These technologies might also in the future allow altering the epigenetic state of the diseases cells and thereby contribute novels tools for the cure of these devastating diseases.

ACKNOWLEDGMENTS

Work in the laboratory of Jörg Tost is supported by grants from the ANR (ANR-13-EPIG-0003-05 and ANR-13-CESA-0011-05), Aviesan/INSERM (EPIG2014-01 and EPIG2014-18), and INCa (PRT-K14-049). Work on inflammatory diseases in the laboratories of C.M.R. and J.T. is supported by a Sirius research award (UCB Pharma S.A.), a Passerelle research award (Pfizer), iCARE (MSD Avenir), and the institutional budget of the CNG. O.F. is supported by a fellowship from the Assistance Publique—Hôpitaux de Paris (AP-HP).

REFERENCES

Abhishek, S., & Palamadai Krishnan, S. (2016). Epidermal differentiation complex: A review on its epigenetic regulation and potential drug targets. *Cell Journal, 18*(1), 1–6.

Absher, D. M., Li, X., Waite, L. L., Gibson, A., Roberts, K., Edberg, J., et al. (2013). Genome-wide DNA methylation analysis of systemic lupus erythematosus reveals persistent hypomethylation of interferon genes and compositional changes to CD4+ T-cell populations. *PLoS Genetics, 9*(8), e1003678. http://dx.doi.org/10.1371/journal.pgen.1003678.

Adams, D., Altucci, L., Antonarakis, S. E., Ballesteros, J., Beck, S., Bird, A., et al. (2012). BLUEPRINT to decode the epigenetic signature written in blood. *Nature Biotechnology, 30*(3), 224–226. http://dx.doi.org/10.1038/nbt.2153.

Adams, A. T., Kennedy, N. A., Hansen, R., Ventham, N. T., O'Leary, K. R., Drummond, H. E., et al. (2014). Two-stage genome-wide methylation profiling in childhood-onset Crohn's disease implicates epigenetic alterations at the VMP1/MIR21 and HLA loci. *Inflammatory Bowel Diseases, 20*(10), 1784–1793. http://dx.doi.org/10.1097/MIB.0000000000000179.

Ahmadi, A., Polyak, S., & Draganov, P. V. (2009). Colorectal cancer surveillance in inflammatory bowel disease: The search continues. *World Journal of Gastroenterology, 15*(1), 61–66.

Ahuja, N., Li, Q., Mohan, A. L., Baylin, S. B., & Issa, J. P. (1998). Aging and DNA methylation in colorectal mucosa and cancer. *Cancer Research, 58*(23), 5489–5494.

Alegria-Torres, J. A., Baccarelli, A., & Bollati, V. (2011). Epigenetics and lifestyle. *Epigenomics, 3*(3), 267–277. http://dx.doi.org/10.2217/epi.11.22.

Allan, R. S. (2015). The epigenetic mechanisms that underlie health and disease. *Immunology and Cell Biology, 93*(3), 224–225. http://dx.doi.org/10.1038/icb.2015.8.

Ansel, K. M., Djuretic, I., Tanasa, B., & Rao, A. (2006). Regulation of Th2 differentiation and Il4 locus accessibility. *Annual Review of Immunology, 24*, 607–656. http://dx.doi.org/10.1146/annurev.immunol.23.021704.115821.

Ansel, K. M., Lee, D. U., & Rao, A. (2003). An epigenetic view of helper T cell differentiation. *Nature Immunology, 4*(7), 616–623. http://dx.doi.org/10.1038/ni0703-616.

Azarschab, P., Porschen, R., Gregor, M., Blin, N., & Holzmann, K. (2002). Epigenetic control of the E-cadherin gene (CDH1) by CpG methylation in colectomy samples of patients with ulcerative colitis. *Genes, Chromosomes & Cancer, 35*(2), 121–126. http://dx.doi.org/10.1002/gcc.10101.

Bai, A. H., Wu, W. K., Xu, L., Wong, S. H., Go, M. Y., Chan, A. W., et al. (2016). Dysregulated lysine acetyltransferase 2B promotes inflammatory bowel disease pathogenesis through transcriptional repression of interleukin-10. *Journal of Crohn's & Colitis, 10*(6), 726–734. http://dx.doi.org/10.1093/ecco-jcc/jjw020.

Balasa, A., Gathungu, G., Kisfali, P., Smith, E. O., Cho, J. H., Melegh, B., et al. (2010). Assessment of DNA methylation at the interferon regulatory factor 5 (IRF5) promoter region in inflammatory bowel diseases. *International Journal of Colorectal Disease, 25*(5), 553–556. http://dx.doi.org/10.1007/s00384-010-0874-0.

Ballestar, E. (2010). Epigenetics lessons from twins: Prospects for autoimmune disease. *Clinical Reviews in Allergy & Immunology, 39*(1), 30–41. http://dx.doi.org/10.1007/s12016-009-8168-4.

Bao, Y., & Cao, X. (2016). Epigenetic control of B cell development and B-cell-related immune disorders. *Clinical Reviews in Allergy & Immunology, 50*(3), 301–311. http://dx.doi.org/10.1007/s12016-015-8494-7.

Barski, A., Cuddapah, S., Cui, K., Roh, T. Y., Schones, D. E., Wang, Z., et al. (2007). High-resolution profiling of histone methylations in the human genome. *Cell, 129*(4), 823–837.

Baylin, S. B., & Jones, P. A. (2011). A decade of exploring the cancer epigenome—Biological and translational implications. *Nature Reviews. Cancer*, *11*(10), 726–734. http://dx.doi.org/10.1038/nrc3130.

Berni Canani, R., Di Costanzo, M., & Leone, L. (2012). The epigenetic effects of butyrate: Potential therapeutic implications for clinical practice. *Clinical Epigenetics*, *4*(1), 4. http://dx.doi.org/10.1186/1868-7083-4-4.

Bibikova, M., Le, J., Barnes, B., Saedinia-Melnyk, S., Zhou, L., Shen, R., et al. (2009). Genome-wide DNA methylation profiling using Infinium((R)) assay. *Epigenomics*, *1*(1), 177–200. http://dx.doi.org/10.2217/epi.09.14.

Bird, A. (2002). DNA methylation patterns and epigenetic memory. *Genes & Development*, *16*(1), 6–21.

Blander, J. M. (2016). Death in the intestinal epithelium-basic biology and implications for inflammatory bowel disease. *The FEBS Journal*, *283*(14), 2720–2730. http://dx.doi.org/10.1111/febs.13771.

Blueprint Consortium. (2016). Quantitative comparison of DNA methylation assays for biomarker development and clinical applications. *Nature Biotechnology*, *34*(7), 726–737. http://dx.doi.org/10.1038/nbt.3605.

Bock, C., Tomazou, E. M., Brinkman, A. B., Muller, F., Simmer, F., Gu, H., et al. (2010). Quantitative comparison of genome-wide DNA methylation mapping technologies. *Nature Biotechnology*, *28*(10), 1106–1114. http://dx.doi.org/10.1038/nbt.1681. nbt.1681 [pii].

Borno, S. T., Fischer, A., Kerick, M., Falth, M., Laible, M., Brase, J. C., et al. (2012). Genome-wide DNA methylation events in TMPRSS2-ERG fusion-negative prostate cancers implicate an EZH2-dependent mechanism with miR-26a hypermethylation. *Cancer Discovery*, *2*(11), 1024–1035. http://dx.doi.org/10.1158/2159-8290.CD-12-0041.

Bowcock, A. M. (2005). The genetics of psoriasis and autoimmunity. *Annual Review of Genomics and Human Genetics*, *6*, 93–122. http://dx.doi.org/10.1146/annurev.genom.6.080604.162324.

Brandtzaeg, P. (2010). Update on mucosal immunoglobulin A in gastrointestinal disease. *Current Opinion in Gastroenterology*, *26*(6), 554–563. http://dx.doi.org/10.1097/MOG.0b013e32833dccf8.

Brandtzaeg, P., Carlsen, H. S., & Halstensen, T. S. (2006). The B-cell system in inflammatory bowel disease. *Advances in Experimental Medicine and Biology*, *579*, 149–167. http://dx.doi.org/10.1007/0-387-33778-4_10.

Brinkman, A. B., Simmer, F., Ma, K., Kaan, A., Zhu, J., & Stunnenberg, H. G. (2010). Whole-genome DNA methylation profiling using MethylCap-seq. *Methods*, *52*(3), 232–236. http://dx.doi.org/10.1016/j.ymeth.2010.06.012. S1046-2023(10)00166-0 [pii].

Campan, M., Moffitt, M., Houshdaran, S., Shen, H., Widschwendter, M., Daxenbichler, G., et al. (2011). Genome-scale screen for DNA methylation-based detection markers for ovarian cancer. *PLoS One*, *6*(12), e28141. http://dx.doi.org/10.1371/journal.pone.0028141. PONE-D-11-08732 [pii].

Cavazza, A., Miccio, A., Romano, O., Petiti, L., Malagoli Tagliazucchi, G., Peano, C., et al. (2016). Dynamic transcriptional and epigenetic regulation of human epidermal Keratinocyte differentiation. *Stem Cell Reports*, *6*(4), 618–632. http://dx.doi.org/10.1016/j.stemcr.2016.03.003.

Chapman, C. G., & Pekow, J. (2015). The emerging role of miRNAs in inflammatory bowel disease: A review. *Therapeutic Advances in Gastroenterology*, *8*(1), 4–22. http://dx.doi.org/10.1177/1756283X14547360.

Chen, M., Chen, Z. Q., Cui, P. G., Yao, X., Li, Y. M., Li, A. S., et al. (2008). The methylation pattern of p16INK4a gene promoter in psoriatic epidermis and its clinical significance. *The British Journal of Dermatology*, *158*(5), 987–993. http://dx.doi.org/10.1111/j.1365-2133.2008.08505.x.

Choi, B., Kim, H. A., Suh, C. H., Byun, H. O., Jung, J. Y., & Sohn, S. (2015). The relevance of miRNA-21 in HSV-induced inflammation in a mouse model. *International Journal of Molecular Sciences, 16*(4), 7413–7427. http://dx.doi.org/10.3390/ijms16047413.

Cohn, S. M., Roth, K. A., Birkenmeier, E. H., & Gordon, J. I. (1991). Temporal and spatial patterns of transgene expression in aging adult mice provide insights about the origins, organization, and differentiation of the intestinal epithelium. *Proceedings of the National Academy of Sciences of the United States of America, 88*(3), 1034–1038.

Cooke, J., Zhang, H., Greger, L., Silva, A. L., Massey, D., Dawson, C., et al. (2012). Mucosal genome-wide methylation changes in inflammatory bowel disease. *Inflammatory Bowel Diseases, 18*(11), 2128–2137. http://dx.doi.org/10.1002/ibd.22942.

Cortese, R., Lu, L., Yu, Y., Ruden, D., & Claud, E. C. (2016). Epigenome–microbiome crosstalk: A potential new paradigm influencing neonatal susceptibility to disease. *Epigenetics, 11*(3), 205–215. http://dx.doi.org/10.1080/15592294.2016.1155011.

Cottrell, S. E., Distler, J., Goodman, N. S., Mooney, S. H., Kluth, A., Olek, A., et al. (2004). A real-time PCR assay for DNA-methylation using methylation-specific blockers. *Nucleic Acids Research, 32*(1), e10.

Couturier-Maillard, A., Secher, T., Rehman, A., Normand, S., De Arcangelis, A., Haesler, R., et al. (2013). NOD2-mediated dysbiosis predisposes mice to transmissible colitis and colorectal cancer. *The Journal of Clinical Investigation, 123*(2), 700–711. http://dx.doi.org/10.1172/JCI62236.

de Zoeten, E. F., Wang, L., Sai, H., Dillmann, W. H., & Hancock, W. W. (2010). Inhibition of HDAC9 increases T regulatory cell function and prevents colitis in mice. *Gastroenterology, 138*(2), 583–594. http://dx.doi.org/10.1053/j.gastro.2009.10.037.

Deng, X., Su, Y., Wu, H., Wu, R., Zhang, P., Dai, Y., et al. (2015). The role of microRNAs in autoimmune diseases with skin involvement. *Scandinavian Journal of Immunology, 81*(3), 153–165. http://dx.doi.org/10.1111/sji.12261.

Djuretic, I. M., Levanon, D., Negreanu, V., Groner, Y., Rao, A., & Ansel, K. M. (2007). Transcription factors T-bet and Runx3 cooperate to activate Ifng and silence Il4 in T helper type 1 cells. *Nature Immunology, 8*(2), 145–153. http://dx.doi.org/10.1038/ni1424.

Eads, C. A., Danenberg, K. D., Kawakami, K., Saltz, L. B., Blake, C., Shibata, D., et al. (2000). MethyLight: A high-throughput assay to measure DNA methylation. *Nucleic Acids Research, 28*(8), e32.

Ehrich, M., Nelson, M. R., Stanssens, P., Zabeau, M., Liloglou, T., Xinarianos, G., et al. (2005). Quantitative high-throughput analysis of DNA methylation patterns by base-specific cleavage and mass spectrometry. *Proceedings of the National Academy of Sciences of the United States of America, 102*(44), 15785–15790.

Ekbom, A., Helmick, C., Zack, M., & Adami, H. O. (1990). Ulcerative colitis and colorectal cancer. A population-based study. *The New England Journal of Medicine, 323*(18), 1228–1233. http://dx.doi.org/10.1056/NEJM199011013231802.

Ekman, A. K., & Enerback, C. (2016). Lack of preclinical support for the efficacy of histone deacetylase inhibitors in the treatment of psoriasis. *The British Journal of Dermatology, 174*(2), 424–426. http://dx.doi.org/10.1111/bjd.14021.

Ellinghaus, D., Bethune, J., Petersen, B. S., & Franke, A. (2015). The genetics of Crohn's disease and ulcerative colitis—Status quo and beyond. *Scandinavian Journal of Gastroenterology, 50*(1), 13–23. http://dx.doi.org/10.3109/00365521.2014.990507.

Farh, K. K., Marson, A., Zhu, J., Kleinewietfeld, M., Housley, W. J., Beik, S., et al. (2015). Genetic and epigenetic fine mapping of causal autoimmune disease variants. *Nature, 518*(7539), 337–343. http://dx.doi.org/10.1038/nature13835.

Fleischer, T., Frigessi, A., Johnson, K. C., Edvardsen, H., Touleimat, N., Klajic, J., et al. (2014). Genome-wide DNA methylation profiles in progression to in situ and invasive carcinoma of the breast with impact on gene transcription and prognosis. *Genome Biology, 15*(8), 435. http://dx.doi.org/10.1186/PREACCEPT-2333349012841587.

Fleisher, A. S., Esteller, M., Harpaz, N., Leytin, A., Rashid, A., Xu, Y., et al. (2000). Microsatellite instability in inflammatory bowel disease-associated neoplastic lesions is associated with hypermethylation and diminished expression of the DNA mismatch repair gene, hMLH1. *Cancer Research, 60*(17), 4864–4868.

Fofanova, T. Y., Petrosino, J. F., & Kellermayer, R. (2016). Microbiome–epigenome interactions and the environmental origins of inflammatory bowel diseases. *Journal of Pediatric Gastroenterology and Nutrition, 62*(2), 208–219. http://dx.doi.org/10.1097/MPG.0000000000000950.

Franke, A., McGovern, D. P., Barrett, J. C., Wang, K., Radford-Smith, G. L., Ahmad, T., et al. (2010). Genome-wide meta-analysis increases to 71 the number of confirmed Crohn's disease susceptibility loci. *Nature Genetics, 42*(12), 1118–1125. http://dx.doi.org/10.1038/ng.717.

Frommer, M., McDonald, L. E., Millar, D. S., Collis, C. M., Watt, F., Grigg, G. W., et al. (1992). A genomic sequencing protocol that yields a positive display of 5-methylcytosine residues in individual DNA strands. *Proceedings of the National Academy of Sciences of the United States of America, 89*(5), 1827–1831.

Gerstein, M. B., Kundaje, A., Hariharan, M., Landt, S. G., Yan, K. K., Cheng, C., et al. (2012). Architecture of the human regulatory network derived from ENCODE data. *Nature, 489*(7414), 91–100. http://dx.doi.org/10.1038/nature11245.

Gervin, K., Vigeland, M. D., Mattingsdal, M., Hammero, M., Nygard, H., Olsen, A. O., et al. (2012). DNA methylation and gene expression changes in monozygotic twins discordant for psoriasis: Identification of epigenetically dysregulated genes. *PLoS Genetics, 8*(1), e1002454. http://dx.doi.org/10.1371/journal.pgen.1002454.

Gilmour, D. S., & Lis, J. T. (1984). Detecting protein–DNA interactions in vivo: Distribution of RNA polymerase on specific bacterial genes. *Proceedings of the National Academy of Sciences of the United States of America, 81*(14), 4275–4279.

Glauben, R., Batra, A., Fedke, I., Zeitz, M., Lehr, H. A., Leoni, F., et al. (2006). Histone hyperacetylation is associated with amelioration of experimental colitis in mice. *Journal of Immunology, 176*(8), 5015–5022.

Globisch, D., Munzel, M., Muller, M., Michalakis, S., Wagner, M., Koch, S., et al. (2010). Tissue distribution of 5-hydroxymethylcytosine and search for active demethylation intermediates. *PLoS One, 5*(12), e15367. http://dx.doi.org/10.1371/journal.pone.0015367.

Gloria, L., Cravo, M., Pinto, A., de Sousa, L. S., Chaves, P., Leitao, C. N., et al. (1996). DNA hypomethylation and proliferative activity are increased in the rectal mucosa of patients with long-standing ulcerative colitis. *Cancer, 78*(11), 2300–2306.

Gonsky, R., Deem, R. L., & Targan, S. R. (2009). Distinct methylation of IFNG in the gut. *Journal of Interferon & Cytokine Research, 29*(7), 407–414. http://dx.doi.org/10.1089/jir.2008.0109.

Gordon, S., & Martinez, F. O. (2010). Alternative activation of macrophages: Mechanism and functions. *Immunity, 32*(5), 593–604. http://dx.doi.org/10.1016/j.immuni.2010.05.007.

Grjibovski, A. M., Olsen, A. O., Magnus, P., & Harris, J. R. (2007). Psoriasis in Norwegian twins: Contribution of genetic and environmental effects. *Journal of the European Academy of Dermatology and Venereology, 21*(10), 1337–1343. http://dx.doi.org/10.1111/j.1468-3083.2007.02268.x.

Gu, X., Nylander, E., Coates, P. J., Fahraeus, R., & Nylander, K. (2015). Correlation between reversal of DNA methylation and clinical symptoms in psoriatic epidermis following narrow-band UVB phototherapy. *The Journal of Investigative Dermatology, 135*(8), 2077–2083. http://dx.doi.org/10.1038/jid.2015.128.

Guo, H., Zhu, P., Guo, F., Li, X., Wu, X., Fan, X., et al. (2015). Profiling DNA methylome landscapes of mammalian cells with single-cell reduced-representation bisulfite sequencing. *Nature Protocols, 10*(5), 645–659. http://dx.doi.org/10.1038/nprot.2015.039.

Guttman-Yassky, E., Suarez-Farinas, M., Chiricozzi, A., Nograles, K. E., Shemer, A., Fuentes-Duculan, J., et al. (2009). Broad defects in epidermal cornification in atopic dermatitis identified through genomic analysis. *The Journal of Allergy and Clinical Immunology*, *124*(6), 1235–1244. http://dx.doi.org/10.1016/j.jaci.2009.09.031. e1258.

Han, J., Park, S. G., Bae, J. B., Choi, J., Lyu, J. M., Park, S. H., et al. (2012). The characteristics of genome-wide DNA methylation in naive CD4+ T cells of patients with psoriasis or atopic dermatitis. *Biochemical and Biophysical Research Communications*, *422*(1), 157–163. http://dx.doi.org/10.1016/j.bbrc.2012.04.128.

Harris, R. A., Nagy-Szakal, D., Mir, S. A., Frank, E., Szigeti, R., Kaplan, J. L., et al. (2014). DNA methylation-associated colonic mucosal immune and defense responses in treatment-naive pediatric ulcerative colitis. *Epigenetics*, *9*(8), 1131–1137. http://dx.doi.org/10.4161/epi.29446.

Harris, R. A., Nagy-Szakal, D., Pedersen, N., Opekun, A., Bronsky, J., Munkholm, P., et al. (2012). Genome-wide peripheral blood leukocyte DNA methylation microarrays identified a single association with inflammatory bowel diseases. *Inflammatory Bowel Diseases*, *18*(12), 2334–2341. http://dx.doi.org/10.1002/ibd.22956.

Harris, R. A., Shah, R., Hollister, E. B., Tronstad, R. R., Hovdenak, N., Szigeti, R., et al. (2016). Colonic mucosal epigenome and microbiome development in children and adolescents. *Journal of Immunology Research*, *2016*. http://dx.doi.org/10.1155/2016/9170162. Art. ID 9170162.

Hasler, R., Feng, Z., Backdahl, L., Spehlmann, M. E., Franke, A., Teschendorff, A., et al. (2012). A functional methylome map of ulcerative colitis. *Genome Research*, *22*(11), 2130–2137. http://dx.doi.org/10.1101/gr.138347.112.

Hawkes, J. E., Nguyen, G. H., Fujita, M., Florell, S. R., Callis Duffin, K., Krueger, G. G., et al. (2016). microRNAs in psoriasis. *The Journal of Investigative Dermatology*, *136*(2), 365–371. http://dx.doi.org/10.1038/JID.2015.409.

Herman, J. G., Graff, J. R., Myohanen, S., Nelkin, B. D., & Baylin, S. B. (1996). Methylation-specific PCR: A novel PCR assay for methylation status of CpG islands. *Proceedings of the National Academy of Sciences of the United States of America*, *93*(18), 9821–9826.

Hoehe, M. R., Kopke, K., Wendel, B., Rohde, K., Flachmeier, C., Kidd, K. K., et al. (2000). Sequence variability and candidate gene analysis in complex disease: Association of mu opioid receptor gene variation with substance dependence. *Human Molecular Genetics*, *9*(19), 2895–2908.

Hoeksema, M. A., Laan, L. C., Postma, J. J., Cummings, R. D., de Winther, M. P., Dijkstra, C. D., et al. (2016). Treatment with Trichuris suis soluble products during monocyte-to-macrophage differentiation reduces inflammatory responses through epigenetic remodeling. *The FASEB Journal*, *30*(8), 2826–2836. http://dx.doi.org/10.1096/fj.201600343R.

Hoeksema, M. A., Stoger, J. L., & de Winther, M. P. (2012). Molecular pathways regulating macrophage polarization: Implications for atherosclerosis. *Current Atherosclerosis Reports*, *14*(3), 254–263. http://dx.doi.org/10.1007/s11883-012-0240-5.

Hou, J. K., Abraham, B., & El-Serag, H. (2011). Dietary intake and risk of developing inflammatory bowel disease: A systematic review of the literature. *The American Journal of Gastroenterology*, *106*(4), 563–573. http://dx.doi.org/10.1038/ajg.2011.44.

Hou, R., Yin, G., An, P., Wang, C., Liu, R., Yang, Y., et al. (2013). DNA methylation of dermal MSCs in psoriasis: Identification of epigenetically dysregulated genes. *Journal of Dermatological Science*, *72*(2), 103–109. http://dx.doi.org/10.1016/j.jdermsci.2013.07.002.

Houseman, E. A., Accomando, W. P., Koestler, D. C., Christensen, B. C., Marsit, C. J., Nelson, H. H., et al. (2012). DNA methylation arrays as surrogate measures of cell mixture distribution. *BMC Bioinformatics*, *13*, 86. http://dx.doi.org/10.1186/1471-2105-13-86.

Houseman, E. A., Kile, M. L., Christiani, D. C., Ince, T. A., Kelsey, K. T., & Marsit, C. J. (2016). Reference-free deconvolution of DNA methylation data and mediation by cell composition effects. *BMC Bioinformatics*, *17*, 259. http://dx.doi.org/10.1186/s12859-016-1140-4.

Hsieh, C. J., Klump, B., Holzmann, K., Borchard, F., Gregor, M., & Porschen, R. (1998). Hypermethylation of the p16INK4a promoter in colectomy specimens of patients with long-standing and extensive ulcerative colitis. *Cancer Research*, *58*(17), 3942–3945.

Hughes, T., Ture-Ozdemir, F., Alibaz-Oner, F., Coit, P., Direskeneli, H., & Sawalha, A. H. (2014). Epigenome-wide scan identifies a treatment-responsive pattern of altered DNA methylation among cytoskeletal remodeling genes in monocytes and CD4+ T cells from patients with Behcet's disease. *Arthritis & Rheumatology*, *66*(6), 1648–1658. http://dx.doi.org/10.1002/art.38409.

Issa, J. P. (2002). Epigenetic variation and human disease. *The Journal of Nutrition*, *132*(8 Suppl.), 2388S–2392S.

Issa, J. P., Ahuja, N., Toyota, M., Bronner, M. P., & Brentnall, T. A. (2001). Accelerated age-related CpG island methylation in ulcerative colitis. *Cancer Research*, *61*(9), 3573–3577.

Jaenisch, R., & Bird, A. (2003). Epigenetic regulation of gene expression: How the genome integrates intrinsic and environmental signals. *Nature Genetics*, *33*(Suppl.), 245–254.

Jiang, Y., & Wang, L. (2016). Role of histone deacetylase 3 in ankylosing spondylitis via negative feedback loop with microRNA-130a and enhancement of tumor necrosis factor-1alpha expression in peripheral blood mononuclear cells. *Molecular Medicine Reports*, *13*(1), 35–40. http://dx.doi.org/10.3892/mmr.2015.4494.

Jung, M., Kadam, S., Xiong, W., Rauch, T. A., Jin, S. G., & Pfeifer, G. P. (2015). MIRA-seq for DNA methylation analysis of CpG islands. *Epigenomics*, 7(5), 695–706. http://dx.doi.org/10.2217/epi.15.33.

Kalla, R., Ventham, N. T., Kennedy, N. A., Quintana, J. F., Nimmo, E. R., Buck, A. H., et al. (2015). MicroRNAs: New players in IBD. *Gut*, *64*(3), 504–517. http://dx.doi.org/10.1136/gutjnl-2014-307891.

Kang, K., Bae, J. H., Han, K., Kim, E. S., Kim, T. O., & Yi, J. M. (2016). A genome-wide methylation approach identifies a new hypermethylated gene panel in ulcerative colitis. *International Journal of Molecular Sciences*, *17*(8), e1291. http://dx.doi.org/10.3390/ijms17081291.

Karatzas, P. S., Mantzaris, G. J., Safioleas, M., & Gazouli, M. (2014). DNA methylation profile of genes involved in inflammation and autoimmunity in inflammatory bowel disease. *Medicine (Baltimore)*, *93*(28), e309. http://dx.doi.org/10.1097/MD.0000000000000309.

Karczewski J., Dobrowolska A., Rychlewska-Hanczewska A. and Adamski Z. (2016). New insights into the role of T cells in pathogenesis of psoriasis and psoriatic arthritis, *Autoimmunity*, 1–16, http://dx.doi.org/10.3109/08916934.2016.1166214, in press.

Katsurano, M., Niwa, T., Yasui, Y., Shigematsu, Y., Yamashita, S., Takeshima, H., et al. (2012). Early-stage formation of an epigenetic field defect in a mouse colitis model, and non-essential roles of T- and B-cells in DNA methylation induction. *Oncogene*, *31*(3), 342–351. http://dx.doi.org/10.1038/onc.2011.241.

Kellermayer, R. (2012). Epigenetics and the developmental origins of inflammatory bowel diseases. *Canadian Journal of Gastroenterology*, *26*(12), 909–915.

Kellermayer, R., Balasa, A., Zhang, W., Lee, S., Mirza, S., Chakravarty, A., et al. (2010). Epigenetic maturation in colonic mucosa continues beyond infancy in mice. *Human Molecular Genetics*, *19*(11), 2168–2176. http://dx.doi.org/10.1093/hmg/ddq095.

Kiernan, R., Bres, V., Ng, R. W., Coudart, M. P., El Messaoudi, S., Sardet, C., et al. (2003). Post-activation turn-off of NF-kappa B-dependent transcription is regulated by acetylation of p65. *The Journal of Biological Chemistry*, *278*(4), 2758–2766. http://dx.doi.org/10.1074/jbc.M209572200.

Koizumi, K., Alonso, S., Miyaki, Y., Okada, S., Ogura, H., Shiiya, N., et al. (2012). Array-based identification of common DNA methylation alterations in ulcerative colitis. *International Journal of Oncology*, *40*(4), 983–994. http://dx.doi.org/10.3892/ijo.2011.1283.

Kominsky, D. J., Keely, S., MacManus, C. F., Glover, L. E., Scully, M., Collins, C. B., et al. (2011). An endogenously anti-inflammatory role for methylation in mucosal inflammation identified through metabolite profiling. *Journal of Immunology*, *186*(11), 6505–6514. http://dx.doi.org/10.4049/jimmunol.1002805.

Korbie, D., Lin, E., Wall, D., Nair, S. S., Stirzaker, C., Clark, S. J., et al. (2015). Multiplex bisulfite PCR resequencing of clinical FFPE DNA. *Clinical Epigenetics*, 7(1), 28. http://dx.doi.org/10.1186/s13148-015-0067-3.

Kraiczy, J., Nayak, K., Ross, A., Raine, T., Mak, T. N., Gasparetto, M., et al. (2016). Assessing DNA methylation in the developing human intestinal epithelium: Potential link to inflammatory bowel disease. *Mucosal Immunology*, *9*(3), 647–658. http://dx.doi.org/10.1038/mi.2015.88.

Kriaucionis, S., & Heintz, N. (2009). The nuclear DNA base 5-hydroxymethylcytosine is present in Purkinje neurons and the brain. *Science*, *324*(5929), 929–930. http://dx.doi.org/10.1126/science.1169786. 1169786 [pii].

Kugathasan, S., & Amre, D. (2006). Inflammatory bowel disease—Environmental modification and genetic determinants. *Pediatric Clinics of North America*, *53*(4), 727–749. http://dx.doi.org/10.1016/j.pcl.2006.05.009.

Ladd-Acosta, C. (2015). Epigenetic signatures as biomarkers of exposure. *Current Environmental Health Reports*, *2*(2), 117–125. http://dx.doi.org/10.1007/s40572-015-0051-2.

Lai, N. S., Chou, J. L., Chen, G. C., Liu, S. Q., Lu, M. C., & Chan, M. W. (2014). Association between cytokines and methylation of SOCS-1 in serum of patients with ankylosing spondylitis. *Molecular Biology Reports*, *41*(6), 3773–3780. http://dx.doi.org/10.1007/s11033-014-3242-2.

Law, J. A., & Jacobsen, S. E. (2010). Establishing, maintaining and modifying DNA methylation patterns in plants and animals. *Nature Reviews. Genetics*, *11*(3), 204–220. http://dx.doi.org/10.1038/nrg2719.

Lee, J., Hong, E. C., Jeong, H., Hwang, J. W., Kim, H., Bae, E. K., et al. (2015). A novel histone deacetylase 6-selective inhibitor suppresses synovial inflammation and joint destruction in a collagen antibody-induced arthritis mouse model. *International Journal of Rheumatic Diseases*, *18*(5), 514–523. http://dx.doi.org/10.1111/1756-185X.12501.

Li, M., Liu, X., Sun, X., Wang, Z., Guo, W., Hu, F., et al. (2013). Therapeutic effects of NK-HDAC-1, a novel histone deacetylase inhibitor, on collagen-induced arthritis through the induction of apoptosis of fibroblast-like synoviocytes. *Inflammation*, *36*(4), 888–896. http://dx.doi.org/10.1007/s10753-013-9616-0.

Li, Q., Suzuki, M., Wendt, J., Patterson, N., Eichten, S. R., Hermanson, P. J., et al. (2015). Post-conversion targeted capture of modified cytosines in mammalian and plant genomes. *Nucleic Acids Research*, *43*(12), e81. http://dx.doi.org/10.1093/nar/gkv244.

Li, Z., Wong, S. H., Shen, J., Chan, M. T., & Wu, W. K. (2016). The role of MicroRNAS in Ankylosing Spondylitis. *Medicine (Baltimore)*, *95*(14), e3325. http://dx.doi.org/10.1097/MD.0000000000003325.

Lin, Z., Hegarty, J. P., Cappel, J. A., Yu, W., Chen, X., Faber, P., et al. (2011). Identification of disease-associated DNA methylation in intestinal tissues from patients with inflammatory bowel disease. *Clinical Genetics*, *80*(1), 59–67. http://dx.doi.org/10.1111/j.1399-0004.2010.01546.x.

Lin, Z., Hegarty, J. P., Yu, W., Cappel, J. A., Chen, X., Faber, P. W., et al. (2012). Identification of disease-associated DNA methylation in B cells from Crohn's disease and ulcerative colitis patients. *Digestive Diseases and Sciences*, *57*(12), 3145–3153. http://dx.doi.org/10.1007/s10620-012-2288-z.

Liu, H. P., Cao, A. T., Feng, T., Li, Q., Zhang, W., Yao, S., et al. (2015). TGF-beta converts Th1 cells into Th17 cells through stimulation of Runx1 expression. *European Journal of Immunology*, *45*(4), 1010–1018. http://dx.doi.org/10.1002/eji.201444726.

Lo Sasso, G., Menzies, K. J., Mottis, A., Piersigilli, A., Perino, A., Yamamoto, H., et al. (2014). SIRT2 deficiency modulates macrophage polarization and susceptibility to experimental colitis. *PLoS One*, *9*(7), e103573. http://dx.doi.org/10.1371/journal.pone.0103573.

Lobaton, T., Azuara, D., Rodriguez-Moranta, F., Loayza, C., Sanjuan, X., de Oca, J., et al. (2014). Relationship between methylation and colonic inflammation in inflammatory bowel disease. *World Journal of Gastroenterology*, *20*(30), 10591–10598. http://dx.doi.org/10.3748/wjg.v20.i30.10591.

Machiels, K., Joossens, M., Sabino, J., De Preter, V., Arijs, I., Eeckhaut, V., et al. (2014). A decrease of the butyrate-producing species Roseburia hominis and Faecalibacterium prausnitzii defines dysbiosis in patients with ulcerative colitis. *Gut*, *63*(8), 1275–1283. http://dx.doi.org/10.1136/gutjnl-2013-304833.

Maeda, T., Guan, J. Z., Higuchi, Y., Oyama, J., & Makino, N. (2009). Aging-related alterations of subtelomeric methylation in sarcoidosis patients. *The Journals of Gerontology. Series A, Biological Sciences and Medical Sciences*, *64*(7), 752–760. http://dx.doi.org/10.1093/gerona/glp049.

Marques-Rocha, J. L., Samblas, M., Milagro, F. I., Bressan, J., Martinez, J. A., & Marti, A. (2015). Noncoding RNAs, cytokines, and inflammation-related diseases. *The FASEB Journal*, *29*(9), 3595–3611. http://dx.doi.org/10.1096/fj.14-260323.

Masser, D. R., Berg, A. S., & Freeman, W. M. (2013). Focused, high accuracy 5-methylcytosine quantitation with base resolution by benchtop next-generation sequencing. *Epigenetics & Chromatin*, *6*(1), 33. http://dx.doi.org/10.1186/1756-8935-6-33.

Meissner, A., Gnirke, A., Bell, G. W., Ramsahoye, B., Lander, E. S., & Jaenisch, R. (2005). Reduced representation bisulfite sequencing for comparative high-resolution DNA methylation analysis. *Nucleic Acids Research*, *33*(18), 5868–5877. http://dx.doi.org/10.1093/nar/gki901.

Miceli-Richard, C., Wang-Renault, S. F., Boudaoud, S., Busato, F., Lallemand, C., Bethune, K., et al. (2016). Overlap between differentially methylated DNA regions in blood B lymphocytes and genetic at-risk loci in primary Sjogren's syndrome. *Annals of the Rheumatic Diseases*, *75*(5), 933–940. http://dx.doi.org/10.1136/annrheumdis-2014-206998.

Mikkelsen, T. S., Ku, M., Jaffe, D. B., Issac, B., Lieberman, E., Giannoukos, G., et al. (2007). Genome-wide maps of chromatin state in pluripotent and lineage-committed cells. *Nature*, *448*(7153), 553–560.

Mischke, M., & Plosch, T. (2016). The gut microbiota and their metabolites: Potential implications for the host epigenome. *Advances in Experimental Medicine and Biology*, *902*, 33–44. http://dx.doi.org/10.1007/978-3-319-31248-4_3.

Miura, F., Enomoto, Y., Dairiki, R., & Ito, T. (2012). Amplification-free whole-genome bisulfite sequencing by post-bisulfite adaptor tagging. *Nucleic Acids Research*, *40*(17), e136. http://dx.doi.org/10.1093/nar/gks454.

Mizoguchi, A., Mizoguchi, E., Takedatsu, H., Blumberg, R. S., & Bhan, A. K. (2002). Chronic intestinal inflammatory condition generates IL-10-producing regulatory B cell subset characterized by CD1d upregulation. *Immunity*, *16*(2), 219–230.

Moran, S., Arribas, C., & Esteller, M. (2016). Validation of a DNA methylation microarray for 850,000 CpG sites of the human genome enriched in enhancer sequences. *Epigenomics*, *8*(3), 389–399. http://dx.doi.org/10.2217/epi.15.114.

Mullican, S. E., Gaddis, C. A., Alenghat, T., Nair, M. G., Giacomin, P. R., Everett, L. J., et al. (2011). Histone deacetylase 3 is an epigenomic brake in macrophage alternative activation. *Genes & Development*, *25*(23), 2480–2488. http://dx.doi.org/10.1101/gad.175950.111.

Nielsen, H. M., & Tost, J. (2012). Epigenetic changes in inflammatory and autoimmune diseases. *Sub-Cellular Biochemistry*, *61*, 455–478. http://dx.doi.org/10.1007/978-94-007-4525-4_20.

Nimmo, E. R., Prendergast, J. G., Aldhous, M. C., Kennedy, N. A., Henderson, P., Drummond, H. E., et al. (2012). Genome-wide methylation profiling in Crohn's disease identifies altered epigenetic regulation of key host defense mechanisms including the Th17 pathway. *Inflammatory Bowel Diseases*, *18*(5), 889–899. http://dx.doi.org/10.1002/ibd.21912.

Noronha, A. M., Liang, Y., Hetzel, J. T., Hasturk, H., Kantarci, A., Stucchi, A., et al. (2009). Hyperactivated B cells in human inflammatory bowel disease. *Journal of Leukocyte Biology*, *86*(4), 1007–1016. http://dx.doi.org/10.1189/jlb.0309203.

Northrop, J. K., Thomas, R. M., Wells, A. D., & Shen, H. (2006). Epigenetic remodeling of the IL-2 and IFN-gamma loci in memory CD8 T cells is influenced by CD4 T cells. *Journal of Immunology*, *177*(2), 1062–1069.

Oakes, C. C., La Salle, S., Robaire, B., & Trasler, J. M. (2006). Evaluation of a quantitative DNA methylation analysis technique using methylation-sensitive/dependent restriction enzymes and real-time PCR. *Epigenetics*, *1*(3), 146–152.

Park, G. T., Han, J., Park, S. G., Kim, S., & Kim, T. Y. (2014). DNA methylation analysis of CD4+ T cells in patients with psoriasis. *Archives of Dermatological Research*, *306*(3), 259–268. http://dx.doi.org/10.1007/s00403-013-1432-8.

Planell, N., Lozano, J. J., Mora-Buch, R., Masamunt, M. C., Jimeno, M., Ordas, I., et al. (2013). Transcriptional analysis of the intestinal mucosa of patients with ulcerative colitis in remission reveals lasting epithelial cell alterations. *Gut*, *62*(7), 967–976. http://dx.doi.org/10.1136/gutjnl-2012-303333.

Qi, J., Hou, S., Zhang, Q., Liao, D., Wei, L., Fang, J., et al. (2013). A functional variant of pre-miRNA-196a2 confers risk for Behcet's disease but not for Vogt–Koyanagi–Harada syndrome or AAU in ankylosing spondylitis. *Human Genetics*, *132*(12), 1395–1404. http://dx.doi.org/10.1007/s00439-013-1346-8.

Qi, J., Yang, Y., Hou, S., Qiao, Y., Wang, Q., Yu, H., et al. (2014). Increased Notch pathway activation in Behcet's disease. *Rheumatology (Oxford)*, *53*(5), 810–820. http://dx.doi.org/10.1093/rheumatology/ket438.

Reilly, C. M., Thomas, M., Gogal, R., Jr., Olgun, S., Santo, A., Sodhi, R., et al. (2008). The histone deacetylase inhibitor trichostatin A upregulates regulatory T cells and modulates autoimmunity in NZB/WF1 mice. *Journal of Autoimmunity*, *31*(2), 123–130. http://dx.doi.org/10.1016/j.jaut.2008.04.020.

Reveille, J. D. (2012). Genetics of spondyloarthritis—Beyond the MHC. *Nature Reviews. Rheumatology*, *8*(5), 296–304. http://dx.doi.org/10.1038/nrrheum.2012.41.

Roadmap Epigenomics, C., Kundaje, A., Meuleman, W., Ernst, J., Bilenky, M., Yen, A., et al. (2015). Integrative analysis of 111 reference human epigenomes. *Nature*, *518*(7539), 317–330. http://dx.doi.org/10.1038/nature14248.

Roberson, E. D., Liu, Y., Ryan, C., Joyce, C. E., Duan, S., Cao, L., et al. (2012). A subset of methylated CpG sites differentiate psoriatic from normal skin. *The Journal of Investigative Dermatology*, *132*(3 Pt. 1), 583–592. http://dx.doi.org/10.1038/jid.2011.348.

Roberts A.R., Vecellio M., Chen L., Ridley A., Cortes A., Knight J.C., et al. (2016). An ankylosing spondylitis-associated genetic variant in the IL23R-IL12RB2 intergenic region modulates enhancer activity and is associated with increased Th1-cell differentiation, *Annals of the Rheumatic Diseases*, http://dx.doi.org/10.1136/annrheumdis-2015-208640, in press.

Ronn, T., & Ling, C. (2015). DNA methylation as a diagnostic and therapeutic target in the battle against Type 2 diabetes. *Epigenomics*, 7(3), 451–460. http://dx.doi.org/10.2217/epi.15.7.

Ruchusatsawat, K., Wongpiyabovorn, J., Shuangshoti, S., Hirankarn, N., & Mutirangura, A. (2006). SHP-1 promoter 2 methylation in normal epithelial tissues and demethylation in psoriasis. *Journal of Molecular Medicine (Berlin, Germany)*, *84*(2), 175–182. http://dx.doi.org/10.1007/s00109-005-0020-6.

Ruzov, A., Tsenkina, Y., Serio, A., Dudnakova, T., Fletcher, J., Bai, Y., et al. (2011). Lineage-specific distribution of high levels of genomic 5-hydroxymethylcytosine in mammalian development. *Cell Research*, *21*(9), 1332–1342. http://dx.doi.org/10.1038/cr.2011.113.

Saito, S., Kato, J., Hiraoka, S., Horii, J., Suzuki, H., Higashi, R., et al. (2011). DNA methylation of colon mucosa in ulcerative colitis patients: Correlation with inflammatory status. *Inflammatory Bowel Diseases*, *17*(9), 1955–1965. http://dx.doi.org/10.1002/ibd.21573.

Samuel N., Wilson G., Lemire M., Id Said B., Lou Y., Li W., et al., (2016). Genome-wide DNA methylation analysis reveals epigenetic dysregulation of MicroRNA-34A in TP53-associated cancer susceptibility, *Journal of Clinical Oncology*, http://dx.doi.org/10.1200/JCO.2016.67.6940, in press.

Sandoval, J., Heyn, H., Moran, S., Serra-Musach, J., Pujana, M. A., Bibikova, M., et al. (2011). Validation of a DNA methylation microarray for 450,000 CpG sites in the human genome. *Epigenetics*, *6*(6), 692–702. doi:16196 [pii].

Santourlidis, S., Graffmann, N., Christ, J., & Uhrberg, M. (2008). Lineage-specific transition of histone signatures in the killer cell Ig-like receptor locus from hematopoietic progenitor to NK cells. *Journal of Immunology*, *180*(1), 418–425.

Sato, F., Harpaz, N., Shibata, D., Xu, Y., Yin, J., Mori, Y., et al. (2002). Hypermethylation of the p14(ARF) gene in ulcerative colitis-associated colorectal carcinogenesis. *Cancer Research*, *62*(4), 1148–1151.

Sato, F., Shibata, D., Harpaz, N., Xu, Y., Yin, J., Mori, Y., et al. (2002). Aberrant methylation of the HPP1 gene in ulcerative colitis-associated colorectal carcinoma. *Cancer Research*, *62*(23), 6820–6822.

Scarpa, M., & Stylianou, E. (2012). Epigenetics: Concepts and relevance to IBD pathogenesis. *Inflammatory Bowel Diseases*, *18*(10), 1982–1996. http://dx.doi.org/10.1002/ibd.22934.

Schlums, H., Cichocki, F., Tesi, B., Theorell, J., Beziat, V., Holmes, T. D., et al. (2015). Cytomegalovirus infection drives adaptive epigenetic diversification of NK cells with altered signaling and effector function. *Immunity*, *42*(3), 443–456. http://dx.doi.org/10.1016/j.immuni.2015.02.008.

Schultz, M. D., He, Y., Whitaker, J. W., Hariharan, M., Mukamel, E. A., Leung, D., et al. (2015). Human body epigenome maps reveal noncanonical DNA methylation variation. *Nature*, *523*(7559), 212–216. http://dx.doi.org/10.1038/nature14465.

Shen, L., Du, J., Xia, Y., Tan, Z., Fu, Y., Yang, Q., et al. (2016). Genome-wide landscape of DNA methylomes and their relationship with mRNA and miRNA transcriptomes in oxidative and glycolytic skeletal muscles. *Scientific Reports*, *6*, 32186. http://dx.doi.org/10.1038/srep32186.

Shen, L., Kondo, Y., Guo, Y., Zhang, J., Zhang, L., Ahmed, S., et al. (2007). Genome-wide profiling of DNA methylation reveals a class of normally methylated CpG island promoters. *PLoS Genetics*, *3*(10), 2023–2036.

Smith, P. J., Levine, A. P., Dunne, J., Guilhamon, P., Turmaine, M., Sewell, G. W., et al. (2014). Mucosal transcriptomics implicates under expression of BRINP3 in the pathogenesis of ulcerative colitis. *Inflammatory Bowel Diseases*, *20*(10), 1802–1812. http://dx.doi.org/10.1097/MIB.0000000000000169.

Subramaniam, R., Mizoguchi, A., & Mizoguchi, E. (2016). Mechanistic roles of epithelial and immune cell signaling during the development of colitis-associated cancer. *Cancer Research Frontiers*, *2*(1), 1–21. http://dx.doi.org/10.17980/2016.1.

Tahara, T., Shibata, T., Nakamura, M., Yamashita, H., Yoshioka, D., Okubo, M., et al. (2009a). Promoter methylation of protease-activated receptor (PAR2) is associated with severe clinical phenotypes of ulcerative colitis (UC). *Clinical and Experimental Medicine*, *9*(2), 125–130. http://dx.doi.org/10.1007/s10238-008-0025-x.

Tahara, T., Shibata, T., Nakamura, M., Yamashita, H., Yoshioka, D., Okubo, M., et al. (2009b). Effect of MDR1 gene promoter methylation in patients with ulcerative colitis. *International Journal of Molecular Medicine*, *23*(4), 521–527.

Tahiliani, M., Koh, K. P., Shen, Y., Pastor, W. A., Bandukwala, H., Brudno, Y., et al. (2009). Conversion of 5-methylcytosine to 5-hydroxymethylcytosine in mammalian DNA by MLL partner TET1. *Science*, *324*(5929), 930–935. http://dx.doi.org/10.1126/science.1170116. 1170116 [pii].

Taiwo, O., Wilson, G. A., Morris, T., Seisenberger, S., Reik, W., Pearce, D., et al. (2012). Methylome analysis using MeDIP-seq with low DNA concentrations. *Nature Protocols*, 7(4), 617–636. http://dx.doi.org/10.1038/nprot.2012.012.

Tost, J. (2008). Methods for the genome-wide and gene-specific analysis of DNA methylation levels and patterns. In J. Tost (Ed.), *Epigenetics* (pp. 63–104). Norwich: Horizon Scientific Press.

Tost, J. (2010). DNA methylation: An introduction to the biology and the disease-associated changes of a promising biomarker. *Molecular Biotechnology*, *44*(1), 71–81. http://dx.doi.org/10.1007/s12033-009-9216-2.

Tost, J. (2016a). Engineering of the epigenome: Synthetic biology to define functional causality and develop innovative therapies. *Epigenomics*, *8*(2), 153–156. http://dx.doi.org/10.2217/epi.15.112.

Tost, J. (2016b). Follow the trace of death: Methylation analysis of cell-free DNA for clinical applications in non-cancerous diseases. *Epigenomics*, *8*, 1169–1172. http://dx.doi.org/10.2217/epi-2016-0080.

Tost, J., & Gut, I. G. (2007). DNA methylation analysis by pyrosequencing. *Nature Protocols*, *2*(9), 2265–2275.

Toussirot, E., Abbas, W., Khan, K. A., Tissot, M., Jeudy, A., Baud, L., et al. (2013). Imbalance between HAT and HDAC activities in the PBMCs of patients with ankylosing spondylitis or rheumatoid arthritis and influence of HDAC inhibitors on TNF alpha production. *PLoS One*, *8*(8), e70939. http://dx.doi.org/10.1371/journal.pone.0070939.

Tovar-Castillo, L. E., Cancino-Diaz, J. C., Garcia-Vazquez, F., Cancino-Gomez, F. G., Leon-Dorantes, G., Blancas-Gonzalez, F., et al. (2007). Under-expression of VHL and over-expression of HDAC-1, HIF-1alpha, LL-37, and IAP-2 in affected skin biopsies of patients with psoriasis. *International Journal of Dermatology*, *46*(3), 239–246. http://dx.doi.org/10.1111/j.1365-4632.2006.02962.x.

Tsaprouni, L. G., Ito, K., Powell, J. J., Adcock, I. M., & Punchard, N. (2011). Differential patterns of histone acetylation in inflammatory bowel diseases. *Journal of Inflammation (London, England)*, *8*(1), 1. http://dx.doi.org/10.1186/1476-9255-8-1.

Tschurtschenthaler, M., Kachroo, P., Heinsen, F. A., Adolph, T. E., Ruhlemann, M. C., Klughammer, J., et al. (2016). Paternal chronic colitis causes epigenetic inheritance of susceptibility to colitis. *Scientific Reports*, *6*, 31640. http://dx.doi.org/10.1038/srep31640.

Ueno, A., Ghosh, A., Hung, D., Li, J., & Jijon, H. (2015). Th17 plasticity and its changes associated with inflammatory bowel disease. *World Journal of Gastroenterology*, *21*(43), 12283–12295. http://dx.doi.org/10.3748/wjg.v21.i43.12283.

Urich, M. A., Nery, J. R., Lister, R., Schmitz, R. J., & Ecker, J. R. (2015). MethylC-seq library preparation for base-resolution whole-genome bisulfite sequencing. *Nature Protocols*, *10*(3), 475–483. http://dx.doi.org/10.1038/nprot.2014.114.

Van den Bossche, J., Neele, A. E., Hoeksema, M. A., & de Winther, M. P. (2014). Macrophage polarization: The epigenetic point of view. *Current Opinion in Lipidology*, *25*(5), 367–373. http://dx.doi.org/10.1097/MOL.0000000000000109.

van Rooij, E. (2011). The art of microRNA research. *Circulation Research*, *108*(2), 219–234. http://dx.doi.org/10.1161/CIRCRESAHA.110.227496.

Vecellio, M., Roberts, A. R., Cohen, C. J., Cortes, A., Knight, J. C., Bowness, P., et al. (2016). The genetic association of RUNX3 with ankylosing spondylitis can be explained by allele-specific effects on IRF4 recruitment that alter gene expression. *Annals of the Rheumatic Diseases*, *75*(8), 1534–1540. http://dx.doi.org/10.1136/annrheumdis-2015-207490.

Vernia, P., Annese, V., Bresci, G., d'Albasio, G., D'Inca, R., Giaccari, S., et al. (2003). Topical butyrate improves efficacy of 5-ASA in refractory distal ulcerative colitis: Results of a multicentre trial. *European Journal of Clinical Investigation*, *33*(3), 244–248.

Voelter-Mahlknecht, S. (2016). Epigenetic associations in relation to cardiovascular prevention and therapeutics. *Clinical Epigenetics*, *8*, 4. http://dx.doi.org/10.1186/s13148-016-0170-0.

Wang, F. Y., Arisawa, T., Tahara, T., Takahama, K., Watanabe, M., Hirata, I., et al. (2008). Aberrant DNA methylation in ulcerative colitis without neoplasia. *Hepato-Gastroenterology*, *55*(81), 62–65.

Wendling, D., Delattre, E., Abbas, W., Guillot, X., Toussirot, E., & Herbein, G. (2015). Sirt-1 activity in PBMC from patients with spondyloarthritis. *Joint, Bone, Spine*, *82*(3), 210–211. http://dx.doi.org/10.1016/j.jbspin.2014.10.002.

Wheeler, J. M., Kim, H. C., Efstathiou, J. A., Ilyas, M., Mortensen, N. J., & Bodmer, W. F. (2001). Hypermethylation of the promoter region of the E-cadherin gene (CDH1) in sporadic and ulcerative colitis associated colorectal cancer. *Gut*, *48*(3), 367–371.

White, C. A., Pone, E. J., Lam, T., Tat, C., Hayama, K. L., Li, G., et al. (2014). Histone deacetylase inhibitors upregulate B cell microRNAs that silence AID and Blimp-1 expression for epigenetic modulation of antibody and autoantibody responses. *Journal of Immunology*, *193*(12), 5933–5950. http://dx.doi.org/10.4049/jimmunol.1401702.

Wilson, T. J., Jobim, M., Jobim, L. F., Portela, P., Salim, P. H., Rosito, M. A., et al. (2010). Study of killer immunoglobulin-like receptor genes and human leukocyte antigens class I ligands in a Caucasian Brazilian population with Crohn's disease and ulcerative colitis. *Human Immunology*, *71*(3), 293–297. http://dx.doi.org/10.1016/j.humimm.2009.12.006.

Wojdacz, T. K., Dobrovic, A., & Hansen, L. L. (2008). Methylation-sensitive high-resolution melting. *Nature Protocols*, *3*(12), 1903–1908. http://dx.doi.org/10.1038/nprot.2008.191. nprot.2008.191 [pii].

Xia, J., & Zhang, W. (2014). MicroRNAs in normal and psoriatic skin. *Physiological Genomics*, *46*(4), 113–122. http://dx.doi.org/10.1152/physiolgenomics.00157.2013.

Yadav, P. K., Chen, C., & Liu, Z. (2011). Potential role of NK cells in the pathogenesis of inflammatory bowel disease. *Journal of Biomedicine & Biotechnology*, *2011*, 348530. http://dx.doi.org/10.1155/2011/348530.

Yamamoto, F., & Yamamoto, M. (2004). A DNA microarray-based methylation-sensitive (MS)-AFLP hybridization method for genetic and epigenetic analyses. *Molecular Genetics and Genomics*, *271*(6), 678–686.

Yu, L., Chibnik, L. B., Srivastava, G. P., Pochet, N., Yang, J., Xu, J., et al. (2015). Association of brain DNA methylation in SORL1, ABCA7, HLA-DRB5, SLC24A4, and BIN1 with pathological diagnosis of Alzheimer disease. *Journal of the American Medical Association Neurology*, *72*(1), 15–24. http://dx.doi.org/10.1001/jamaneurol.2014.3049.

Yuksel, S., Kucukazman, S. O., Karatas, G. S., Ozturk, M. A., Prombhul, S., & Hirankarn, N. (2016). Methylation status of Alu and LINE-1 interspersed repetitive sequences in Behcet's disease patients. *BioMed Research International*, *2016*, 1393089. http://dx.doi.org/10.1155/2016/1393089.

Zan, H., & Casali, P. (2015). Epigenetics of peripheral B-cell differentiation and the antibody response. *Frontiers in Immunology*, *6*, 631. http://dx.doi.org/10.3389/fimmu.2015.00631.

Zawada, A. M., Schneider, J. S., Michel, A. I., Rogacev, K. S., Hummel, B., Krezdorn, N., et al. (2016). DNA methylation profiling reveals differences in the 3 human monocyte subsets and identifies uremia to induce DNA methylation changes during differentiation. *Epigenetics*, *11*(4), 259–272. http://dx.doi.org/10.1080/15592294.2016.1158363.

Zhang, T., Cooper, S., & Brockdorff, N. (2015). The interplay of histone modifications—Writers that read. *EMBO Reports*, *16*(11), 1467–1481. http://dx.doi.org/10.15252/embr.201540945.

Zhang, P., Su, Y., Chen, H., Zhao, M., & Lu, Q. (2010). Abnormal DNA methylation in skin lesions and PBMCs of patients with psoriasis vulgaris. *Journal of Dermatological Science*, *60*(1), 40–42. http://dx.doi.org/10.1016/j.jdermsci.2010.07.011.

Zhang, P., Su, Y., Zhao, M., Huang, W., & Lu, Q. (2011). Abnormal histone modifications in PBMCs from patients with psoriasis vulgaris. *European Journal of Dermatology*, *21*(4), 552–557. http://dx.doi.org/10.1684/ejd.2011.1383.

Zhang, K., Zhang, R., Li, X., Yin, G., & Niu, X. (2009). Promoter methylation status of p15 and p21 genes in HPP-CFCs of bone marrow of patients with psoriasis. *European Journal of Dermatology*, *19*(2), 141–146. http://dx.doi.org/10.1684/ejd.2008.0618.

Zhang, P., Zhao, M., Liang, G., Yin, G., Huang, D., Su, F., et al. (2013). Whole-genome DNA methylation in skin lesions from patients with psoriasis vulgaris. *Journal of Autoimmunity*, *41*, 17–24. http://dx.doi.org/10.1016/j.jaut.2013.01.001.

Zhang, Q., Zhao, K., Shen, Q., Han, Y., Gu, Y., Li, X., et al. (2015). Tet2 is required to resolve inflammation by recruiting Hdac2 to specifically repress IL-6. *Nature*, *525*(7569), 389–393. http://dx.doi.org/10.1038/nature15252.

Zhou, F., Wang, W., Shen, C., Li, H., Zuo, X., Zheng, X., et al. (2016). Epigenome-wide association analysis identified nine skin DNA methylation loci for psoriasis. *The Journal of Investigative Dermatology*, *136*(4), 779–787. http://dx.doi.org/10.1016/j.jid.2015.12.029.

CHAPTER SEVEN

Regulation of Cellular Immune Responses in Sepsis by Histone Modifications

W.F. Carson IV[1], S.L. Kunkel
University of Michigan Medical School, Ann Arbor, MI, United States
[1]Corresponding author: e-mail address: wfcarson@umich.edu

Contents

Abstract

Severe sepsis, septic shock, and related inflammatory syndromes are driven by the aberrant expression of proinflammatory mediators by immune cells. During the acute phase of sepsis, overexpression of chemokines and cytokines drives physiological stress leading to organ failure and mortality. Following recovery from sepsis, the immune system exhibits profound immunosuppression, evidenced by an inability to produce the same proinflammatory mediators that are required for normal responses to infectious microorganisms. Gene expression in inflammatory responses is influenced by the transcriptional accessibility of the chromatin, with histone posttranslational modifications determining whether inflammatory gene loci are set to transcriptionally active, repressed, or poised states. Experimental evidence indicates that histone modifications play a central role in governing the cytokine storm of severe sepsis, and that aberrant chromatin modifications induced during the acute phase of sepsis may mediate chronic immunosuppression in sepsis survivors. This review will focus on the role of histone modifications in governing immune responses in severe sepsis, with an emphasis on specific leukocyte subsets and the histone modifications observed in these cells during chronic stages of sepsis. Additionally, the expression and function of chromatin-modifying enzymes (CMEs) will be discussed in the context of severe sepsis, as potential mediators of epigenetic regulation of gene expression in sepsis responses. In summary, this review will argue for the use of chromatin modifications and CME expression in leukocytes as potential biomarkers of immunosuppression in patients with severe sepsis.

Advances in Protein Chemistry and Structural Biology, Volume 106
ISSN 1876-1623
http://dx.doi.org/10.1016/bs.apcsb.2016.08.004

1. INTRODUCTION

An argument can be made to consider sepsis as a chronic inflammatory disorder rather than simply an acute inflammatory event. Of course, this is not meant to downplay the life-threatening syndrome driven by the "cytokine storm" induced in response to exposure to pathogen- and damage-associated molecular patterns (PAMPs and DAMPs, respectively) (Wiersinga, Leopold, Cranendonk, & van der Poll, 2014). Mortality due to severe sepsis, septic shock, and related syndromes remains high despite advances in treatment strategies and improved knowledge of the underlying mechanisms of disease (Kempker & Martin, 2016). However, the long-term consequences of severe sepsis remain underappreciated in regards to the standard of care normally prescribed to survivors. Over the past 2 decades, extensive epidemiological evidence in human patient populations, coupled with basic research into the functionality of postseptic leukocytes from both human patients and experimental animal models, has described a persistent immunosuppressive state with multiple causative factors (Bermejo-Martin et al., 2016; Delano & Ward, 2016; Gotts & Matthay, 2016). Survivors of severe sepsis exhibit increases in all-cause mortality as compared to the healthy age-matched population for months, if not years following recovery from the acute inflammatory event (Perl, Dvorak, Hwang, & Wenzel, 1995; Quartin, Schein, Kett, & Peduzzi, 1997). These patients also exhibit increases in a wide range of negative health outcomes, from as minor as chronic fatigue and decreased quality of life (Maley & Mikkelsen, 2016), to increases in the occurrence of cancer (Jones et al., 2015) and secondary infections (Ortego et al., 2015). Critically, the time frame for postseptic immunosuppression remains difficult to accurately identify; in fact, much of the literature from experimental animal models of sepsis suggests that certain aspects of postseptic immunosuppression may persist far beyond the resolution of acute inflammation (Wen, Dou, Hogaboam, & Kunkel, 2008). Taken together, these results suggest that following recovery from severe sepsis, a patient's immune system may be irrevocably altered (Fig. 1).

One of the classic phenotypes observed in both postseptic patients and experimental animal models of sepsis is a widespread apoptotic event affecting leukocytes throughout both lymphatic and nonlymphatic tissues (Hotchkiss et al., 2001; Le Tulzo et al., 2002; Tinsley et al., 2003; Wang, Huang, Lin, & Lei, 1994). This apoptotic event is mechanistically driven by activation-induced cell death responses mediated by overproduction of

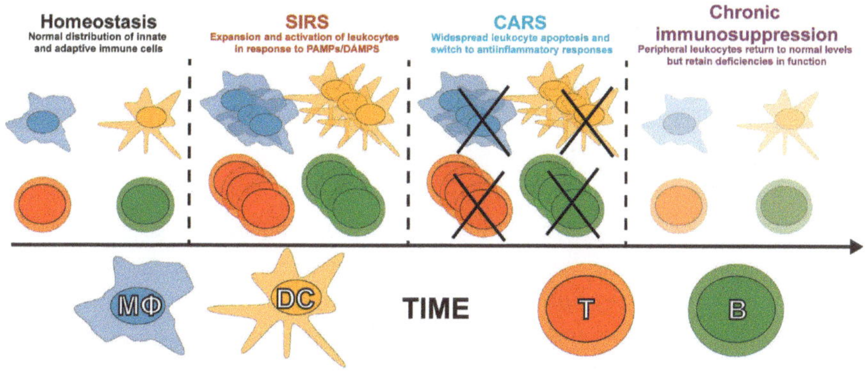

Fig. 1 The progression from initial activation to immune suppression in leukocytes during sepsis. Following the initial exposure to high levels of PAMPs/DAMPs, both innate and adaptive immune cells become activated, proliferate and begin to produce proinflammatory mediators (SIRS). When the magnitude of the inflammatory response crosses over to a "cytokine storm," leukocytes respond via widespread apoptosis and a switch to antiinflammatory responses (CARS). Following resolution of sepsis, total numbers of peripheral leukocytes return to similar levels as found prior to sepsis; however, these cells retain multifactorial defects in activation and effector function (chronic immunosuppression).

proinflammatory mediators such as tumor necrosis factor alpha (TNFα) (Wesche-Soldato, Swan, Chung, & Ayala, 2007). During the acute phase of sepsis, leukocyte apoptosis can be observed in peripheral blood, spleen, thymus, lymph nodes, and bone marrow. This widespread immune cell death has a direct negative effect on host immune defense, with leukopenia leaving the septic patient susceptible to secondary infections (Drewry et al., 2014). This raises the question as to the teleology of the apoptotic event seen in sepsis, as it appears to contribute significantly to poor outcomes (i.e., increased morbidity and mortality). However, widespread leukocyte apoptosis does serve two important functions in the context of acute sepsis. First, it helps limit overproduction of proinflammatory cytokines and chemokines, albeit in a very blunt manner, by eliminating the immune cells responsible for prolonging the cytokine storm (Hotchkiss et al., 1999). Second, leukocyte apoptosis is preferable to leukocyte necrosis and pyroptosis, both of which can perpetuate the cytokine storm of sepsis through the release of DAMPs that can trigger downstream inflammatory responses (Cavassani et al., 2008, 2013).

It is therefore tempting to attribute postseptic immunosuppression directly to the leukopenia resulting from sepsis-induced apoptosis. However, there are two important observations attesting to mechanisms other than cell death for the initiation and maintenance of postseptic immunosuppression. First, the compensatory antiinflammatory response syndrome (CARS) is characterized by a switch from the production of proinflammatory mediators (such as TNFα) to antiinflammatory (e.g., IL-10) and immune-skewing cytokines (e.g., IL-4) (Mack, McCarter, Naama, Calvano, & Daly, 1996; Osuchowski, Welch, Siddiqui, & Remick, 2006; Xu, Lou, Ren, Miao, & Deng, 2012; Zedler, Bone, Baue, von Donnersmarck, & Faist, 1999). This upregulation of signaling mediators strongly suggests that sepsis-induced leukocyte apoptosis does not result in the complete ablation of cellular immunity. Second, as time progresses and patients/experimental animals begin to recover from the acute inflammatory event, the total cellularity of immune tissues and peripheral organs begins to return to levels observed prior to the sepsis-inducing event (Unsinger, Kazama, McDonough, Hotchkiss, & Ferguson, 2009). If the postseptic immunosuppression phenotype was driven primarily by leukopenia, it would be expected that as the total numbers of peripheral leukocytes returned to normal levels, so too would immune function and immunity against opportunistic infections. However, susceptibility to opportunistic infections can be observed in postseptic human patients and experimental animals long past recovery from the acute inflammatory event and the return of peripheral leukocytes. These observations argue for molecular mechanisms driving postseptic immunosuppression that can both persist in leukocytes long after the resolution of inflammation, and also have a negative impact on cell activation and effector function in response to subsequent pathogen challenge.

2. HISTONE MODIFICATIONS: POTENT TOOLS FOR EPIGENETIC CONTROL

As with all cells of the body, immune cell functions are governed by the expression of genes leading to changes in cell phenotype. Gene expression regulation can be directly affected by changes to DNA accessibility, with tightly packaged DNA inaccessible to transcription factors, and unpackaged DNA readily available for transcription (Turner, 2014). The physical packaging of DNA can be regulated through direct chemical modification of DNA (DNA methylation), or through the posttranslational modification of the unstructured tails of the core histone components (H2A, H2B, H3, and H4). DNA methylation is thought to regulate gene expression in a binary fashion, whereby the addition of methyl groups to CpG dinucleotides results in chromatin condensation and silencing of gene expression (Razin & Cedar, 1994). In contrast, the end result of histone modifications depends on the type of chemical modification (e.g., methylation, acetylation, phosphorylation, etc.), the specific amino acid residue(s) modified on each histone subunit tail, as well as the sum total of modifications located on all histone subunit tails within a particular nucleosome complex (Ng & Cheung, 2016). This increased level of complexity allows for the fine-tuning of gene expression patterns by histone modifications to regulate DNA accessibility. From the perspective of histone modifications, genes can be found in three distinct states—transcriptionally active, repressed, or "poised." Active genes are often characterized by increased histone acetylation on multiple subunits, as well as increases in permissive histone methylation events (e.g., histone H3 lysine 4 trimethylation, H3K4me3) (Vermeulen & Timmers, 2010). Repressed genes are often characterized by a dearth of histone acetylation, along with increases in repressive histone methylation events (e.g., histone H3 lysine 27 trimethylation, H3K27me3) (Grossniklaus & Paro, 2014). "Poised" genes are balanced between both active and repressive modifications, resulting in a gene without bias toward expression or silencing (Puri, Gala, Mishra, & Dhawan, 2015). This allows the cell to rapidly pattern histones in poised genes toward expression or repression depending on the immediate needs dictated by the local microenvironment.

Histone posttranslational modifications are an important mechanism for epigenetic regulation of gene expression. As multipotent stem cells differentiate into the various germ layers of a developing multicellular organism, histone modifications help exert epigenetic control over gene expression, driving differentiation of cells and tissues (Rasmussen, 2014). In a similar fashion, immune cells utilize histone modifications to guide gene expression decisions during hematopoiesis, during the initiation of inflammatory responses, and during differentiation into various effector and/or regulatory lineages (Lim, Li, Holloway, & Rao, 2013; Winter & Amit, 2014). Histone modifications are made by chromatin-modifying enzymes (CMEs), which work in concert with cofactors such as scaffolding proteins and transcription factors to interact with nucleosomes and add or remove posttranslational modifications from histone tails. As a result, gene expression patterns can be regulated through CME-directed changes in chromatin dynamics, allowing for discrete epigenetic control over the cellular immune response to specific pathogens.

A growing body of research describes an important link between the microenvironment of an immune cell, the up- or downregulation of

CME expression and function, and changes in chromatin modifications guiding epigenetic regulation of gene expression. CMEs appear to fill an important niche between receptor-mediated signal transduction and transcription factor/DNA binding, through their ability to remodel chromatin, stabilize genes into transcriptionally active or silent states, and interact directly with transcription factors, cofactors, and other DNA-binding proteins. Importantly, specific CMEs appear to be governed by distinct extracellular signals, leading to links between qualitatively different inflammatory responses, and the presence or absence of particular histone modifications in immune cells. For example, STAT1 and IRF7 activation in response to Type I interferon stimulation results in the upregulation of the histone 3 lysine 9 methyltransferase Setdb2 in macrophages, resulting in increased repressive H3K9me3 remodeling events, and suppression of proinflammatory functions (Kroetz et al., 2015). The connections between extracellular signals and immune responses also create commonalities across leukocyte subsets. For example, proinflammatory cytokines such as IL-12 (STAT4) and TNFα (NF-kB) can drive expression of MLL1 in CD4+ T cells and macrophages (respectively), leading to the further promotion of T-helper type-1/M1-type inflammatory responses by these cells (Schaller et al., 2015; Wang et al., 2012). In this fashion, expression and function of MLL1 can be seen as a common epigenetic mechanism for the promotion of inflammation aimed at the clearance of microbial pathogens.

Histone modifications are utilized by immune cells to stabilize gene expression patterns throughout development, from early hematopoiesis to the final differentiation to terminal effector lineages. However, if these processes are aberrantly regulated, the resulting changes in gene expression patterns can serve to drive immunopathology. Such is the case with the immunosuppression observed following recovery from severe sepsis. As indicated previously, following recovery from sepsis, the widespread apoptosis of immune cells appears to subside, and total numbers of leukocytes will return to normal/baseline levels in the peripheral blood and lymphoid organs. Their general distribution also returns to levels observed in healthy human patients or control animals, although modulations can occur in the distribution of functional subsets (such as effector vs regulatory CD4+ T cells) (Monneret et al., 2003; Venet et al., 2009, 2004). However, the function of these immune cells is drastically altered postsepsis, with immune cells of all lineages suffering from significant impairments in activation and effector function. Innate immune cells such as macrophages and dendritic cells exhibit reduced production of proinflammatory mediators, decreased phagocytic capacity, and a reduction in their ability to process and present antigen to T cells (Benjamim, Hogaboam, & Kunkel, 2004; Xiu & Jeschke, 2013). Adaptive immune cells suffer from similar impairments in cytokine and chemokine production, along with significant difficulties in committing to specific effector lineages (Cabrera-Perez, Condotta, Badovinac, & Griffith, 2014). For example, postseptic CD4+ T cells are often observed to have a T_H2 bias, with increased production of IL-4 at the expense of IFNγ production (Ayala, Deol, Lehman, Herdon, & Chaudry, 1994; Mack et al., 1996). These CD4+ cells can also exhibit a more complex "confusion" phenotype, whereby they fail to properly commit to either T_H1, T_H2 or T_H17 functional lineages characterized by the production of specific lineage-specific inflammatory mediators (IFNg, IL-4, or IL-17, respectively)—instead, these cells respond to lineage commitment instructions by producing a mixture of multiple cytokines, leading to an unfocused inflammatory response that alters immune responses in peripheral tissues (Carson et al., 2010; Carson, Ito, et al., 2011; Mukherjee, Allen, Lukacs, Kunkel, & Carson, 2012). While these deficiencies in effector function are due to multiple complex factors, there is an increasing body of literature implicating aberrant histone modifications and expression/function of CMEs as being central to both the development and persistence of cellular immunosuppression phenotypes postsepsis (Carson, Cavassani, Dou, & Kunkel, 2011).

3. HISTONE MODIFICATIONS AND POSTSEPTIC INNATE IMMUNE RESPONSES

Innate immune cells are the primary causative agents of the pathophysiology of severe sepsis. Cells of the innate immune system can respond rapidly to PAMPs and DAMPs in their microenvironment through their expression of pattern recognition receptors on the cell surface (e.g., TLR2 and TLR4), within endocytic vesicles (e.g., TLR3 and TLR9) and in the cytosol (e.g., NOD2 and RIG-I) (Bentzer, Russell, & Walley, 2015; Kang, Kim, Cho, & Lee, 2015). Binding of these receptors by their cognate molecules results in rapid activation and production of proinflammatory mediators. Additionally, innate immune cells can be rapidly recruited in large numbers to sites of inflammation via differentiation of progenitor cells in the bone marrow and direct chemotaxis to inflamed tissues. This is in contrast to adaptive immune cells that require further maturation steps in specialized immune organs (such as the thymus) before they can participate in inflammatory responses. This rapid and intense inflammatory response driven by innate immune cells, especially neutrophils, monocytes, and monocyte-derived dendritic cells and macrophages, initiates the inflammatory signaling cascade that ultimately results in the "cytokine storm" of severe sepsis.

Of course, despite their role as central mediators of immunopathology in severe inflammatory responses, innate immunity remains a critical first line of defense against pathogens. When innate immune cells function at the small scale rather than systemically—in the mucosa, in barrier tissues such as the skin, or in response to small numbers of infectious agents—these cells can rapidly phagocytose and kill pathogens well before an infection can take root and spread (Scott, Henri, & Guilliams, 2014). Innate immune sentinels at barrier surfaces can normally clear microorganisms efficiently without initiating systemic inflammatory processes, serving to limit tissue damage while promoting tolerogenic responses to nonpathogenic stimuli (Allavena, Chieppa, Monti, & Piemonti, 2004; Divangahi, King, & Pernet, 2015). In the context of severe sepsis, the immune system is faced with a difficult decision regarding innate immunity: if these cells are suppressed or otherwise removed from the periphery (i.e., induced to apoptose), the host may be able to limit the overproduction of proinflammatory mediators, but at the cost of increasing susceptibility to secondary infections. Indeed, one of the primary causes of mortality in septic patients in intensive care is due to superimposed nosocomial infections, which are able to take root due to the loss of innate immune function as a result of sepsis-induced leukocyte apoptosis and the switch from SIRS to CARS (Adib-Conquy & Cavaillon, 2009; Parlato & Cavaillon, 2015).

As mentioned previously, while immune cells ultimately return to the same level of total numbers as observed in healthy individuals following recovery from severe sepsis, they manifest deficiencies in activation potential and effector function that preclude their ability to protect the postseptic patient from future pathogens. In the case of innate immunity, these deficiencies primarily manifest as reductions in the production of proinflammatory mediators such as chemokines as cytokines (Deng et al., 2006) and an impaired ability to process and present antigen to cells of the adaptive immune system (Monneret et al., 2004; Venet et al., 2007). The molecular mechanisms governing the refractory nature of innate immune responses postsepsis are myriad in nature, including impairments at all stages of innate immune activation (Wang & Deng, 2008). In all cases, however, the root issue for innate immune cells postsepsis is an inability to express their genes in an appropriate fashion, either in regards to kinetics, intensity or a combination of both factors. Histone modifications, with their ability to govern gene expression through regulating chromatin accessibility, play a central role in governing innate immune dysfunction postsepsis.

In regards to neutrophil function in severe sepsis, the study of histone modifications is of particular interest due to the unique link between the antimicrobial functions of these cells and the posttranslational modification of their histone subunits. As with other immune cells, neutrophils utilize histone modifications to stabilize gene loci into transcriptionally active/repressed states in response to activating signals (Iida, Watanabe-Fukunaga, Nagata, & Fukunaga, 2008). Unique to neutrophils is the use of DNA/histone complexes to trap microbes in extracellular spaces. These DNA/histone complexes, known as neutrophil extracellular traps (NETs), are produced by activated neutrophils in response to microbial stimulus (i.e., lipopolysaccharide, LPS) (Neeli, Dwivedi, Khan, & Radic, 2009) or inflammatory mediators (i.e., IL-8) (Marcos et al., 2010). NETs serve to sequester and inactive microbial pathogens, and in concert with antimicrobial proteins such as neutrophil elastase, serve to destroy pathogens in extracellular spaces (Brinkmann et al., 2004). Prior to the production of NETs, neutrophils modify their histones through conversion of histone subunit arginase residues to citrulline residues, a process known as citrullination (Wang et al., 2009). Citrullination of histone subunits is a required initial step for the production of NETs by activated neutrophils, requiring the functional activity of specific CMEs known as peptidylarginine deiminases (PADs) (Anzilotti, Pratesi, Tommasi, & Migliorini, 2010). PAD-dependent citrullination of histones is also an important molecular mechanism for epigenetic regulation, depending on the posttranslational state of the original arginine residue prior to citrullination. For unmodified arginine residues, citrullination can help to drive gene expression through its ability to lower the binding strength of the DNA/histone complex (Zhang et al., 2012). However, in the case of monomethylated arginine, which is normally an activating epigenetic mark, citrullination can result in the loss of activating monomethylation, leading to an overall decrease in gene expression (Cuthbert et al., 2004).

Interestingly, despite their apparent function as antimicrobial defense mechanisms, the role of NETs in sepsis remains unclear. NETosis is clearly induced in septic responses, and extracellular NETs can bind, sequester and kill sepsis-inducing microorganisms, especially bacteria (Kaplan & Radic, 2012). However, NETs can also serve to drive uncontrolled inflammation in a similar fashion as secreted chemokines and cytokines. Histone proteins themselves have been shown to be proinflammatory, and their presence in tissues induces chemotaxis of innate immune cells in a chemokine-dependent manner (Westman et al., 2015). Additionally, DNA in the NETs can potentially stimulate TLRs, especially if the DNA is enriched in unmethylated CpG dinucleotides, a pattern recognized by TLR9 (Huang et al., 2011). Compounding the issue is the connection between NETosis and cell death, as neutrophils must die to release their DNA/histone complexes (Vanden Berghe, Linkermann, Jouan-Lanhouet, Walczak, & Vandenabeele, 2014). The resulting neutropenia induced in part by enhanced NETosis during sepsis may serve to drive postseptic immunosuppression, and in concert with the proinflammatory signals initiated by NET components, may exacerbate postseptic immunosuppression. It is intriguing then to hypothesize that NETosis is ultimately a net negative in severe inflammatory processes such as sepsis. Indeed, studies of sepsis responses in PAD4-deficient animals, which have significant defects in NETosis due to the loss of the PAD4 histone arginine deiminase, indicate no significant modulation in bacteremia following loss of NETosis (Martinod et al., 2015). Additionally, disease severity was reduced in PAD4-deficient animals in response to endotoxin-induced shock, suggesting that in sterile or non-infectious sepsis responses, the presence of NETs may strictly serve to drive inflammatory processes leading to physiological stress and increased mortality (Martinod et al., 2015).

In contrast to the extensive body of literature dealing with neutrophil responses during the acute phase of sepsis, there are few studies tracking the long-term negative effects of sepsis on neutrophil functions. The best described phenomenon for neutrophil dysfunction in sepsis is so-called "neutrophil paralysis," an inability of neutrophils to migrate to sites of infection in response to chemoattractants (Alves-Filho, Spiller, & Cunha, 2010). Neutrophil paralysis is a critical mechanism underlying susceptibility to secondary infections in acute sepsis; however, the persistence of neutrophil paralysis in long-term survivors of severe sepsis remains understudied. It would be of interest to determine if the histone citrullination induced in neutrophils during sepsis had long-term epigenetic effects on gene expression, especially in regards to those gene families responsible for governing neutrophil chemotaxis (such as chemokine receptors). Despite these limitations, investigations into neutrophil NETosis in sepsis provide evidence that sepsis directly induces posttranslational modifications to histones in immune cells. This finding provides context for the more extensive body of research dealing with chromatin modifications in monocytes, macrophages, and dendritic cells postsepsis.

Antigen-presenting myeloid cells (including monocytes, macrophages, and dendritic cells) have been shown in numerous studies to develop significant functional defects in both humans and experimental animals following severe sepsis. Upon rechallenge either in vivo with a secondary infection or in vitro with microbes or microbial products (i.e., TLR ligands), macrophages, and dendritic cells exhibit similar functional deficiencies, such as suppressed production of proinflammatory mediators (Munoz et al., 1991). Interestingly, these deficiencies are not necessarily due to concurrent increased in antiinflammatory gene expression, such as IL-10 (Wen, Dou, et al., 2008). Rather, the deficiencies in activation observed in macrophages and dendritic cells postsepsis appears to be cell intrinsic, in that these cells exhibit decreased proinflammatory cytokine production even in optimized culture conditions (Flohe et al., 2006; Huang, Venet, Chung, Lomas-Neira, & Ayala, 2007; Muthu et al., 2008). This repression of effector function is correlated with a switch in the balance of histone modifications away from transcriptionally active chromatin (H3 and H4 acetylation, along with H3K4me3) and toward transcriptionally silent chromatin (reductions in H3 and H4 acetylation and increases in H3K27me3, for example) (Lyn-Kew et al., 2010; Wen, Dou, et al., 2008; Wen, Schaller, Dou, Hogaboam, & Kunkel, 2008).

Following recovery from sepsis, both macrophages and dendritic cells produce significantly less proinflammatory mediators as compared to the same cells from healthy age-matched patients and control animals. This includes reductions in early response cytokines such as TNFα, inflammasome cytokines such as IL-1β, and adaptive immune-instructing cytokines such as IL-12 (Wen, Hogaboam, Gauldie, & Kunkel, 2006; Xu et al., 2012). These suppressed cytokine responses mimic the phenotype of "LPS tolerance," whereby LPS-stimulated myeloid cells (and macrophages in particular) lose their ability to respond to secondary LPS challenge (Biswas & Lopez-Collazo, 2009). Analysis of the promoter regions of these genes via chromatin immunoprecipitation (ChIP) reveals a shift toward repressive histone patterning. This includes significant decreases in H3K4me3, as well as H3 and H4 acetylation (Rios, de Lima, Moretti, & Soriano, 2016; Schaafsma et al., 2015; Yoza, Hu, & McCall, 2002). In macrophages, a switchover to repressive histone modifications can also occur in promoters of antimicrobial genes, such as is the case with reduced AcH4 and H3K4me3 in the promoter of Nos2, the enzyme responsible for the generation of nitric oxide species

(Lyn-Kew et al., 2010). There is also growing evidence for histone modifications to serve as molecular intermediaries for driving immunosuppression following sepsis. For example, the NF-kB repressor protein TNFAIP3/A20 is upregulated in postseptic macrophages exhibiting LPS tolerance, resulting in the suppression of a broad range of proinflammatory cytokines (such as TNFα, IL-6, and IFN-β) (Li et al., 2013). A20 expression is itself epigenetically regulated in LPS tolerant cells, as the promoter of A20 becomes enriched in H3K4me3, stabilizing transcriptional accessibility (Li et al., 2013). In this instance, the histone modifications induced following severe sepsis can still mediate immune suppression despite being characteristic of active chromatin, as the genes driven by the increase in H3K4me3 serve to negatively regulate proinflammatory signal transduction.

Interestingly, these changes in histone modifications can persist far beyond recovery from sepsis, suggesting that these epigenetic marks set in innate immune cells may serve to regulate subsequent inflammatory responses well into the future (Benjamim et al., 2004). For example, aberrant histone methylation patterns and concurrent reductions in inflammatory cytokine expression can be observed in spleen-resident dendritic cells at 6 weeks postsepsis, long after the resolution of acute inflammation (Wen, Dou, et al., 2008). This long-term epigenetic dysregulation is characterized by significant decreases in activating H3K4me3 and significant increases in repressive H3K27me3 at specific sites within the IL-12p35 and -p40 promoters. The persistent dysregulation in cytokine expression driven by epigenetic changes has measurable functional consequences. For example, secondary challenge of fully recovered postseptic mice with the opportunistic fungal pathogen *Aspergillus fumigatus* results in significant fungal outgrowth and mortality, whereas healthy mice that did not undergo sepsis fully clear the fungal challenge (Benjamim, Hogaboam, Lukacs, & Kunkel, 2003; Benjamim, Lundy, Lukacs, Hogaboam, & Kunkel, 2005). This susceptibility to *A. fumigatus* infection is in large part due to decreased IL-12 production by lung-resident dendritic cells, and can only be reversed through the adoptive transfer of healthy dendritic cells into the airways of postseptic mice, cells which presumably do not bear the aberrant histone methylation marks that suppress IL-12 production in postseptic dendritic cells (Benjamim et al., 2005).

These modulations in histone posttranslational modifications observed in immune cells (both innate and adaptive, as will be discussed) suggest that CME expression and function is also modulated in immune cells during sepsis. Aside from the direct effects on PAD expression and function as discussed previously during sepsis-induced NETosis, studies in experimental animal models have implicated an important role for histone deacetylases (HDACs) in the acute phase of sepsis (Ciarlo, Savva, & Roger, 2013). HDACs remove acetyl groups from histone tails, which drives transcriptional repression of gene expression when this deacetylation occurs in promoter or enhancer regions. Most experimental data regarding the role of HDAC expression and function in sepsis comes indirectly through the use of histone deacetylase inhibitors (HDACi) to either blunt the inflammatory response during the acute phase, or attempt to modulate negative outcomes of sepsis in survivors (i.e., reversal of LPS tolerance in macrophages). The results of these studies can appear strikingly incongruous. For example, HDACi treatment can serve to inhibit macrophage functions in vitro, including pathogen killing responses (Mombelli et al., 2011). This would suggest that HDACi treatment would exacerbate sepsis responses, as it would limit the ability of macrophages to clear the microorganisms initiating the cytokine storm of sepsis. However, experimental models of sepsis often show improved outcome measures in response to HDACi treatment, including decreases in mortality rate and improvements in other negative outcomes of sepsis, such as coagulopathy (Zhao et al., 2014). This raises the question as to whether HDACi treatment is primarily serving to limit expression of proinflammatory mediators (a beneficial outcome) or serving to block molecular pathways leading to phagocytosis and killing of microorganisms (a negative outcome).

In addition, histone methyltransferase expression is also modulated in innate immune cells in response to sepsis and during the transition to tolerant phenotypes. For example, LPS tolerance induction in macrophages leads to overexpression of the histone methyltransferase EHMT2/G9a, which adds repressive H3K9me2 to promoter regions of TNFα, resulting in transcriptional repression (Yoshida et al., 2015). EHMT2 also works in concert with the Heterochromatin Protein 1 (HP1) family to recruit DNA methyltransferase 3a/3b (Dnmt3a/3b) to the TNFα promoter, resulting in increased DNA methylation and further suppression of TNFα expression (El Gazzar et al., 2008). The interaction between EHMT2 and Dnmt3a/3b is important from the perspective of epigenetics as DNA methylation patterns are known to be transferred from parent to daughter cells though well-defined molecular mechanisms (Tammen, Friso, & Choi, 2013). Therefore, EHMT2-directed methylation of promoters has the potential to persist in macrophage populations, which is essential for the persistence of the LPS tolerance phenotype both in vitro and in vivo. In regards to histone methyltransferases and demethylases, studies of splenic DCs from postseptic animals identified long-term changes in CME binding to the IL-12a and IL-12b promoters. This includes significant decreases in binding of Mixed-Lineage Leukemia (MLL) complex proteins WDR5 and RbBP5, with concurrent increases in Polycomb Repressive Complex 2 (PRC2) components EED and SUZ12 (Wen, Dou, et al., 2008). Importantly, these functional changes in CME and CME cofactor binding can be observed in primary tissue-resident innate immune cells long after the resolution of inflammation, providing further evidence that sepsis induces long-term changes in chromatin packaging with significant effects on gene expression.

At present, there is a strong body of literature correlating histone modifications with aberrant innate immune responses following sepsis in vivo, and in vitro using standard models of innate immune tolerance induction (LPS tolerance). These histone modifications have been shown to persist for long periods of time in peripheral innate immune cells following sepsis, and they appear to play an important mechanistic role in the decreased effector functions of these cells during subsequent pathogen challenge (Fig. 2). However, there remains a paucity of data studying the long-term expression patterns of CMEs in innate immune cells following sepsis. Recent studies have begun to investigate the dynamic regulation of CMEs during the acute phase of sepsis, and experimental animal models have identified numerous enzymes that may help to drive pathogenic inflammatory responses. For example, expression of the histone deacetylase SIRT1 is induced in human sepsis patient samples (Liu, Vachharajani, Yoza, & McCall, 2012; Liu, Yoza, El Gazzar, Vachharajani, & McCall, 2011) and innate immune cells during the development of LPS tolerance (Liu et al., 2011). Pharmacological inhibition of SIRT1 can also protect against sepsis-induced mortality during the acute phase (Lugrin et al., 2013) and may block the development of innate immune paralysis in survivors (Vachharajani et al., 2014). Whether SIRT1 expression remains upregulated in cells and tissues following recovery from sepsis, and in immune cells in particular, remains to be studied. Additionally, aside from functional studies of histone methyltransferase and demethylase binding to proinflammatory gene promoters of primary DCs following recovery from sepsis, there have been few studies tracking the expression/function of these CMEs in either animal models or human patients. This knowledge gap in the primary literature will be discussed at the close of this chapter, and in summary with findings from studies in adaptive immunity, a case will be made for the use of CME expression in peripheral leukocytes as possible biomarkers for the development of postseptic immunosuppression.

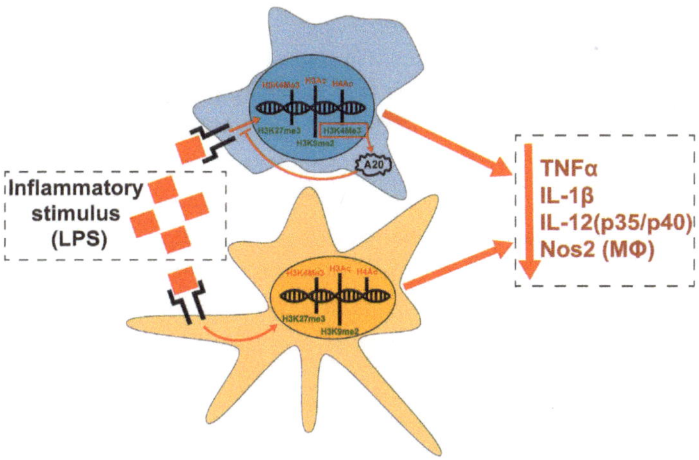

Fig. 2 A shift toward repressive histone modifications impairs proinflammatory gene expression in postseptic innate immune cells. When postseptic macrophages and dendritic cells encounter subsequent inflammatory stimuli (e.g., LPS), they are tasked with expressing proinflammatory mediators in the context of aberrant histone modification patterns, including decreases (*red*) in activating marks (H3K4me3, H3Ac, and H4Ac) and increases (*green*) in repressive marks (H3K27me3 and H3K9me2). This shift toward repressive histone modifications results in decreased expression in a wide range of proinflammatory genes, including cytokines and antimicrobial enzymes. When increases in activating histone modifications are observed in postseptic innate immune cells (i.e., H3K4me3), they promote the expression of negative regulators of inflammatory signal transduction (such as A20 and its ability to repress LPS-dependent signals), further dampening inflammatory responses by innate immunity.

4. HISTONE MODIFICATIONS AND POSTSEPTIC ADAPTIVE IMMUNE RESPONSES

Similar to the deficiencies observed in postseptic innate immune cells, adaptive immune responses are also impacted following recovery from sepsis. However, unlike innate immune responses that appear broadly dampened, adaptive immune responses exhibit complex immunomodulatory phenotypes that speak to their roles as master regulators of the cellular immune response. Adaptive immune responses are uniquely tailored to each specific pathogen challenge, and these inflammatory subtypes promote only those innate immune responses that best control and clear the invading pathogen. For example, bacterial infections often stimulate T-helper type 1 responses (T_H1), characterized by the production of IFNγ leading to the activation of macrophage killing functions (Oestreich & Weinmann, 2012). In contrast, extracellular parasites stimulate T-helper type 2 responses (T_H2) characterized by the production of IL-5 and the mobilization of eosinophils from the bone marrow (Kon & Kay, 1999). When these responses are aberrantly regulated, and the quality of the inflammatory process is not appropriate for the particular infection, negative outcomes such as immunopathology and increased infection severity can result.

Following recovery from sepsis, the decision-making capabilities of adaptive immunity are impaired, leaving the host particularly susceptible to infections. This impaired ability to commit to tailored inflammatory responses was classically identified in the literature as a "T_H2 bias" (Mack et al., 1996; O'Sullivan et al., 1995). Peripheral blood leukocytes from postseptic patients often exhibit increased production of T_H2 cytokines such as IL-4 (Spolarics et al., 2003). However, this phenotype is much more complex than simply a bias in a binary decision tree. When tasked with committing to either the T_H1 or T_H2 functional lineages in vivo, postseptic CD4+ T cells exhibit an intermediate phenotype, where cells will make inappropriate cytokines in response to the cytokine milieu, such as production of both IL-4 and IFNγ in cultures that only drive IL-4 production in control T cells (Carson et al., 2010). Postseptic CD4+ T cells also exhibit a bias toward overproduction of IL-17, the phenotypic cytokine produced by T_H17 cells, even in inflammatory conditions that do not normally promote T_H17 conditions; this pathogenic overproduction of IL-17 can exacerbate immunopathology during secondary viral infections (Mukherjee et al., 2012). The T cell "confusion" phenotype can also be observed in adoptive transfer models utilizing T_H-subtype specific inflammatory models. In these, postseptic T cells maintain their inability to commit to functional lineages even when transferred into lymphopenic hosts that have never been through a severe septic episode, and should therefore maintain functionality of their innate immune system (Carson, Ito, et al., 2011). Taken together, these findings provide evidence that postseptic CD4+ T cells suffer from cell-intrinsic defects in their ability to commit to functional lineages. As the decision-making processes of immune cells can be regulated through epigenetics, these studies provide strong evidence for the role of aberrant histone modifications and CME expression/function in CD4+ T cells following sepsis.

Significant modulations in both histone methylation and acetylation can be observed in postseptic CD4+ T cells. For example, changes in histone methylation track with aberrant T_H-subtype lineage commitment in postseptic T cells cultured in vitro. Increased levels of repressive H3K27me3 can be observed in the promoters of both IFNγ and GATA3, indicating that increased repressive histone methylation may be participating in the inability of these cells to lock in specific gene loci into transcriptionally accessible states (Carson et al., 2010). Interestingly, these same promoter sites do not show any modulations in activating H3K4me3, suggesting that as a consequence of sepsis, CD4+ T cell specific genes are changed from "poised" to transcriptionally repressed states (Carson et al., 2010). This switch to epigenetic repression for T_H-specific gene loci may explain in part the "confusion" phenotype displayed by postseptic CD4+ T cells both in vitro and in vivo.

Interestingly, when postseptic CD4+ T cells do exhibit modulations in activating histone modifications, these occur in the promoter regions of genes involved with driving suppressive activity. In addition to deficiencies in T_H-subtype polarization, postseptic CD4+ T cells exhibit an increased propensity to develop suppressive phenotypes (Venet et al., 2008). Regulatory CD4+ T cells (T_{reg}) are essential adaptive immune cells for the maintenance of tissue homeostasis and the resolution of inflammation (Panduro, Benoist, & Mathis, 2016; Rodriguez-Perea, Arcia, Rueda, & Velilla, 2016). While the specific mechanisms utilized by T_{regs} to suppress inflammatory responses remain controversial, experimental evidence points to important roles for antiinflammatory cytokine production, cell–cell interactions, and the ability of T_{regs} to significantly alter the local microenvironment (bind and sequester IL-2, for example) (von Boehmer, 2005). T_{regs} arise from two specific progenitor populations: "natural" (n)T_{regs} arise from thymocyte progenitors, and exit the thymus as self-reactive T cells with immunosuppressive phenotypes (Schmetterer, Neunkirchner, & Pickl, 2012), whereas "inducible" (i)Tregs exit the thymus as traditional, naïve T cells and then switch to a suppressive phenotype in response to cytokine stimulus and/or cell–cell interactions in peripheral tissues (Dons, Raimondi, Cooper, & Thomson, 2012). nTregs are thought to be critically important for protection against the development of autoimmune disease through their ability to restrict pathogenic self-antigen responses by effector T cells in the periphery (Langier, Sade, & Kivity, 2010). iTregs are important for the resolution of inflammation following pathogen clearance, and the conversion of immature T cells into iTregs is critical for the development and maintenance of tolerance at barrier surfaces (Pabst & Mowat, 2012; Peters & Sacks, 2006).

Following recovery from sepsis, the distribution of T_{regs} increases in both human patients and experimental animal models (Jiang, Yao, & Sheng, 2012). There is some controversy about whether this increase is due to an expansion of T_{regs} during the acute phase of sepsis as a consequence of the switch from SIRS to CARS (Monneret et al., 2003), or whether this is simply a relative increase in T_{regs} due to the widespread apoptosis of effector T cells during SIRS (Venet et al., 2004). In either case, peripheral T_{reg} populations are clearly increased following recovery from sepsis, and this increase in T_{regs} is a central component of postseptic immunosuppression. For example, enhanced T_{reg} populations following recovery from sepsis can enhance susceptibility to secondary infections (Huo et al., 2016; Molinaro et al., 2015), as well as suppressing tumor surveillance in experimental animal models (Cavassani et al., 2010). Interestingly, it appears that the contribution of T_{regs} to the suppressive phenotype of severe sepsis survivors is due to both the survival/expansion of previously committed nT_{regs}, but also to the preferential differentiation of naïve peripheral T cells into iT_{regs} in inflammatory contexts. As with the effector T cell deficiencies discussed earlier, histone modifications play a central role in the preferential induction of T_{reg} phenotypes in postseptic CD4+ T cells.

When naïve CD4+ CD25− effector T cells from postseptic mice are analyzed, no significant changes in histone methylation are observed in the promoter region of the key T_{reg} transcription factor Foxp3 (Cavassani et al., 2010). This is counter to the results observed with other T-helper lineage subtypes, such as the enhanced suppressive H3K27me3 levels observed in the GATA3 locus (Carson et al., 2010). However, significant increases in permissive histone acetylation are observed in the Foxp3 locus in these cells, especially at lysine 9 of the tail of histone 3 (H3K9ac). This increase in H3K9ac in CD4+ CD25− T cells correlates with enhanced Foxp3 expression and increased suppressive activity both in vitro and in vivo when these effector cells are skewed toward the T_{reg} lineage. Importantly, these CD4+ CD25− T cells also exhibit increased expression of the histone acetyltransferase Kat2a/GCN5L2, and this enhanced expression is observable at time points following resolution of sepsis-induced inflammation (Cavassani et al., 2010). These results provide evidence for persistent modulation in CME expression in T cells as a consequence of severe sepsis.

Interestingly, this increase in H3K9ac observed at the Foxp3 gene loci in CD4+ T cells might not necessarily reflect the effects of sepsis and related syndromes on the histone acetylation landscape in CD4+ T cells. As mentioned previously, postseptic CD4+ T cells tend to exhibit increases in repressive histone modifications following sepsis; this is counter to the findings regarding increased iT_{reg} functionality, which relies on increases in activating histone acetylation. However, when the genome-wide H3K9ac landscape is analyzed in human patients with chronic organ failure syndromes that closely mimic sepsis, a different pattern of histone acetylation emerges. When human peripheral blood CD4+ T cells are analyzed from patients suffering from acute-on-chronic liver failure, a disease that manifests similar immunosuppression as severe sepsis, a global decrease in H3K9ac levels are observed (Jin et al., 2015). Specific genes that exhibited decreases in activating H3K9ac included transcription factors such as IRF3, CEBPE, and ATF4, among others, suggesting that decreased histone acetylation in CD4+ T cells in liver failure patients is particularly targeted toward repression of transcription factor expression.

At present, there is no clear mechanistic explanation for the seemingly disparate effects of sepsis on H3K9ac levels in CD4+ cells, namely the genome-wide decrease with selective increases in loci critical for T_{reg} function. One possible explanation may be due to the differential expression of CMEs in postseptic CD4+ T cells. For example, expression and function of the SIRT family of deacetylases appear linked with inflammatory responses in sepsis syndromes. SIRT1 and SIRT6 have both been shown to play important roles in mediating monocyte responses to LPS (Liu et al., 2012), and modulation in SIRT1 expression can have direct effects on survival in animal models of sepsis (Vachharajani et al., 2014). Additionally, pharmacological inhibition of SIRT2 in experimental models of sepsis can improve survival rates while decreasing the expression of proinflammatory cytokines such as IL-6 and TNFγ (Zhao et al., 2015). It is difficult to state conclusively that the decrease in H3K9ac observed in immunoparalyzed CD4+ T cells is directly due to enhanced SIRT expression/function, as SIRTs have not yet been widely studied in postseptic CD4+ T cells. Additionally, SIRTs serve as deacetylases to a wide range of protein substrates and are not limited specifically to histones (Carafa et al., 2016). However, it is intriguing to postulate that differential specificity for SIRT1, 2, and 6 compared to Kat2a may explain how global H3K9ac levels can decrease in CD4+ T cells while specific gene loci such as Foxp3 become enriched in this activating epigenetic mark.

Taken together, these studies indicate an important role for modulations in histone methylation and acetylation in governing the switch to immunosuppressive phenotypes in postseptic adaptive immunity. In contrast to innate immune cells that primarily exhibit modulations toward repressive chromatin states in proinflammatory gene loci (such as decreases in H3K4me3 with increases in H3K27me3), adaptive immune cells exhibit targeted epigenetic changes in both proinflammatory mediators and transcription factors that predispose peripheral adaptive immune cells to mature

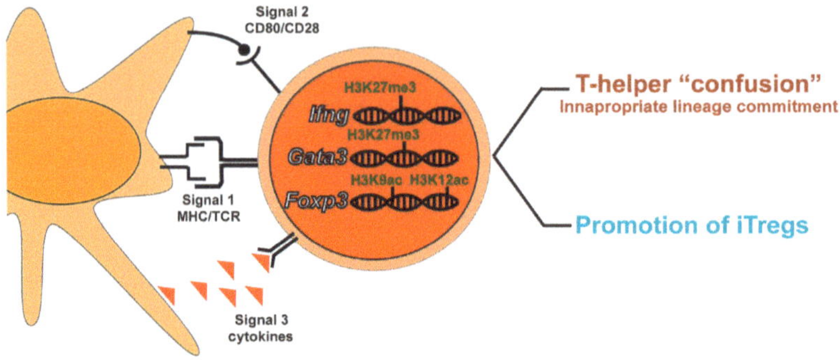

Fig. 3 Changes in histone methylation and acetylation in postseptic CD4+ T cells promote suppressive phenotypes at the expense of inflammatory effector cell lineages. Increases in repressive H3K27me3 in the promoter regions of T_H1 (*Ifng*) and T_H2 (*Gata3*) gene loci inhibit lineage commitment to effector cell phenotypes, whereas increases in permissive histone acetylation (H3K9ac and H3K12ac) bias the activation of CD4+ T cells toward the iTreg lineage.

toward immunosuppressive lineages. This is accomplished through global decreases in activating H3K9ac coupled with targeted increases of H3K27me3 at promoter regions of lineage-specific gene loci. In addition, increases in activating H3 acetylation (including H3K9ac and H4K12ac) in the promoter region of the transcription factor Foxp3, leading to the development of regulatory T cell functions (Fig. 3). Concurrent changes in CME expression help to drive and possibly maintain these marks beyond the resolution of sepsis-induced inflammation, providing one possible mechanism for the persistence of chromatin modifications in immune cells postsepsis.

5. CONCLUSIONS AND FUTURE PERSPECTIVES

Chromatin dynamics serve to regulate leukocyte gene expression through all stages of differentiation, activation, and commitment to effector phenotypes. As immune cells proceed toward terminal differentiation, the expression and function of CMEs leads to changes in histone modifications, with subsequent effects on lineage-specific gene expression. When CMEs are aberrantly regulated and histone modifications are improperly propagated, the resulting aberrant gene expression can significantly interfere with the proper functionality of cellular immunity. The studies summarized in this chapter highlight the importance of histone modifications in the initiation and maintenance of immunosuppressive phenotypes following recovery from severe sepsis. However, despite the experimental evidence currently available, there still remain significant gaps in the literature concerning the effects of severe sepsis on histone modifications in immune cells.

As mentioned previously, histone modifications in neutrophils postsepsis are surprisingly understudied, especially considering the importance of histone citrullination in the production of antimicrobial NETs. Histone citrullination is considered to be a repressive epigenetic mark and expression of the protein deiminase PAD4 can drive suppression of gene expression in target cells. As PAD4 expression is an essential first step in NETosis, it is intriguing to postulate that PAD4 expression may be increased in neutrophil populations postsepsis as a consequence of sepsis-induced NET production. Subsequent histone citrullination that then occurs due to the overexpression of PAD4 may inhibit neutrophil responses to subsequent infections due to repressed proinflammatory gene expression. This mechanism could have a significant impact not just on postseptic immunosuppression, but also on the acute phase of sepsis/SIRS, as experimental models suggest that innate immune cells in animals destined to succumb to sepsis-induced mortality can exhibit suppressive phenotypes as early as 6 h following the induction of experimental sepsis (cecal ligation and puncture) (Chiswick, Mella, Bernardo, & Remick, 2015). Further studies are required to determine if genome-wide histone citrullination is modulated in postseptic neutrophils, if this citrullination persists past the resolution of sepsis-induced inflammation, and if this histone citrullination drives aberrant epigenetic control of proinflammatory genes in neutrophils postsepsis.

In regards to monocytic innate immune cells, such as macrophages and dendritic cells, the experimental literature is measurably more robust. Decreases in activating histone methylation and acetylation with concurrent increases in repressive histone methylation are correlated with the suppression of proinflammatory cytokines and chemokines from both cell populations during secondary infection responses. In some cases, the CMEs mediating these changes in histone modifications have been identified, as is the case with EHMT2 and cofactors of the MLL1 and PRC2 chromatin remodeling complexes. However, there are two important facets of CME expression in postseptic myeloid cell populations that remain understudied. First, are the changes in histone modifications observed in these cell populations due to modulations in CME expression during the acute phase of sepsis, or are the expression pattern of CMEs permanently altered in these cells postsepsis? If it is the former, how are these changes in histone modifications maintained or propagated to daughter cells without concurrent expression of the required CMEs? If it is the latter, then what are the molecular mechanisms governing persistent under/overexpression of CMEs in myeloid populations postsepsis? Second, are these changes in CME expression and/or histone modifications limited specifically to peripheral myeloid cell populations, or are these changes also occurring in bone marrow-resident progenitor cells? If it is the former, than the same question applies: how are these epigenetic modifications persisting in peripheral myeloid populations if the bone marrow is constantly reseeding the blood with new monocytes that presumably do not contain these epigenetic "lesions?" If it is the latter, than this would presumably explain the persistence of sepsis-induced histone modifications in myeloid cells, but then a mechanistic explanation is required for the ability of severe systemic inflammation to make measurable changes to the epigenome of progenitor cells in the bone marrow.

Many of these same questions remain for the study of adaptive immune responses postsepsis. In regards to CME expression, there have been few studies linking changes in histone modifications to up- or downregulation of CMEs in postseptic CD4+ T cells. Of the work that has been published—for example, the upregulation of Kat2a in postseptic CD4+ T cells that exhibit a bias toward T_{reg} differentiation—the results are strictly correlative, and do not provide direct experimental evidence for induced or persistent expression of specific CMEs in maintaining histone modifications in postseptic cells. In addition, little is understood about the developmental stages of CD4+ T cells that are most affected by severe sepsis in regards to changes in CME expression and histone modifications. As with innate immunity, there remains questions about whether changes in CMEs or histone modifications occur in peripheral CD4+ T cells that survive the widespread apoptotic event of acute sepsis, or if these changes occur in bone marrow-resident lymphocyte progenitor cells that then give rise to peripheral CD4+ T cells with altered epigenetic landscapes. Adding to the complexity is the requirement for adaptive immune cells to mature through genomic editing and T cell/B cell receptor generation and expression, either in the thymus for T cells or in the bone marrow and spleen for B cells. It is therefore possible that the changes in CME expression and/or histone modifications observed in these adaptive immune cells postsepsis may have been driven by changes to the microenvironment of these specialized tissues.

In a broader sense, studies of histone modifications and postseptic adaptive immunity need to be expanded to encompass more than strictly CD4+ T cell responses. Cytotoxic CD8+ T cells also exhibit deficiencies in activation and effector function postsepsis, and many of these functional changes mimic the proliferation and cytokine production defects observed in CD4+ T cells (Condotta, Rai, James, Griffith, & Badovinac, 2013). This evidence argues strongly for similar mechanisms—i.e., histone modifications—to be governing CD8+ T cell dysfunction postsepsis. However, at the present time, there remains a paucity of published data investigating this possibility. Additionally, B cells have been shown to have significant defects postsepsis, especially in regards to the antibody response (Mohr et al., 2012). As the production of antibodies is inexorably linked with chromatin dynamics, including histone modifications (Zan & Casali, 2015), it is a strong hypothesis to propose that dysregulated CME expression following sepsis could be a major mechanistic driver of B cell dysfunction. As with studies in CD8+ T cells, however, the available literature regarding CMEs and histone modifications in postseptic B cells remains limited.

Another important area of investigation for all aspect of postsepsis cellular immunity is the lack of clear mechanistic links between the acute inflammatory event and the subsequent modulation of CME expression and downstream histone modification events. In a broad sense, the current working model of sepsis-induced immunosuppression argues that the physiological stress of acute sepsis induces epigenetic changes in immune cells, either through the modulation of CME expression, the differential survival of epigenetically altered immune cells during sepsis-induced apoptotic responses, or both. This paradigm still contains a "black box" of a mechanism to explain exactly how sepsis results in modulations to histone modifications, and more importantly, why sepsis-induced inflammation drives these changes when other inflammatory responses fail to do so. While there remains much to be investigated regarding this phenomenon in sepsis, there may be clues to be taken from basic immunology regarding the specific mechanisms of postseptic changes in histone modifications. As studies into the epigenetics of cellular immune responses become more robust, a pattern has started to emerge linking specific extracellular signals to the expression of particular CMEs, suggesting that CMEs and their corresponding histone modifications might be compartmentalized along with PAMPs, DAMPs, and cytokines as effector cell-specific factors guiding cellular immune responses (Table 1). For example, human monocyte-derived macrophages exhibit unique patterns of CME expression when induced to M1 or M2 phenotypes with exogenous cytokines; M1 macrophages preferentially express MLL1 in response to the M1-promoting cytokine IFNγ, while M2 macrophages express JMJD3 in response to the M2-promoting cytokine IL-4 (Kittan et al., 2013). MLL1 expression in macrophages also plays a key role in response to NF-kB dependent signal transduction (i.e., TNFα), providing additional evidence for the importance of MLL1 in mediating M1 responses (Wang et al., 2012). Type I interferons, especially IFNβ, have been shown to induce expression of the histone methyltransferase Setdb2 in macrophages, leading to epigenetic repression of proinflammatory genes through increased H3K9me3 (Kroetz et al., 2015; Schliehe et al., 2015). In T cells, differential CME expression has been linked to the generation of specific T-helper cell lineages, in particular IL-12-dependent modulation of MLL1 function in T_H1 cells (Schaller et al., 2015), and TGFβ-dependent modulation of the histone methyltransferase SMYD3 and its function in iTreg cells (Nagata et al., 2015).

The hallmark of severe sepsis and sepsis-related immunopathologies, from trauma to burn to ischemia/reperfusion injury, is the overproduction of proinflammatory mediators in response to high levels of PAMPs and DAMPs. More importantly, many of the cytokines already identified as having cognate CMEs are also important players in the cytokine storm of sepsis (e.g., Type I IFN and Setdb2, IL-12 and MLL1, etc.). Given the important link between cytokines and CME expression in immune cells, and the specificity of CMEs for specific histone modifications, it is tempting to hypothesize that it is the cytokine storm of sepsis that specifically alters the histone epigenetic landscape in cellular immunity, leading to sepsis-induced immunoparalysis. In such a hypothetical model, both the intensity and

Table 1 A Selection of Chromatin-Modifying Enzymes, Their Associated Histone Marks and Identified Cytokine/Signal Transduction Pathways Associated with Enzyme Expression and Function

CME	Histone Modification	Activating Pathway
MLL1	H3K4me3	IFNγ/STAT1 (Kittan et al., 2013)
		IL-12/STAT4 (Schaller et al., 2015)
		TNFα/NF-kB (Wang et al., 2012)
JMJD3	H3K27me3	LPS/NF-kB (De Santa et al., 2009)
		IL-4/STAT6 (Ishii et al., 2009; Kittan et al., 2013)
Setdb2	H3K9me3	IFNβ/STAT1/IRF7 (Kroetz et al., 2015; Schliehe et al., 2015)
SMYD3	H3K4me3	TGFβ1/Smad3 (Nagata et al., 2015)

duration of the cytokine-dependent inflammatory signals would drive aberrant, possibly pathologic expression of CMEs in immune cells that survive the initial apoptotic event, and perhaps even bone marrow-resident progenitor cells. This modulation in CME expression would be drastically different from expression changes seen during more classic pathogen-induced responses based mainly on the intensity of the cytokine storm induced during sepsis. Of course, such a model of sepsis-induced epigenetic paralysis requires a mechanistic explanation for the ability of the cytokine storm to set permanent histone modification changes in a fashion not seen in more classical single pathogen infections. Additionally, many CMEs have been shown to have multiple extracellular factors that drive their expression. Sometimes, these factors work in concert to support the production of particular cell lineages (for example, TNFα and IL-12 signals both leading to MLL1 expression and M1/T_H1 lineage commitment). In other cases, seemingly disparate responses can work through similar CMEs, complicating the links between inflammatory factors and specific lineages (for example, expression of JMJD3 is driven by both LPS to promote M1 responses, and IL-4 to promote M2 responses) (De Santa et al., 2009; Ishii et al., 2009). Despite these challenges, studies to date for both basic immunology and the pathophysiology of sepsis continue to point to the cytokine storm, and/or exposure to high levels of PAMPs/DAMPs, as the driving factors behind modulations in histone modifications following sepsis.

Looking forward to translational and clinical applications, it may be possible to utilize CME expression and histone modification patterns as biomarkers of sepsis-induced immunosuppression. At present, there is no widely acceptable methodology for the diagnosis of postseptic immunosuppression, with studies that identify the immunosuppression phenotype based on large-scale population studies or patient cohorts. On a per-patient level, there remains no consistently accurate and technically feasible methodology to diagnose postseptic cellular immune dysfunction outside of a secondary infection, save for costly, and complex in vitro cell biology experiments. In experimental animal models, the best predictor of postseptic immunosuppression is often the mortality rate within the experimental cohort, with only severe sepsis models ($>$40% mortality) exhibiting phenotypes consistent with postseptic immunosuppression (Benjamim et al., 2004). These give clues as to the status of the cellular immune response within the experimental group, but do not accurately predict the immune function of each individual animal. Despite the many avenues of investigation still left to address, the research summarized herein provides evidence for an epigenetic signature in peripheral immune cells that may signal postseptic immunosuppression. This would include overexpression of HDACs and repressive histone demethylases, for example, and would involve a switch away from activating H3K4me3 and H3/H4ac toward enriched H3K27me3 in the promoters of proinflammatory genes and lineage-specific transcription factors. It may be possible to screen peripheral blood samples from patients who have recovered from severe sepsis or related immunopathologies for modulated expression of CMEs involved with making these specific histone modifications. While direct diagnostic assessment of the histone modification landscape in peripheral human blood samples may be prohibitively challenging at present (issues relating to cost, availability of screening technology, etc.), there remains hope for the future as high-throughout epigenetic sequencing technologies (such as ChIP-on-chip and ChIP-seq) become more widespread, user friendly and cost effective. The ultimate goal would be to provide clinicians with an epigenetic toolkit from which they could actively identify survivors of severe sepsis who are at high risk for negative outcomes in response to secondary infections, allowing for proactive rather than reactive treatment regimens. As the armamentarium of pharmacological inhibitors of CME function continue to expand, encompassing broad-spectrum HDAC inhibitors, inhibitors of specific HDACs, and the first generation of histone methyltransferase and demethylase inhibitors, it is intriguing to predict that in the future, the prophylaxis regimen of survivors of severe sepsis may include CME inhibitors to help break postseptic immunosuppression.

6. SUMMARY

Survivors of life-threatening immunopathologies such as severe sepsis often go on to develop severe immunosuppression, characterized by an inability of the cellular immune response to activate effectively in response to secondary pathogens. This immunosuppression is driven by an inability of both innate and adaptive immunity to produce key proinflammatory mediators and lineage-specific transcription factors, leading to improper immune responses that fail to provide immunity. Aberrant histone modifications help to drive this suppressed gene expression by locking inflammatory gene loci into transcriptionally inaccessible states, through the removal of activating histone methylation and acetylation, and/or the increase in repressive histone methylation. These histone modifications are driven by the expression and function of CMEs, which are themselves regulated by inflammatory signals (such as PAMPs, DAMPs, and cytokines) that are found at high concentration during the acute phase of severe sepsis. It is believed that the persistence of these histone modifications, either in peripheral immune cells, in innate and adaptive immune progenitor cells in the bone marrow, or a combination of the two, helps to explain how this immunoparalysis can last for long periods of time after the resolution of sepsis-induced inflammation. While much still remains to be elucidated regarding the link between sepsis, CMEs and histone modifications, studies in both animal models and human patients have provided convincing evidence of the importance of histone modifications to exert epigenetic control over postseptic immune cells to drive sepsis-induced immunosuppression.

REFERENCES

Adib-Conquy, M., & Cavaillon, J. M. (2009). Compensatory anti-inflammatory response syndrome. *Thrombosis and Haemostasis*, *101*(1), 36–47. doi: 09010036 [pii].

Allavena, P., Chieppa, M., Monti, P., & Piemonti, L. (2004). From pattern recognition receptor to regulator of homeostasis: The double-faced macrophage mannose receptor. *Critical Reviews in Immunology*, *24*(3), 179–192.

Alves-Filho, J. C., Spiller, F., & Cunha, F. Q. (2010). Neutrophil paralysis in sepsis. *Shock*, *34*(Suppl. 1), 15–21. http://dx.doi.org/10.1097/SHK.0b013e3181e7e61b.

Anzilotti, C., Pratesi, F., Tommasi, C., & Migliorini, P. (2010). Peptidylarginine deiminase 4 and citrullination in health and disease. *Autoimmunity Reviews*, *9*(3), 158–160. http://dx.doi.org/10.1016/j.autrev.2009.06.002.

Ayala, A., Deol, Z. K., Lehman, D. L., Herdon, C. D., & Chaudry, I. H. (1994). Polymicrobial sepsis but not low-dose endotoxin infusion causes decreased splenocyte IL-2/IFN-gamma release while increasing IL-4/IL-10 production. *The Journal of Surgical Research*, *56*(6), 579–585. http://dx.doi.org/10.1006/jsre.1994.1092.

Benjamim, C. F., Hogaboam, C. M., & Kunkel, S. L. (2004). The chronic consequences of severe sepsis. *Journal of Leukocyte Biology*, *75*(3), 408–412. http://dx.doi.org/10.1189/jlb.0503214.

Benjamim, C. F., Hogaboam, C. M., Lukacs, N. W., & Kunkel, S. L. (2003). Septic mice are susceptible to pulmonary aspergillosis. *The American Journal of Pathology*, *163*(6), 2605–2617.

Benjamim, C. F., Lundy, S. K., Lukacs, N. W., Hogaboam, C. M., & Kunkel, S. L. (2005). Reversal of long-term sepsis-induced immunosuppression by dendritic cells. *Blood*, *105*(9), 3588–3595. http://dx.doi.org/10.1182/blood-2004-08-3251.

Bentzer, P., Russell, J. A., & Walley, K. R. (2015). Advances in sepsis research. *Clinics in Chest Medicine*, *36*(3), 521–530. http://dx.doi.org/10.1016/j.ccm.2015.05.009.

Bermejo-Martin, J. F., Andaluz-Ojeda, D., Almansa, R., Gandia, F., Gomez-Herreras, J. I., Gomez-Sanchez, E., … Tamayo, E. (2016). Defining immunological dysfunction in sepsis: A requisite tool for precision medicine. *The Journal of Infection*, *72*(5), 525–536. http://dx.doi.org/10.1016/j.jinf.2016.01.010.

Biswas, S. K., & Lopez-Collazo, E. (2009). Endotoxin tolerance: New mechanisms, molecules and clinical significance. *Trends in Immunology*, *30*(10), 475–487. http://dx.doi.org/10.1016/j.it.2009.07.009.

Brinkmann, V., Reichard, U., Goosmann, C., Fauler, B., Uhlemann, Y., Weiss, D. S., … Zychlinsky, A. (2004). Neutrophil extracellular traps kill bacteria. *Science*, *303*(5663), 1532–1535. http://dx.doi.org/10.1126/science.1092385.

Cabrera-Perez, J., Condotta, S. A., Badovinac, V. P., & Griffith, T. S. (2014). Impact of sepsis on CD4 T cell immunity. *Journal of Leukocyte Biology*, *96*(5), 767–777. http://dx.doi.org/10.1189/jlb.5MR0114-067R.

Carafa, V., Rotili, D., Forgione, M., Cuomo, F., Serretiello, E., Hailu, G. S., … Altucci, L. (2016). Sirtuin functions and modulation: From chemistry to the clinic. *Clinical Epigenetics*, *8*, 61. http://dx.doi.org/10.1186/s13148-016-0224-3.

Carson, W. F., IV, Cavassani, K. A., Dou, Y., & Kunkel, S. L. (2011). Epigenetic regulation of immune cell functions during post-septic immunosuppression. *Epigenetics*, *6*(3), 273–283.

Carson, W. F., IV, Cavassani, K. A., Ito, T., Schaller, M., Ishii, M., Dou, Y., & Kunkel, S. L. (2010). Impaired CD4+ T-cell proliferation and effector function correlates with repressive histone methylation events in a mouse model of severe sepsis. *European Journal of Immunology*, *40*(4), 998–1010.

Carson, W. F., IV, Ito, T., Schaller, M., Cavassani, K. A., Chensue, S. W., & Kunkel, S. L. (2011). Dysregulated cytokine expression by CD4+ T cells from post-septic mice modulates both Th1 and Th2-mediated granulomatous lung inflammation. *PLoS One*, *6*(5), e20385. http://dx.doi.org/10.1371/journal.pone.0020385.

Cavassani, K. A., Carson, W. F. I. V., Moreira, A. P., Wen, H., Schaller, M. A., Ishii, M., … Kunkel, S. L. (2010). The post sepsis-induced expansion and enhanced function of regulatory T cells create an environment to potentiate tumor growth. *Blood*, *115*(22), 4403–4411. http://dx.doi.org/10.1182/blood-2009-09-241083.

Cavassani, K. A., Ishii, M., Wen, H., Schaller, M. A., Lincoln, P. M., Lukacs, N. W., … Kunkel, S. L. (2008). TLR3 is an endogenous sensor of tissue necrosis during acute inflammatory events. *The Journal of Experimental Medicine*, *205*(11), 2609–2621. http://dx.doi.org/10.1084/jem.20081370.

Cavassani, K. A., Moreira, A. P., Habiel, D., Ito, T., Coelho, A. L., Allen, R. M., … Kunkel, S. L. (2013). Toll like receptor 3 plays a critical role in the progression and severity of acetaminophen-induced hepatotoxicity. *PLoS One*, *8*(6), e65899. http://dx.doi.org/10.1371/journal.pone.0065899.

Chiswick, E. L., Mella, J. R., Bernardo, J., & Remick, D. G. (2015). Acute-phase deaths from murine polymicrobial sepsis are characterized by innate immune suppression rather than exhaustion. *The Journal of Immunology*, *195*(8), 3793–3802. http://dx.doi.org/10.4049/jimmunol.1500874.

Ciarlo, E., Savva, A., & Roger, T. (2013). Epigenetics in sepsis: Targeting histone deacetylases. *International Journal of Antimicrobial Agents*, *42*(Suppl.), S8–S12. http://dx.doi.org/10.1016/j.ijantimicag.2013.04.004.

Condotta, S. A., Rai, D., James, B. R., Griffith, T. S., & Badovinac, V. P. (2013). Sustained and incomplete recovery of naive CD8+ T cell precursors after sepsis contributes to impaired CD8+ T cell responses to infection. *The Journal of Immunology*, *190*(5), 1991–2000. http://dx.doi.org/10.4049/jimmunol.1202379.

Cuthbert, G. L., Daujat, S., Snowden, A. W., Erdjument-Bromage, H., Hagiwara, T., Yamada, M., … Kouzarides, T. (2004). Histone deimination antagonizes arginine methylation. *Cell*, *118*(5), 545–553. http://dx.doi.org/10.1016/j.cell.2004.08.020.

De Santa, F., Narang, V., Yap, Z. H., Tusi, B. K., Burgold, T., Austenaa, L., … Natoli, G. (2009). Jmjd3 contributes to the control of gene expression in LPS-activated macrophages. *The EMBO Journal*, *28*(21), 3341–3352. http://dx.doi.org/10.1038/emboj.2009.271.

Delano, M. J., & Ward, P. A. (2016). Sepsis-induced immune dysfunction: Can immune therapies reduce mortality? *The Journal of Clinical Investigation*, *126*(1), 23–31. http://dx.doi.org/10.1172/JCI82224.

Deng, J. C., Cheng, G., Newstead, M. W., Zeng, X., Kobayashi, K., Flavell, R. A., & Standiford, T. J. (2006). Sepsis-induced suppression of lung innate immunity is mediated by IRAK-M. *The Journal of Clinical Investigation*, *116*(9), 2532–2542. http://dx.doi.org/10.1172/JCI28054.

Divangahi, M., King, I. L., & Pernet, E. (2015). Alveolar macrophages and type I IFN in airway homeostasis and immunity. *Trends in Immunology*, *36*(5), 307–314. http://dx.doi.org/10.1016/j.it.2015.03.005.

Dons, E. M., Raimondi, G., Cooper, D. K., & Thomson, A. W. (2012). Induced regulatory T cells: Mechanisms of conversion and suppressive potential. *Human Immunology*, *73*(4), 328–334. http://dx.doi.org/10.1016/j.humimm.2011.12.011.

Drewry, A. M., Samra, N., Skrupky, L. P., Fuller, B. M., Compton, S. M., & Hotchkiss, R. S. (2014). Persistent lymphopenia after diagnosis of sepsis predicts mortality. *Shock*, *42*(5), 383–391. http://dx.doi.org/10.1097/SHK.0000000000000234.

El Gazzar, M., Yoza, B. K., Chen, X., Hu, J., Hawkins, G. A., & McCall, C. E. (2008). G9a and HP1 couple histone and DNA methylation to TNFalpha transcription silencing during endotoxin tolerance. *The Journal of Biological Chemistry*, *283*(47), 32198–32208. http://dx.doi.org/10.1074/jbc.M803446200.

Flohe, S. B., Agrawal, H., Schmitz, D., Gertz, M., Flohe, S., & Schade, F. U. (2006). Dendritic cells during polymicrobial sepsis rapidly mature but fail to initiate a protective Th1-type immune response. *Journal of Leukocyte Biology*, *79*(3), 473–481. http://dx.doi.org/10.1189/jlb.0705413.

Gotts, J. E., & Matthay, M. A. (2016). Sepsis: Pathophysiology and clinical management. *BMJ*, *353*, i1585. http://dx.doi.org/10.1136/bmj.i1585.

Grossniklaus, U., & Paro, R. (2014). Transcriptional silencing by polycomb-group proteins. *Cold Spring Harbor Perspectives in Biology*, *6*(11), a019331. http://dx.doi.org/10.1101/cshperspect.a019331.

Hotchkiss, R. S., Swanson, P. E., Freeman, B. D., Tinsley, K. W., Cobb, J. P., Matuschak, G. M., … Karl, I. E. (1999). Apoptotic cell death in patients with sepsis, shock, and multiple organ dysfunction. *Critical Care Medicine*, *27*(7), 1230–1251.

Hotchkiss, R. S., Tinsley, K. W., Swanson, P. E., Schmieg, R. E., Jr., Hui, J. J., Chang, K. C., … Karl, I. E. (2001). Sepsis-induced apoptosis causes progressive profound depletion of B and CD4+ T lymphocytes in humans. *The Journal of Immunology*, *166*(11), 6952–6963.

Huang, H., Evankovich, J., Yan, W., Nace, G., Zhang, L., Ross, M., … Tsung, A. (2011). Endogenous histones function as alarmins in sterile inflammatory liver injury through Toll-like receptor 9 in mice. *Hepatology*, *54*(3), 999–1008. http://dx.doi.org/10.1002/hep.24501.

Huang, X., Venet, F., Chung, C. S., Lomas-Neira, J., & Ayala, A. (2007). Changes in dendritic cell function in the immune response to sepsis. Cell- & tissue-based therapy. *Expert Opinion on Biological Therapy*, 7(7), 929–938. http://dx.doi.org/10.1517/14712598.7.7.929.

Huo, R., Wang, L., Wang, X., Zhao, Y., Wang, Y., Zhao, X., … Huang, Y. (2016). Removal of regulatory T cells prevents secondary chronic infection but increases the mortality of subsequent sub-acute infection in sepsis mice. *Oncotarget*, 7(10), 10962–10975. http://dx.doi.org/10.18632/oncotarget.7682.

Iida, S., Watanabe-Fukunaga, R., Nagata, S., & Fukunaga, R. (2008). Essential role of C/EBPalpha in G-CSF-induced transcriptional activation and chromatin modification of myeloid-specific genes. *Genes to Cells*, *13*(4), 313–327. http://dx.doi.org/10.1111/j.1365-2443.2008.01173.x.

Ishii, M., Wen, H., Corsa, C. A., Liu, T., Coelho, A. L., Allen, R. M., … Kunkel, S. L. (2009). Epigenetic regulation of the alternatively activated macrophage phenotype. *Blood*, *114*(15), 3244–3254. http://dx.doi.org/10.1182/blood-2009-04-217620.

Jiang, L. N., Yao, Y. M., & Sheng, Z. Y. (2012). The role of regulatory T cells in the pathogenesis of sepsis and its clinical implication. *Journal of Interferon & Cytokine Research*, *32*(8), 341–349. http://dx.doi.org/10.1089/jir.2011.0080.

Jin, L., Wang, K., Liu, H., Chen, T., Yang, Y., Ma, X., … He, Y. (2015). Genomewide histone H3 lysine 9 acetylation profiling in CD4+ T cells revealed endoplasmic reticulum stress deficiency in patients with acute-on-chronic liver failure. *Scandinavian Journal of Immunology*, *82*(5), 452–459. http://dx.doi.org/10.1111/sji.12341.

Jones, T. K., Fuchs, B. D., Small, D. S., Halpern, S. D., Hanish, A., Umscheid, C. A., … Mikkelsen, M. E. (2015). Post-acute care use and hospital readmission after sepsis. *Annals of the American Thoracic Society*, *12*(6), 904–913. http://dx.doi.org/10.1513/AnnalsATS.201411-504OC.

Kang, J. W., Kim, S. J., Cho, H. I., & Lee, S. M. (2015). DAMPs activating innate immune responses in sepsis. *Ageing Research Reviews*, *24*(Pt. A), 54–65. http://dx.doi.org/10.1016/j.arr.2015.03.003.

Kaplan, M. J., & Radic, M. (2012). Neutrophil extracellular traps: Double-edged swords of innate immunity. *The Journal of Immunology*, *189*(6), 2689–2695. http://dx.doi.org/10.4049/jimmunol.1201719.

Kempker, J. A., & Martin, G. S. (2016). The changing epidemiology and definitions of sepsis. *Clinics in Chest Medicine*, *37*(2), 165–179. http://dx.doi.org/10.1016/j.ccm.2016.01.002.

Kittan, N. A., Allen, R. M., Dhaliwal, A., Cavassani, K. A., Schaller, M., Gallagher, K. A., … Hogaboam, C. M. (2013). Cytokine induced phenotypic and epigenetic signatures are key to establishing specific macrophage phenotypes. *PLoS One*, *8*(10), e78045. http://dx.doi.org/10.1371/journal.pone.0078045.

Kon, O. M., & Kay, A. B. (1999). T cells and chronic asthma. *International Archives of Allergy and Immunology*, *118*(2–4), 133–135. doi: 24049.

Kroetz, D. N., Allen, R. M., Schaller, M. A., Cavallaro, C., Ito, T., & Kunkel, S. L. (2015). Type I interferon induced epigenetic regulation of macrophages suppresses innate and adaptive immunity in acute respiratory viral infection. *PLoS Pathogens*, *11*(12), e1005338. http://dx.doi.org/10.1371/journal.ppat.1005338.

Langier, S., Sade, K., & Kivity, S. (2010). Regulatory T cells: The suppressor arm of the immune system. *Autoimmunity Reviews*, *10*(2), 112–115. http://dx.doi.org/10.1016/j.autrev.2010.08.013.

Le Tulzo, Y., Pangault, C., Gacouin, A., Guilloux, V., Tribut, O., Amiot, L., … Drenou, B. (2002). Early circulating lymphocyte apoptosis in human septic shock is associated with poor outcome. *Shock*, *18*(6), 487–494.

Li, Y., Zhang, P., Wang, C., Han, C., Meng, J., Liu, X., … Cao, X. (2013). Immune responsive gene 1 (IRG1) promotes endotoxin tolerance by increasing A20 expression in macrophages through reactive oxygen species. *The Journal of Biological Chemistry*, *288*(23), 16225–16234. http://dx.doi.org/10.1074/jbc.M113.454538.

Lim, P. S., Li, J., Holloway, A. F., & Rao, S. (2013). Epigenetic regulation of inducible gene expression in the immune system. *Immunology*, *139*(3), 285–293. http://dx.doi.org/10.1111/imm.12100.

Liu, T. F., Vachharajani, V. T., Yoza, B. K., & McCall, C. E. (2012). NAD+-dependent sirtuin 1 and 6 proteins coordinate a switch from glucose to fatty acid oxidation during the acute inflammatory response. *The Journal of Biological Chemistry*, *287*(31), 25758–25769. http://dx.doi.org/10.1074/jbc.M112.362343.

Liu, T. F., Yoza, B. K., El Gazzar, M., Vachharajani, V. T., & McCall, C. E. (2011). NAD+-dependent SIRT1 deacetylase participates in epigenetic reprogramming during endotoxin tolerance. *The Journal of Biological Chemistry*, *286*(11), 9856–9864. http://dx.doi.org/10.1074/jbc.M110.196790.

Lugrin, J., Ciarlo, E., Santos, A., Grandmaison, G., dos Santos, I., Le Roy, D., & Roger, T. (2013). The sirtuin inhibitor cambinol impairs MAPK signaling, inhibits inflammatory and innate immune responses and protects from septic shock. *Biochimica et Biophysica Acta*, *1833*(6), 1498–1510. http://dx.doi.org/10.1016/j.bbamcr.2013.03.004.

Lyn-Kew, K., Rich, E., Zeng, X., Wen, H., Kunkel, S. L., Newstead, M. W., … Standiford, T. J. (2010). IRAK-M regulates chromatin remodeling in lung macrophages during experimental sepsis. *PLoS One*, *5*(6), e11145. http://dx.doi.org/10.1371/journal.pone.0011145.

Mack, V. E., McCarter, M. D., Naama, H. A., Calvano, S. E., & Daly, J. M. (1996). Dominance of T-helper 2-type cytokines after severe injury. *Archives of Surgery*, *131*(12), 1303–1308. discussion 1308–1309.

Maley, J. H., & Mikkelsen, M. E. (2016). Short-term gains with long-term consequences: the evolving story of sepsis survivorship. *Clinics in Chest Medicine*, *37*(2), 367–380. http://dx.doi.org/10.1016/j.ccm.2016.01.017.

Marcos, V., Zhou, Z., Yildirim, A. O., Bohla, A., Hector, A., Vitkov, L., … Hartl, D. (2010). CXCR2 mediates NADPH oxidase-independent neutrophil extracellular trap formation in cystic fibrosis airway inflammation. *Nature Medicine*, *16*(9), 1018–1023. http://dx.doi.org/10.1038/nm.2209.

Martinod, K., Fuchs, T. A., Zitomersky, N. L., Wong, S. L., Demers, M., Gallant, M., … Wagner, D. D. (2015). PAD4-deficiency does not affect bacteremia in polymicrobial sepsis and ameliorates endotoxemic shock. *Blood*, *125*(12), 1948–1956. http://dx.doi.org/10.1182/blood-2014-07-587709.

Mohr, A., Polz, J., Martin, E. M., Griessl, S., Kammler, A., Potschke, C., … Mannel, D. N. (2012). Sepsis leads to a reduced antigen-specific primary antibody response. *European Journal of Immunology*, *42*(2), 341–352. http://dx.doi.org/10.1002/eji.201141692.

Molinaro, R., Pecli, C., Guilherme, R. F., Alves-Filho, J. C., Cunha, F. Q., Canetti, C., … Benjamim, C. F. (2015). CCR4 controls the suppressive effects of regulatory T cells on early and late events during severe sepsis. *PLoS One*, *10*(7), e0133227. http://dx.doi.org/10.1371/journal.pone.0133227.

Mombelli, M., Lugrin, J., Rubino, I., Chanson, A. L., Giddey, M., Calandra, T., & Roger, T. (2011). Histone deacetylase inhibitors impair antibacterial defenses of macrophages. *The Journal of Infectious Diseases*, *204*(9), 1367–1374. http://dx.doi.org/10.1093/infdis/jir553.

Monneret, G., Debard, A. L., Venet, F., Bohe, J., Hequet, O., Bienvenu, J., & Lepape, A. (2003). Marked elevation of human circulating CD4+CD25+ regulatory T cells in sepsis-induced immunoparalysis. *Critical Care Medicine*, *31*(7), 2068–2071.

Monneret, G., Finck, M. E., Venet, F., Debard, A. L., Bohe, J., Bienvenu, J., & Lepape, A. (2004). The anti-inflammatory response dominates after septic shock: Association of low monocyte HLA-DR expression and high interleukin-10 concentration. *Immunology Letters*, *95*(2), 193–198. http://dx.doi.org/10.1016/j.imlet.2004.07.009.

Mukherjee, S., Allen, R. M., Lukacs, N. W., Kunkel, S. L., & Carson, W. F., IV (2012). STAT3-mediated IL-17 production by postseptic T cells exacerbates viral immunopathology of the lung. *Shock*, *38*(5), 515–523. http://dx.doi.org/10.1097/SHK.0b013e31826f862c.

Munoz, C., Carlet, J., Fitting, C., Misset, B., Bleriot, J. P., & Cavaillon, J. M. (1991). Dysregulation of in vitro cytokine production by monocytes during sepsis. *The Journal of Clinical Investigation*, *88*(5), 1747–1754. http://dx.doi.org/10.1172/JCI115493.

Muthu, K., He, L. K., Melstrom, K., Szilagyi, A., Gamelli, R. L., & Shankar, R. (2008). Perturbed bone marrow monocyte development following burn injury and sepsis promote hyporesponsive monocytes. *Journal of Burn Care & Research*, *29*(1), 12–21.

Nagata, D. E., Ting, H. A., Cavassani, K. A., Schaller, M. A., Mukherjee, S., Ptaschinski, C., … Lukacs, N. W. (2015). Epigenetic control of Foxp3 by SMYD3 H3K4 histone methyltransferase controls iTreg development and regulates pathogenic T-cell responses during pulmonary viral infection. *Mucosal Immunology*, *8*(5), 1131–1143. http://dx.doi.org/10.1038/mi.2015.4.

Neeli, I., Dwivedi, N., Khan, S., & Radic, M. (2009). Regulation of extracellular chromatin release from neutrophils. *Journal of Innate Immunity*, *1*(3), 194–201. http://dx.doi.org/10.1159/000206974.

Ng, M. K., & Cheung, P. (2016). A brief histone in time: Understanding the combinatorial functions of histone PTMs in the nucleosome context. *Biochemistry and Cell Biology*, *94*(1), 33–42. http://dx.doi.org/10.1139/bcb-2015-0031.

Oestreich, K. J., & Weinmann, A. S. (2012). Transcriptional mechanisms that regulate T helper 1 cell differentiation. *Current Opinion in Immunology*, *24*(2), 191–195. http://dx.doi.org/10.1016/j.coi.2011.12.004.

Ortego, A., Gaieski, D. F., Fuchs, B. D., Jones, T., Halpern, S. D., Small, D. S., … Mikkelsen, M. E. (2015). Hospital-based acute care use in survivors of septic shock. *Critical Care Medicine*, *43*(4), 729–737. http://dx.doi.org/10.1097/CCM.0000000000000693.

Osuchowski, M. F., Welch, K., Siddiqui, J., & Remick, D. G. (2006). Circulating cytokine/inhibitor profiles reshape the understanding of the SIRS/CARS continuum in sepsis and predict mortality. *The Journal of Immunology*, *177*(3), 1967–1974.

O'Sullivan, S. T., Lederer, J. A., Horgan, A. F., Chin, D. H., Mannick, J. A., & Rodrick, M. L. (1995). Major injury leads to predominance of the T helper-2 lymphocyte phenotype and diminished interleukin-12 production associated with decreased resistance to infection. *Annals of Surgery*, *222*(4), 482–490. discussion 490–492.

Pabst, O., & Mowat, A. M. (2012). Oral tolerance to food protein. *Mucosal Immunology*, *5*(3), 232–239. http://dx.doi.org/10.1038/mi.2012.4.

Panduro, M., Benoist, C., & Mathis, D. (2016). Tissue Tregs. *Annual Review of Immunology*, *34*, 609–633. http://dx.doi.org/10.1146/annurev-immunol-032712-095948.

Parlato, M., & Cavaillon, J. M. (2015). Host response biomarkers in the diagnosis of sepsis: A general overview. *Methods in Molecular Biology*, *1237*, 149–211. http://dx.doi.org/10.1007/978-1-4939-1776-1_15.

Perl, T. M., Dvorak, L., Hwang, T., & Wenzel, R. P. (1995). Long-term survival and function after suspected gram-negative sepsis. *JAMA*, *274*(4), 338–345.

Peters, N., & Sacks, D. (2006). Immune privilege in sites of chronic infection: Leishmania and regulatory T cells. *Immunological Reviews*, *213*, 159–179. http://dx.doi.org/10.1111/j.1600-065X.2006.00432.x.

Puri, D., Gala, H., Mishra, R., & Dhawan, J. (2015). High-wire act: The poised genome and cellular memory. *The FEBS Journal*, *282*(9), 1675–1691. http://dx.doi.org/10.1111/febs.13165.

Quartin, A. A., Schein, R. M., Kett, D. H., & Peduzzi, P. N. (1997). Magnitude and duration of the effect of sepsis on survival. Department of Veterans Affairs Systemic Sepsis Cooperative Studies Group. *JAMA*, *277*(13), 1058–1063.

Rasmussen, T. P. (2014). The epigenetics of early development: Inferences from stem cells. *Molecular Reproduction and Development*, *81*(2), 194–201. http://dx.doi.org/10.1002/mrd.22269.

Razin, A., & Cedar, H. (1994). DNA methylation and genomic imprinting. *Cell*, *77*(4), 473–476.

Rios, E. C., de Lima, T. M., Moretti, A. I., & Soriano, F. G. (2016). The role of nitric oxide in the epigenetic regulation of THP-1 induced by lipopolysaccharide. *Life Sciences*, *147*, 110–116. http://dx.doi.org/10.1016/j.lfs.2016.01.041.

Rodriguez-Perea, A. L., Arcia, E. D., Rueda, C. M., & Velilla, P. A. (2016). Phenotypic characterization of regulatory T cells in humans and rodents. *Clinical and Experimental Immunology*, *185*(3), 281–291. http://dx.doi.org/10.1111/cei.12804.

Schaafsma, W., Zhang, X., van Zomeren, K. C., Jacobs, S., Georgieva, P. B., Wolf, S. A., … Eggen, B. J. (2015). Long-lasting pro-inflammatory suppression of microglia by LPS-preconditioning is mediated by RelB-dependent epigenetic silencing. *Brain, Behavior, and Immunity*, *48*, 205–221. http://dx.doi.org/10.1016/j.bbi.2015.03.013.

Schaller, M., Ito, T., Allen, R. M., Kroetz, D., Kittan, N., Ptaschinski, C., … Kunkel, S. L. (2015). Epigenetic regulation of IL-12-dependent T cell proliferation. *Journal of Leukocyte Biology*, *98*(4), 601–613. http://dx.doi.org/10.1189/jlb.1A0814-375RR.

Schliehe, C., Flynn, E. K., Vilagos, B., Richson, U., Swaminathan, S., Bosnjak, B., ... Bergthaler, A. (2015). The methyltransferase Setdb2 mediates virus-induced susceptibility to bacterial superinfection. *Nature Immunology*, *16*(1), 67–74. http://dx.doi.org/10.1038/ni.3046.

Schmetterer, K. G., Neunkirchner, A., & Pickl, W. F. (2012). Naturally occurring regulatory T cells: Markers, mechanisms, and manipulation. *The FASEB Journal*, *26*(6), 2253–2276. http://dx.doi.org/10.1096/fj.11-193672.

Scott, C. L., Henri, S., & Guilliams, M. (2014). Mononuclear phagocytes of the intestine, the skin, and the lung. *Immunological Reviews*, *262*(1), 9–24. http://dx.doi.org/10.1111/imr.12220.

Spolarics, Z., Siddiqi, M., Siegel, J. H., Garcia, Z. C., Stein, D. S., Denny, T., & Deitch, E. A. (2003). Depressed interleukin-12-producing activity by monocytes correlates with adverse clinical course and a shift toward Th2-type lymphocyte pattern in severely injured male trauma patients. *Critical Care Medicine*, *31*(6), 1722–1729. http://dx.doi.org/10.1097/01.CCM.0000063579.43470.AA.

Tammen, S. A., Friso, S., & Choi, S. W. (2013). Epigenetics: The link between nature and nurture. *Molecular Aspects of Medicine*, *34*(4), 753–764. http://dx.doi.org/10.1016/j.mam.2012.07.018.

Tinsley, K. W., Grayson, M. H., Swanson, P. E., Drewry, A. M., Chang, K. C., Karl, I. E., & Hotchkiss, R. S. (2003). Sepsis induces apoptosis and profound depletion of splenic interdigitating and follicular dendritic cells. *The Journal of Immunology*, *171*(2), 909–914.

Turner, B. M. (2014). Nucleosome signalling; An evolving concept. *Biochimica et Biophysica Acta*, *1839*(8), 623–626. http://dx.doi.org/10.1016/j.bbagrm.2014.01.001.

Unsinger, J., Kazama, H., McDonough, J. S., Hotchkiss, R. S., & Ferguson, T. A. (2009). Differential lymphopenia-induced homeostatic proliferation for CD4+ and CD8+ T cells following septic injury. *Journal of Leukocyte Biology*, *85*(3), 382–390.

Vachharajani, V. T., Liu, T., Brown, C. M., Wang, X., Buechler, N. L., Wells, J. D., ... McCall, C. E. (2014). SIRT1 inhibition during the hypoinflammatory phenotype of sepsis enhances immunity and improves outcome. *Journal of Leukocyte Biology*, *96*(5), 785–796. http://dx.doi.org/10.1189/jlb.3MA0114-034RR.

Vanden Berghe, T., Linkermann, A., Jouan-Lanhouet, S., Walczak, H., & Vandenabeele, P. (2014). Regulated necrosis: The expanding network of non-apoptotic cell death pathways. *Nature Reviews. Molecular Cell Biology*, *15*(2), 135–147. http://dx.doi.org/10.1038/nrm3737.

Venet, F., Chung, C. S., Kherouf, H., Geeraert, A., Malcus, C., Poitevin, F., ... Monneret, G. (2009). Increased circulating regulatory T cells (CD4(+)CD25 (+) CD127 (−)) contribute to lymphocyte anergy in septic shock patients. *Intensive Care Medicine*, *35*(4), 678–686.

Venet, F., Chung, C. S., Monneret, G., Huang, X., Horner, B., Garber, M., & Ayala, A. (2008). Regulatory T cell populations in sepsis and trauma. *Journal of Leukocyte Biology*, *83*(3), 523–535.

Venet, F., Pachot, A., Debard, A. L., Bohe, J., Bienvenu, J., Lepape, A., & Monneret, G. (2004). Increased percentage of CD4+CD25+ regulatory T cells during septic shock is due to the decrease of CD4+CD25− lymphocytes. *Critical Care Medicine*, *32*(11), 2329–2331.

Venet, F., Tissot, S., Debard, A. L., Faudot, C., Crampe, C., Pachot, A., ... Monneret, G. (2007). Decreased monocyte human leukocyte antigen-DR expression after severe burn injury: Correlation with severity and secondary septic shock. *Critical Care Medicine*, *35*(8), 1910–1917. http://dx.doi.org/10.1097/01.CCM.0000275271.77350.B6.

Vermeulen, M., & Timmers, H. T. (2010). Grasping trimethylation of histone H3 at lysine 4. *Epigenomics*, *2*(3), 395–406. http://dx.doi.org/10.2217/epi.10.11.

von Boehmer, H. (2005). Mechanisms of suppression by suppressor T cells. *Nature Immunology, 6*(4), 338–344. http://dx.doi.org/10.1038/ni1180.
Wang, T. S., & Deng, J. C. (2008). Molecular and cellular aspects of sepsis-induced immunosuppression. *Journal of Molecular Medicine (Berlin, Germany), 86*(5), 495–506. http://dx.doi.org/10.1007/s00109-007-0300-4.
Wang, S. D., Huang, K. J., Lin, Y. S., & Lei, H. Y. (1994). Sepsis-induced apoptosis of the thymocytes in mice. *The Journal of Immunology, 152*(10), 5014–5021.
Wang, Y., Li, M., Stadler, S., Correll, S., Li, P., Wang, D., … Coonrod, S. A. (2009). Histone hypercitrullination mediates chromatin decondensation and neutrophil extracellular trap formation. *The Journal of Cell Biology, 184*(2), 205–213. http://dx.doi.org/10.1083/jcb.200806072.
Wang, X., Zhu, K., Li, S., Liao, Y., Du, R., Zhang, X., … Wu, M. (2012). MLL1, a H3K4 methyltransferase, regulates the TNFalpha-stimulated activation of genes downstream of NF-kappaB. *Journal of Cell Science, 125*(Pt. 17), 4058–4066. http://dx.doi.org/10.1242/jcs.103531.
Wen, H., Dou, Y., Hogaboam, C. M., & Kunkel, S. L. (2008). Epigenetic regulation of dendritic cell-derived interleukin-12 facilitates immunosuppression after a severe innate immune response. *Blood, 111*(4), 1797–1804. http://dx.doi.org/10.1182/blood-2007-08-106443.
Wen, H., Hogaboam, C. M., Gauldie, J., & Kunkel, S. L. (2006). Severe sepsis exacerbates cell-mediated immunity in the lung due to an altered dendritic cell cytokine profile. *The American Journal of Pathology, 168*(6), 1940–1950.
Wen, H., Schaller, M. A., Dou, Y., Hogaboam, C. M., & Kunkel, S. L. (2008). Dendritic cells at the interface of innate and acquired immunity: The role for epigenetic changes. *Journal of Leukocyte Biology, 83*(3), 439–446. http://dx.doi.org/10.1189/jlb.0607357.
Wesche-Soldato, D. E., Swan, R. Z., Chung, C. S., & Ayala, A. (2007). The apoptotic pathway as a therapeutic target in sepsis. *Current Drug Targets, 8*(4), 493–500.
Westman, J., Papareddy, P., Dahlgren, M. W., Chakrakodi, B., Norrby-Teglund, A., Smeds, E., … Herwald, H. (2015). Extracellular histones induce chemokine production in whole blood ex vivo and leukocyte recruitment in vivo. *PLoS Pathogens, 11*(12), e1005319. http://dx.doi.org/10.1371/journal.ppat.1005319.
Wiersinga, W. J., Leopold, S. J., Cranendonk, D. R., & van der Poll, T. (2014). Host innate immune responses to sepsis. *Virulence, 5*(1), 36–44. http://dx.doi.org/10.4161/viru.25436.
Winter, D. R., & Amit, I. (2014). The role of chromatin dynamics in immune cell development. *Immunological Reviews, 261*(1), 9–22. http://dx.doi.org/10.1111/imr.12200.
Xiu, F., & Jeschke, M. G. (2013). Perturbed mononuclear phagocyte system in severely burned and septic patients. *Shock, 40*(2), 81–88. http://dx.doi.org/10.1097/SHK.0b013e318299f774.
Xu, P. B., Lou, J. S., Ren, Y., Miao, C. H., & Deng, X. M. (2012). Gene expression profiling reveals the defining features of monocytes from septic patients with compensatory anti-inflammatory response syndrome. *The Journal of Infection, 65*(5), 380–391. http://dx.doi.org/10.1016/j.jinf.2012.08.001.
Yoshida, K., Maekawa, T., Zhu, Y., Renard-Guillet, C., Chatton, B., Inoue, K., … Ishii, S. (2015). The transcription factor ATF7 mediates lipopolysaccharide-induced epigenetic changes in macrophages involved in innate immunological memory. *Nature Immunology, 16*(10), 1034–1043. http://dx.doi.org/10.1038/ni.3257.
Yoza, B. K., Hu, J. Y., & McCall, C. E. (2002). Inhibition of histone deacetylation enhances endotoxin-stimulated transcription but does not reverse endotoxin tolerance. *Journal of Endotoxin Research, 8*(2), 109–114. http://dx.doi.org/10.1179/096805102125000245.

Zan, H., & Casali, P. (2015). Epigenetics of peripheral B-cell differentiation and the antibody response. *Frontiers in Immunology*, *6*, 631. http://dx.doi.org/10.3389/fimmu.2015.00631.

Zedler, S., Bone, R. C., Baue, A. E., von Donnersmarck, G. H., & Faist, E. (1999). T-cell reactivity and its predictive role in immunosuppression after burns. *Critical Care Medicine*, *27*(1), 66–72.

Zhang, X., Bolt, M., Guertin, M. J., Chen, W., Zhang, S., Cherrington, B. D., … Coonrod, S. A. (2012). Peptidylarginine deiminase 2-catalyzed histone H3 arginine 26 citrullination facilitates estrogen receptor alpha target gene activation. *Proceedings of the National Academy of Sciences of the United States of America*, *109*(33), 13331–13336. http://dx.doi.org/10.1073/pnas.1203280109.

Zhao, T., Alam, H. B., Liu, B., Bronson, R. T., Nikolian, V. C., Wu, E., … Li, Y. (2015). Selective inhibition of SIRT2 improves outcomes in a lethal septic model. *Current Molecular Medicine*, *15*(7), 634–641.

Zhao, T., Li, Y., Liu, B., Wu, E., Sillesen, M., Velmahos, G. C., … Alam, H. B. (2014). Histone deacetylase inhibitor treatment attenuates coagulation imbalance in a lethal murine model of sepsis. *Surgery*, *156*(2), 214–220. http://dx.doi.org/10.1016/j.surg.2014.04.022.

CHAPTER EIGHT

A New Molecular Mechanism Underlying the Antitumor Effect of DNA Methylation Inhibitors via an Antiviral Immune Response

Y. Saito[1], T. Nakaoka, H. Saito
Division of Pharmacotherapeutics, Keio University Faculty of Pharmacy, Minato-ku, Japan
[1]Corresponding author: e-mail address: saito-ys@pha.keio.ac.jp

Contents

Abstract

Chromatin remodeling mediated by DNA methylation and histone modifications play critical roles in the transcriptional regulation of protein-coding genes, noncoding RNAs such as microRNAs, and endogenous retroviruses (ERVs). Many studies have shown that aberrant DNA methylation and histone modifications are associated with the initiation and progression of various malignancies. Epigenetic silencing of tumor suppressor genes in cancer is generally mediated by DNA hypermethylation of CpG island promoters and histone modifications such as histone deacetylation, methylation of histone H3 lysine 9 (H3K9), and trimethylation of H3K27. Chromatin-modifying drugs such as DNA methylation inhibitors and histone deacetylase inhibitors have clinical promise for cancer therapy. However, details of the mechanisms responsible for the antitumor effects of these drugs have been unclear.

Recently, a new molecular mechanism for the antitumor effect of DNA methylation inhibitors has been proposed: induction of interferon-responsive genes via double-stranded RNAs derived from ERVs. We have also confirmed the same effect of DNA demethylation using a 3D culture system for stem cells known as organoid culture.

Advances in Protein Chemistry and Structural Biology, Volume 106
ISSN 1876-1623
http://dx.doi.org/10.1016/bs.apcsb.2016.08.005

Our findings indicated that DNA demethylation suppresses the proliferation of cancer-initiating cells by inducing an antiviral response, including activation of interferon-responsive genes. Treatment with DNA methylation inhibitors to activate a growth-inhibiting immune response may be an effective therapeutic approach for malignant disorders.

1. INTRODUCTION

Epigenetic alterations are acquired modifications of chromatin DNA or histone proteins resulting from methylation and/or acetylation, which regulate downstream gene expression. Aberrant DNA methylation and histone modifications can be induced by aging, chronic inflammation, and viral infection. Many studies have shown that aberrant epigenetic alterations play critical roles in the initiation and progression of various malignancies (Baylin & Jones, 2011; Gal-Yam, Saito, Egger, & Jones, 2008; Jones & Baylin, 2007). Epigenetic silencing of tumor suppressor genes in cancer is generally mediated by DNA hypermethylation of CpG island promoters and histone modifications such as histone deacetylation, methylation of histone H3 lysine 9 (H3K9), and trimethylation of H3K27 (Baylin & Jones, 2011). We have reported that microRNAs (miRNAs), small noncoding RNAs that regulate various target genes, are also under epigenetic control in cancer cells (Hibino et al., 2014; Saito, Friedman, et al., 2009; Saito et al., 2006; Saito, Suzuki, et al., 2009).

Aberrant DNA methylation and histone modifications at CpG island promoters of tumor suppressor genes are frequently observed in various types of cancers (Baylin & Jones, 2011; Gal-Yam et al., 2008; Jones & Baylin, 2007). Chromatin-modifying drugs such as DNA methylation inhibitors and histone deacetylase (HDAC) inhibitors can reactivate tumor suppressor genes and miRNAs, and have great clinical promise for cancer therapy (Kelly, De Carvalho, & Jones, 2010). Inhibitors of DNA methylation and histone deacetylation can work synergistically to suppress the growth of cancer cell lines both in vitro and in vivo (Cameron, Bachman, Myohanen, Herman, & Baylin, 1999; Zahnow et al., 2016). By inducing the expression of tumor suppressor genes, epigenetic treatment inhibits not only the growth but also possibly the invasiveness and metastatic potential of cancer. The DNA methylation inhibitor 5-aza-2′-deoxycytidine (5-Aza-CdR), an analog of cytidine, has been widely studied and was recently approved for the treatment of myelodysplastic syndrome (Kelly et al., 2010; Yoo & Jones,

2006). The HDAC inhibitor suberoylanilide hydroxamic acid (SAHA) has also been approved for patients with cutaneous T-cell lymphoma (Kelly et al., 2010). Other epigenetic drugs are also being assessed in clinical trials. However, how these epigenetic drugs work has remained unclear.

Recently, a new molecular mechanism for the antitumor effect of DNA methylation inhibitors has been proposed: induction of interferon-responsive genes via double-stranded RNAs (dsRNAs) derived from endogenous retroviruses (ERVs) (Chiappinelli et al., 2015; Roulois et al., 2015). We have also confirmed the same effect of DNA demethylation using a 3D culture system for stem cells known as organoid culture (Saito et al., 2016). The available data indicate that DNA demethylation suppresses the proliferation of cancer-initiating cells (CICs) by inducing an antiviral response including activation of interferon-responsive genes. Here, we review a new molecular mechanism underlying the antitumor effect of DNA methylation inhibitors that involves induction of an antiviral immune response.

2. CHROMATIN REMODELING MEDIATED BY DNA METHYLATION AND HISTONE MODIFICATION

Epigenetic alterations such as DNA methylation and histone modifications play critical roles in chromatin remodeling, which regulates gene expression in both mammalian development and human diseases. Alteration of DNA methylation is one of the most consistent epigenetic changes in human cancers (Jones & Baylin, 2002). DNA methylation patterns are generated and heritably maintained by three active DNA methyltransferases: DNMT1, DNMT3A, and DNMT3B. The de novo methyltransferases DNMT3A and DNMT3B act independently of replication and show equal preference for both unmethylated and hemimethylated DNA, whereas the maintenance DNA methyltransferase DNMT1 acts during replication and preferentially methylates hemimethylated DNA (Okano, Xie, & Li, 1998). DNA hypermethylation occurs mainly in short CpG-rich DNA stretches called "CpG islands." DNA methylation can lead to gene silencing by either preventing or promoting the recruitment of regulatory proteins to DNA. Alternatively, it can provide binding sites for methyl-binding domain proteins, which can mediate gene repression through interactions with HDACs. This phenomenon of aberrant promoter CpG island hypermethylation has been associated with stabilization of transcriptional repression and loss of gene function, and occurs fundamentally in tumor suppressor genes. In contrast, DNA hypomethylation is associated mainly with loss of DNA methylation in genome-wide regions. It has been reported that DNA hypomethylation occurs in several tumor types, such as colorectal and gastric cancers, involving many gene-poor genomic areas such as repetitive elements, retrotransposons, and introns, where it leads to genomic instability (Eden, Gaudet, Waghmare, & Jaenisch, 2003; Suzuki et al., 2006).

Histone modifications also regulate the transcription of genes that play important roles in cellular processes. The N-terminal tails of histones can undergo a variety of posttranslational modifications including methylation and acetylation of specific residues. Unlike DNA methylation, histone modifications can lead to either activation or repression, depending on the residues that are modified and the type of modifications present. For example, trimethylation of lysine 4 on histone H3 (H3K4me3) is enriched at transcriptionally active gene promoters (Liang et al., 2004), whereas di- and trimethylation of H3K9 and trimethylation of H3K27 are present at gene promoters that are transcriptionally repressed (Jenuwein & Allis, 2001; Kouzarides, 2007).

Active and inactive chromatin structures are shown in Fig. 1. Transcriptionally active chromatin structure is characterized by acetylation of histone H3 and trimethylation of H3K4. Epigenetic gene silencing is generally mediated by two distinct histone modifications: methylation of H3K9 and trimethylation of H3K27. The polycomb repressive complex 2 (PRC2) mediates epigenetic gene silencing through trimethylation of H3K27. Methylation of H3K9 works in combination with DNA methylation, whereas trimethylation of H3K27 works independently of DNA methylation (Kondo et al., 2008). HDAC induces deacetylation of histone H3 in both of these pathways of epigenetic gene silencing.

Yao et al. investigated histone modifications in the promoter regions of p16 (INK4a) during differentiation of embryonic stem cell-hepatoma hybrid cells (Yao et al., 2010). Trimethylation of H3K27 at the p16 (INK4a) promoter region, occurring in the early phase of p16 (INK4a) silencing, was followed by dimethylation of H3K9 at later stages. During the induced differentiation of hepatomas, di- and trimethylation of H3K4 were maintained at high levels during the silencing of p16 (INK4a), suggesting that H3K4 methylation events did not cause the silencing of p16INK4a. These findings indicated that the enrichment of H3K27 trimethylation, independent of H3K9 methylation and DNA methylation, was an early event in the silencing of p16 (INK4a) during tumor development. This histone modification

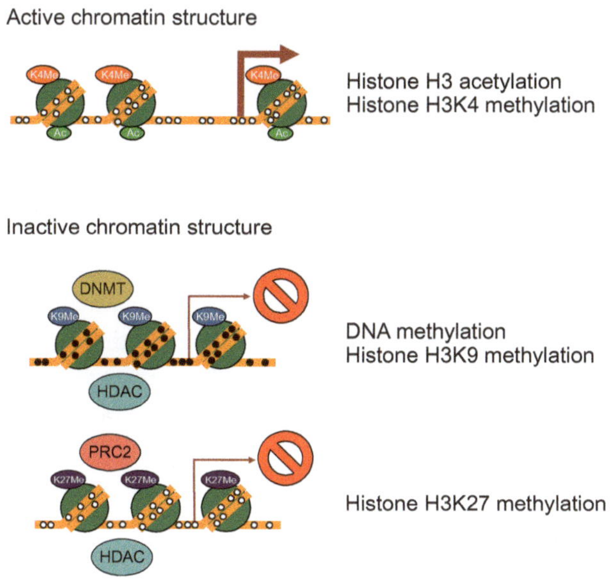

Fig. 1 Active and inactive chromatin structures. Transcriptionally active chromatin in normal cells is characterized by acetylation of histone H3 (Ac) and trimethylation of H3K4 (K4Me). Epigenetic gene silencing is generally mediated by two distinct histone modifications: methylation of H3K9 (K9Me) and trimethylation of H3K27 (K27Me). PRC2 mediates epigenetic gene silencing by trimethylating H3K27. Methylation of H3K9 works in combination with DNA methylation induced by DNMTs, whereas trimethylation of H3K27 works independently of DNA methylation. HDAC induces deacetylation of histone H3 in both of the pathways of epigenetic silencing. *Clear circle*, unmethylated DNA; *solid circle*, methylated DNA.

pattern may be a heritable marker for epigenetic silencing of p16 (INK4a) during the development of liver cancers (Yao et al., 2010).

3. ACTIVATION OF TUMOR-SUPPRESSOR miRNAs BY CHROMATIN-MODIFYING DRUGS IN CANCER CELLS

miRNAs are small noncoding RNAs that function as endogenous silencers of various target genes and control cell proliferation and differentiation. Misexpression of cancer-related miRNAs leads to the initiation and progression of cancer by modulating their target oncogenes or tumor suppressor genes (Calin & Croce, 2006; Garzon, Calin, & Croce, 2009). We have discovered that ~5% of human miRNAs are upregulated more than threefold by treatment of T24 bladder cancer cells with the DNA demethylating agent 5-Aza-CdR and the HDAC inhibitor 4-phenylbutyric acid (PBA). In particular, *miR-127*, which is embedded in a CpG island, is markedly induced by a decrease in DNA methylation levels and an increase in active histone marks around the promoter region of the *miR-127* gene. In addition, activation of *miR-127* by epigenetic treatment induces down-regulation of its target oncogene *BCL6* (Saito et al., 2006). We have also demonstrated that treatment of gastric cancer cells with 5-Aza-CdR and PBA induces activation of *miR-512-5p*, which is located at Alu repeats on chromosome 19. Activation of *miR-512-5p* by epigenetic treatment induces suppression of *MCL1*, resulting in apoptosis of gastric cancer cells (Saito, Suzuki, et al., 2009). These observations indicate that chromatin remodeling by epigenetic treatment can directly activate miRNA expression and that activation of silenced tumor-suppressor miRNAs could be a novel therapeutic approach for human cancers.

Polycomb repressive complex (PRC1 and PRC2)-mediated epigenetic regulation is critical for maintaining cellular homeostasis. PRC2 mediates epigenetic gene silencing through trimethylation of histone H3K27 and is known to aberrantly silence tumor suppressor genes in cancer. Enhancer of zeste homolog 2 (EZH2), which is the catalytic subunit of PRC2, enhances tumorigenesis and is commonly overexpressed in several types of cancer. We have recently reported miRNA expression profiles in gastric and liver cancer cells treated with SAHA and 3-deazaneplanocin A (DZNep) (Hibino et al., 2014). We confirmed that SAHA and DZNep suppressed EZH2 expression in AGS and HepG2 cells and inhibited their proliferation. *miR-1246* was commonly upregulated in cancer cells by treatment with SAHA and DZNep. *miR-302a* and *miR-4448* were markedly upregulated by treatment with SAHA and DZNep, respectively. DYRK1A, CDK2, BMI-1, and Girdin, which are targets of *miR-1246*, *miR-302a*, and *miR-4448*, were suppressed by treatment with SAHA and DZNep, leading to apoptosis, cell cycle arrest, and reduced migration of AGS and HepG2 cells. Thus, EZH2 inhibitors such as SAHA and DZNep exert multiple anti-cancer effects through activation of tumor-suppressor miRNAs.

4. INDUCTION OF IMMUNE RESPONSE BY CHROMATIN-MODIFYING DRUGS IN CANCER CELLS

Previous studies have demonstrated that 5-Aza-CdR treatment can induce the expression of interferon-responsive genes and ERVs (Karpf et al., 2004, 1999); expression of the STAT1, 2, and 3 genes was significantly upregulated by DNA methylation inhibitors in colon cancer cells and suggested that induction of the interferon-response pathway was a major cellular response to 5-Aza-CdR. Interferon regulatory factors (IRFs) are transcription factors known to play key roles in innate and adaptive immune responses, cell growth, apoptosis, and development. A study using gastric cancer cell lines has shown that expression of IRF4, IRF5, and IRF8 was frequently suppressed by DNA hypermethylation of their promoter regions (Yamashita et al., 2010). Treatment of these gastric cancer cell lines with 5-Aza-CdR restored the expression of these genes. Interferon-γ-induced expression of IRF8 was also enhanced by 5-Aza-CdR. Moreover, treatment of gastric cancer cells with 5-Aza-CdR enhanced the suppressive effects of interferon-α, interferon-β, and interferon-γ on cell growth.

Tsai et al. have demonstrated that transient low doses of 5-Aza-CdR and 5-azacytidine (5-Aza-CR) exert durable antitumor effects on hematological and epithelial tumor cells (Tsai et al., 2012). A study using 63 cancer cell lines (breast, colorectal, and ovarian) has shown that low doses of 5-Aza-CR upregulated genes associated with immunomodulatory processing and presentation, and cytokines/chemokines in several types of cancer (Li et al., 2014). Thus, the major activity of DNA methylation inhibitors seems to be induction of immune responses in cancer cells.

In recent clinical trials for nonsmall cell lung cancer (NSCLC), patients showed remarkably robust and durable responses to immune checkpoint blockade therapy after first receiving 5-Aza-CR (Juergens et al., 2011; Wrangle et al., 2013). 5-Aza-CR upregulates genes and pathways related to both innate and adaptive immunity and genes related to immune evasion in several types of NSCLC. In addition, 5-Aza-CR upregulates PD-L1 transcripts and protein, a key ligand-mediator of immune tolerance. These findings indicate that epigenetic therapy combined with blockade of immune checkpoints such as the PD-1/PD-L1 pathway may increase the immune response of cancer cells.

Type III interferon (IFN-λ) exhibits potent antiviral activity similar to IFN-α/β. The IFN-λ receptor is restricted to cells of epithelial origin, whereas the IFN-α/β receptor is expressed ubiquitously. Ding et al. have demonstrated that the HDAC repression machinery mediates transcriptional silencing of the unique IFN-λ receptor subunit (IFNLR1) in a cell-type-specific manner. HDAC inhibitors increase receptor expression and restore sensitivity to IFN-λ in previously nonresponsive cells, thereby enhancing protection against viral pathogens. Thus, the type III interferon response induced by HDAC inhibitors may be an antitumor strategy (Ding, Khoury-Hanold, Iwasaki, & Robek, 2014).

5. ACTIVATION OF ERVs BY CHROMATIN REMODELING

Although it has become increasingly apparent that DNA methylation inhibitors upregulate the expression of interferon-responsive genes, details of the mechanism by which DNA demethylation induces an immune response have been unclear. Recent studies have yielded new data showing that DNA methylation inhibitors upregulate ERVs in cancer cells to induce a growth-inhibiting immune response (Chiappinelli et al., 2015; Roulois et al., 2015). Different ERV gene families constitute about 8% of the human genome and are considered to be long terminal repeat (LTR) retrotransposons, in contrast to non-LTR retrotransposons such as LINE and SINE. The expression of ERVs could produce nucleic acids or proteins with viral signatures, which activate the innate immune response (Hurst & Magiorkinis, 2015). Most human endogenous retroviruses (HERVs) are defective on account of multiple stop codons within essential viral genes, whereas the human endogenous retrovirus-K (HERV-K) family shows evidence of recent activity within the human genome. Members of the HERV-K family often harbor open reading frames for the *gag*, *pol*, and *env* genes (Bannert & Kurth, 2004).

Overexpression of HERV *env* genes has been observed in endometrial carcinomas, with a significant correlation between HERV expression levels and tumor stage (Strissel et al., 2012). Elevated expression of HERV-K has been found in melanomas, and detection of HERV-K-specific antibodies in melanoma patients has been found to correlate with reduced survival (Stengel, Fiebig, Kurth, & Denner, 2010). These findings indicate that HERVs are overexpressed in several invasive carcinomas and could serve as a new molecular marker in cancer patients.

In 1981, Groudine et al. demonstrated that transcriptionally inactive endogenous retroviral genes can be reactivated by the DNA methylation inhibitor 5-Aza-CR (Groudine, Eisenman, & Weintraub, 1981). HERV-K Gag protein is expressed in teratocarcinoma cell lines and human testicular tumors, and demethylation upon treatment with 5-Aza-CR results in overexpression of Gag, suggesting that endogenous retroviral elements are subject to regulation through the methylation of CpG dinucleotides (Gotzinger, Sauter, Roemer, & Mueller-Lantzsch, 1996). A higher level of DNA methylation at LTR was observed in melanoma cells not expressing HERV-K relative to cells expressing HERV-K. Treatment of melanoma cell lines with 5-Aza-CdR resulted in increased levels of HERV-K expression in cells that did not previously express it, indicating that increased HERV-K expression in melanomas may be due to increased promoter activity and demethylation of the 5′-LTR (Stengel et al., 2010). It has been reported that expression of HERVs is regulated by CpG methylation of 5′ LTRs in germ cell tumors and endometrial carcinoma (Lavie, Kitova, Maldener, Meese, & Mayer, 2005; Strissel et al., 2012). These findings indicate that CpG methylation at 5′ LTRs directly regulates the transcriptional activity of ERVs.

Recent studies have shown that histone modifications also play critical roles in the transcriptional regulation of ERVs. An important modification for initiating repressive chromatin formation is trimethylation of histone H3K9 by two major histone methyltransferases. SUV39H1/2 complexes maintain this modification at constitutive heterochromatin and most LINEs, which together encompass the majority of H3K9 trimethylation in chromatin (Collins, Kyle, Egawa, Shinkai, & Oltz, 2015). In contrast, the histone methyltransferase SETDB1 is targeted to ERVs, which must be silenced to maintain genome and transcriptome integrity (Matsui et al., 2010). Repressive histone methylation catalyzed by the enzyme SETDB1 is required for germline development and silencing of H3K9 trimethylation-marked ERVs in primordial germ cells (Liu et al., 2014). SETDB1 is also reported to be critical for suppression of specific ERV families and exogenous retroviruses in committed B-lineage cells from adult mice (Collins et al., 2015). The profile of ERV activation in SETDB1-deficient B cells is distinct from that observed in corresponding embryonic tissues, despite the loss of repressive chromatin modifications at all ERVs. On loss of SETDB1, ERVs are activated in a lineage-specific manner depending on the set of transcription factors available to target proviral regulatory elements. These observations reveal that the histone methyltransferase SETDB1 plays an essential role in ERV silencing and genome stability in both germ cells and somatic cells.

6. INHIBITION OF DNA METHYLATION SUPPRESSES CANCER CELLS BY INDUCING AN ANTIVIRAL IMMUNE RESPONSE

So far, the major mechanism responsible for the antitumor effect of DNA methylation inhibitors has been considered to be reactivation of silenced tumor suppressor genes by aberrant DNA methylation and/or histone modification of their CpG island promoters.

Recently, however, two groups, Roulois et al. and Chiappinelli et al., have demonstrated that DNA demethylating agents can induce a cell-autonomous immune activation response by stimulating the expression of dsRNAs derived from ERVs (Chiappinelli et al., 2015; Roulois et al., 2015). It was recently reported that ERV RNAs can trigger signaling by cytosolic pattern-recognition receptors and activate MAVS in mammals (Zeng et al., 2014). Therefore, ERVs can be the source of 5-Aza-CdR-induced dsRNAs and a trigger for the antiviral response. This may represent a paradigm shift in our understanding of the antitumor effect of DNA demethylating agents.

In ovarian cancer cells, DNA methyltransferase inhibitors (DNMTis) trigger cytosolic sensing of dsRNA, causing a type I interferon response and apoptosis (Chiappinelli et al., 2015). Knockdown of the dsRNA sensors TLR3 and MAVS markedly reduces this interferon response, and blocking of interferon β or its receptor abrogates this effect. DNMTis induce ERV demethylation and expression, helping to trigger the dsRNA response. Thus, inhibition of DNA methylation causes an interferon response in cancer via dsRNAs derived from ERVs. Moreover, in melanoma patients receiving immune checkpoint therapy, it has been shown that a high viral defense signature in tumors is significantly associated with a durable clinical response, and in a preclinical melanoma model, DNMTi treatment has been shown to sensitize cells to anti-CTLA4 therapy. Treatment with DNA methylation inhibitors to activate a growth-inhibiting immune response may therefore be an effective therapeutic approach for human malignancies.

Roulois et al. have demonstrated that low-dose 5-Aza-CdR targets colorectal CICs by inducing viral mimicry (Roulois et al., 2015). This effect is associated with induction of dsRNAs derived from endogenous retroviral elements, activation of the MDA5/MAVS RNA recognition pathway, and downstream activation of IRF7. Disruption of virus recognition pathways, by individually knocking down MDA5, MAVS, or IRF7, inhibits the ability of 5-Aza-CdR to target colorectal CICs and significantly decreases its long-term growth effects. Induction of dsRNAs derived from ERVs causes activation of the MDA5 RNA recognition receptor, followed by downstream activation of MAVS, IRF7, and type III IFNs and upregulation of interferon-responsive genes, independent of their promoter demethylation. These findings indicate that DNA-demethylating agents target colorectal cancer cells through induction of viral mimicry by endogenous transcripts.

Fig. 2 shows the molecular pathway of the antiviral immune response induced by inhibition of DNA methylation. In a normal state, the

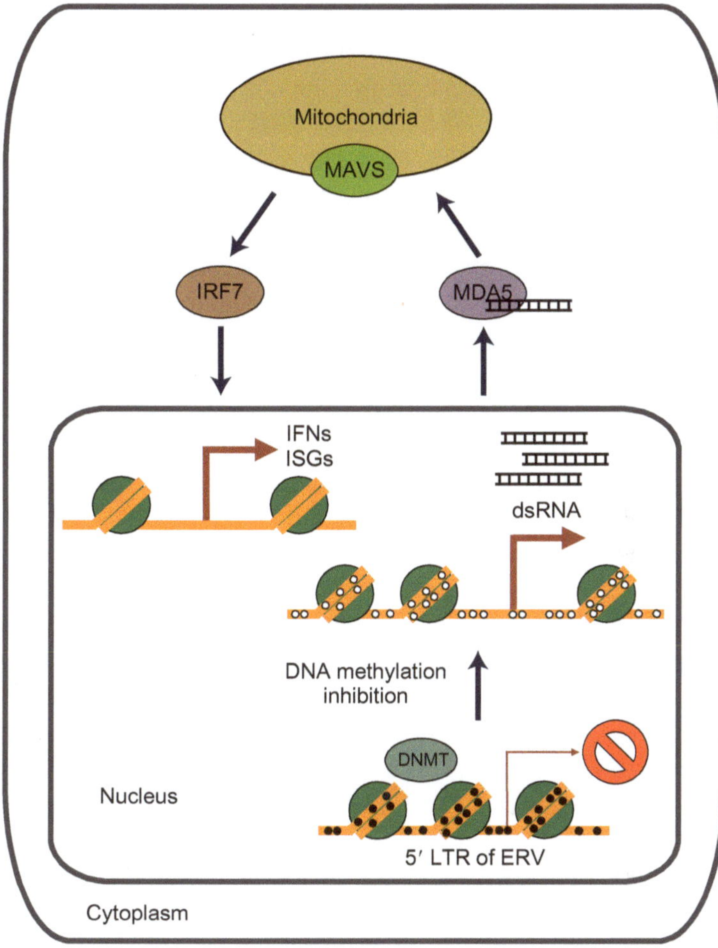

Fig. 2 Molecular pathway of the antiviral immune response induced by inhibition of DNA methylation In a normal state, the 5′ LTR sequences of ERVs are heavily methylated and expression of ERVs is silenced. When DNA methylation at the 5′ LTR sequences is inhibited, expression of ERVs is induced. MDA5 recognizes nucleic acids associated with viral infections such as dsRNAs. The CARDs of MDA5 can recruit the signaling adaptor protein MAVS, which resides in the outer mitochondrial membrane. MAVS induces signaling cascades that result in nuclear translocation of IRF7 and activation of IFNs and ISGs as an antiviral immune response.

5′ LTR sequences of ERVs are heavily methylated and expression of ERVs is silenced. When DNA methylation of the 5′ LTR sequences is inhibited, expression of ERVs is induced. MDA5 is a pattern-recognition receptor ubiquitously expressed in the cytoplasm of most human cells. MDA5 recognizes nucleic acids associated with viral infections such as dsRNAs and has two amino-terminal caspase recruitment domains (CARDs). The CARDs of MDA5, upon activation, can recruit the signaling adaptor protein MAVS, which resides in the outer mitochondrial membrane. MAVS induces signaling cascades that result in nuclear translocation of IRF7 and activation of IFNs and interferon-stimulated genes (ISGs) as an antiviral immune response (Barbalat, Ewald, Mouchess, & Barton, 2011).

Previous studies have demonstrated that DNA methylation inhibitors such as 5-Aza-CdR and zebularine suppress intestinal tumors in $Apc^{min/+}$ (Min) mice (Laird et al., 1995; Yoo et al., 2008). However, the molecular mechanism underlying the in vivo antitumor effect of DNA methylation inhibitors remains to be clarified. To investigate the antitumor effect of DNA methylation inhibitors in an in vitro tumor model, we established organoids derived from intestinal tumors of Min mice (Saito et al., 2016). A newly developed 3D culture system allows Lgr5-positive stem cells to form budding cyst-like structures (organoids) that resemble the properties of the original tissues (Sato et al., 2009). This type of 3D culture uses serum-free medium that includes only specifically defined factors such as R-spondin 1 (Rspo1), epidermal growth factor, and noggin. Rspo1 has been identified as a ligand for Lgr5 and an essential factor for activation of the Wnt signaling pathway (Carmon, Gong, Lin, Thomas, & Liu, 2011; de Lau et al., 2011). Tumor tissues contain many stromal cells such as fibroblasts. Since the organoid culture system allows expansion of Lgr5-positive stem cells into cyst-like structures with properties resembling the original tissues, it is possible to observe the biological signatures of the tumor cells alone, to the exclusion of stromal cells. In addition, a recent study has clearly demonstrated that tumor organoids closely recapitulate the properties of the original tumors (van de Wetering et al., 2015), suggesting that patient-derived organoids may allow drug screening and personalized cancer therapy. Thus, organoids derived from tumors of Min mice are considered to be a desirable model for the initial stage of intestinal tumorigenesis according to the adenoma–carcinoma sequence.

We have subjected organoids derived from intestinal tumors of *Min* mice to treatment with 5-Aza-CdR and knockdown of *Dnmt1*. Our results revealed that pharmacologic and genetic DNA demethylation induced by 5-Aza-CdR treatment and *Dnmt1* knockdown suppressed the proliferation of intestinal tumor organoids (Saito et al., 2016). Expression analyses of the tumor organoids after 5-Aza-CdR treatment and *Dnmt1* knockdown revealed that interferon-responsive genes including *Oas1*, *Irf7*, *Isg15 Rig1*, and *Mda5* were activated by DNA demethylation. Gene ontology and pathway analyses clearly demonstrated that these genes activated by DNA demethylation are involved in the antiviral response. We then examined the expression levels of ERVs in intestinal tumor organoids after 5-Aza-CdR treatment and *Dnmt1* knockdown. Expression of murine ERVs such as IAP-1, MLV, LV30-2, and MuRRS was significantly increased in the organoids after 5-Aza-CdR treatment and knockdown. Our results strongly support the hypothesis that the major antitumor effect of DNA-demethylating agents is induction of an antiviral response via activation of dsRNAs derived from ERVs (Chiappinelli et al., 2015; Roulois et al., 2015). These findings indicate that inhibition of DNA methylation suppresses intestinal tumor organoids by inducing a growth-inhibitory immune response, including activation of interferon-responsive genes, in response to ERVs. Treatment with DNA methylation inhibitors to activate a growth-inhibiting immune response could be an effective therapeutic approach for malignant disorders.

ACKNOWLEDGMENTS

This work was supported by a Grant-in-Aid for Scientific Research B (26290049) from the Japan Society for Promotion of Science (to Y.S.).

Conflict of interest: None of the authors have any conflicts of interest to declare.

REFERENCES

Bannert, N., & Kurth, R. (2004). Retroelements and the human genome: New perspectives on an old relation. *Proceedings of the National Academy of Sciences of the United States of America, 101*(Suppl. 2), 14572–14579. http://dx.doi.org/10.1073/pnas.0404838101.

Barbalat, R., Ewald, S. E., Mouchess, M. L., & Barton, G. M. (2011). Nucleic acid recognition by the innate immune system. *Annual Review of Immunology, 29*, 185–214. http://dx.doi.org/10.1146/annurev-immunol-031210-101340.

Baylin, S. B., & Jones, P. A. (2011). A decade of exploring the cancer epigenome—Biological and translational implications. *Nature Reviews. Cancer, 11*(10), 726–734. http://dx.doi.org/10.1038/nrc3130.

Calin, G. A., & Croce, C. M. (2006). Microrna signatures in human cancers. *Nature Reviews. Cancer, 6*(11), 857–866.

Cameron, E. E., Bachman, K. E., Myohanen, S., Herman, J. G., & Baylin, S. B. (1999). Synergy of demethylation and histone deacetylase inhibition in the re-expression of genes silenced in cancer. *Nature Genetics, 21*(1), 103–107. http://dx.doi.org/10.1038/5047.

Carmon, K. S., Gong, X., Lin, Q., Thomas, A., & Liu, Q. (2011). R-spondins function as ligands of the orphan receptors lgr4 and lgr5 to regulate wnt/beta-catenin signaling. *Proceedings of the National Academy of Sciences of the United States of America, 108*(28), 11452–11457. http://dx.doi.org/10.1073/pnas.1106083108.

Chiappinelli, K. B., Strissel, P. L., Desrichard, A., Li, H., Henke, C., Akman, B., et al. (2015). Inhibiting DNA methylation causes an interferon response in cancer via dsRNA including endogenous retroviruses. *Cell, 162*(5), 974–986. http://dx.doi.org/10.1016/j.cell.2015.07.011.

Collins, P. L., Kyle, K. E., Egawa, T., Shinkai, Y., & Oltz, E. M. (2015). The histone methyltransferase setdb1 represses endogenous and exogenous retroviruses in b lymphocytes. *Proceedings of the National Academy of Sciences of the United States of America*, *112*(27), 8367–8372. http://dx.doi.org/10.1073/pnas.1422187112.

de Lau, W., Barker, N., Low, T. Y., Koo, B. K., Li, V. S., Teunissen, H., et al. (2011). Lgr5 homologues associate with wnt receptors and mediate r-spondin signalling. *Nature*, *476*(7360), 293–297. http://dx.doi.org/10.1038/nature10337.

Ding, S., Khoury-Hanold, W., Iwasaki, A., & Robek, M. D. (2014). Epigenetic reprogramming of the type III interferon response potentiates antiviral activity and suppresses tumor growth. *PLoS Biology*, *12*(1), e1001758. http://dx.doi.org/10.1371/journal.pbio.1001758.

Eden, A., Gaudet, F., Waghmare, A., & Jaenisch, R. (2003). Chromosomal instability and tumors promoted by DNA hypomethylation. *Science*, *300*(5618), 455. http://dx.doi.org/10.1126/science.1083557.

Gal-Yam, E. N., Saito, Y., Egger, G., & Jones, P. A. (2008). Cancer epigenetics: Modifications, screening, and therapy. *Annual Review of Medicine*, *59*, 267–280. http://dx.doi.org/10.1146/annurev.med.59.061606.095816.

Garzon, R., Calin, G. A., & Croce, C. M. (2009). Micrornas in cancer. *Annual Review of Medicine*, *60*, 167–179. http://dx.doi.org/10.1146/annurev.med.59.053006.104707.

Gotzinger, N., Sauter, M., Roemer, K., & Mueller-Lantzsch, N. (1996). Regulation of human endogenous retrovirus-k gag expression in teratocarcinoma cell lines and human tumours. *The Journal of General Virology*, 77(Pt. 12), 2983–2990. http://dx.doi.org/10.1099/0022-1317-77-12-2983.

Groudine, M., Eisenman, R., & Weintraub, H. (1981). Chromatin structure of endogenous retroviral genes and activation by an inhibitor of DNA methylation. *Nature*, *292*(5821), 311–317.

Hibino, S., Saito, Y., Muramatsu, T., Otani, A., Kasai, Y., Kimura, M., et al. (2014). Inhibitors of enhancer of zeste homolog 2 (ezh2) activate tumor-suppressor microRNAs in human cancer cells. *Oncogenesis*, *3*, e104. http://dx.doi.org/10.1038/oncsis.2014.17.

Hurst, T. P., & Magiorkinis, G. (2015). Activation of the innate immune response by endogenous retroviruses. *The Journal of General Virology*, *96*(Pt. 6), 1207–1218. http://dx.doi.org/10.1099/jgv.0.000017. 0.1099/jgv.0.000017.

Jenuwein, T., & Allis, C. D. (2001). Translating the histone code. *Science*, *293*(5532), 1074–1080. http://dx.doi.org/10.1126/science.1063127.

Jones, P. A., & Baylin, S. B. (2002). The fundamental role of epigenetic events in cancer. *Nature Reviews. Genetics*, *3*(6), 415–428. http://dx.doi.org/10.1038/nrg816.

Jones, P. A., & Baylin, S. B. (2007). The epigenomics of cancer. *Cell*, *128*(4), 683–692. http://dx.doi.org/10.1016/j.cell.2007.01.029.

Juergens, R. A., Wrangle, J., Vendetti, F. P., Murphy, S. C., Zhao, M., Coleman, B., et al. (2011). Combination epigenetic therapy has efficacy in patients with refractory advanced non-small cell lung cancer. *Cancer Discovery*, *1*(7), 598–607. http://dx.doi.org/10.1158/2159-8290.CD-11-0214.

Karpf, A. R., Lasek, A. W., Ririe, T. O., Hanks, A. N., Grossman, D., & Jones, D. A. (2004). Limited gene activation in tumor and normal epithelial cells treated with the DNA methyltransferase inhibitor 5-aza-2'-deoxycytidine. *Molecular Pharmacology*, *65*(1), 18–27. http://dx.doi.org/10.1124/mol.65.1.18.

Karpf, A. R., Peterson, P. W., Rawlins, J. T., Dalley, B. K., Yang, Q., Albertsen, H., et al. (1999). Inhibition of DNA methyltransferase stimulates the expression of signal transducer and activator of transcription 1, 2, and 3 genes in colon tumor cells. *Proceedings of the National Academy of Sciences of the United States of America*, *96*(24), 14007–14012.

Kelly, T. K., De Carvalho, D. D., & Jones, P. A. (2010). Epigenetic modifications as therapeutic targets. *Nature Biotechnology*, *28*(10), 1069–1078. http://dx.doi.org/10.1038/nbt.1678.

Kondo, Y., Shen, L., Cheng, A. S., Ahmed, S., Boumber, Y., Charo, C., et al. (2008). Gene silencing in cancer by histone h3 lysine 27 trimethylation independent of promoter DNA methylation. *Nature Genetics*, *40*(6), 741–750. http://dx.doi.org/10.1038/ng.159.

Kouzarides, T. (2007). Chromatin modifications and their function. *Cell*, *128*(4), 693–705. http://dx.doi.org/10.1016/j.cell.2007.02.005.

Laird, P. W., Jackson-Grusby, L., Fazeli, A., Dickinson, S. L., Jung, W. E., Li, E., et al. (1995). Suppression of intestinal neoplasia by DNA hypomethylation. *Cell*, *81*(2), 197–205.

Lavie, L., Kitova, M., Maldener, E., Meese, E., & Mayer, J. (2005). Cpg methylation directly regulates transcriptional activity of the human endogenous retrovirus family herv-k(hml-2). *Journal of Virology*, *79*(2), 876–883. http://dx.doi.org/10.1128/JVI.79.2.876-883.2005.

Li, H., Chiappinelli, K. B., Guzzetta, A. A., Easwaran, H., Yen, R. W., Vatapalli, R., et al. (2014). Immune regulation by low doses of the DNA methyltransferase inhibitor 5-azacitidine in common human epithelial cancers. *Oncotarget*, *5*(3), 587–598. http://dx.doi.org/10.18632/oncotarget.1782.

Liang, G., Lin, J. C., Wei, V., Yoo, C., Cheng, J. C., Nguyen, C. T., et al. (2004). Distinct localization of histone h3 acetylation and h3-k4 methylation to the transcription start sites in the human genome. *Proceedings of the National Academy of Sciences of the United States of America*, *101*(19), 7357–7362. http://dx.doi.org/10.1073/pnas.0401866101.

Liu, S., Brind'Amour, J., Karimi, M. M., Shirane, K., Bogutz, A., Lefebvre, L., et al. (2014). Setdb1 is required for germline development and silencing of h3k9me3-marked endogenous retroviruses in primordial germ cells. *Genes & Development*, *28*(18), 2041–2055. http://dx.doi.org/10.1101/gad.244848.114.

Matsui, T., Leung, D., Miyashita, H., Maksakova, I. A., Miyachi, H., Kimura, H., et al. (2010). Proviral silencing in embryonic stem cells requires the histone methyltransferase eset. *Nature*, *464*(7290), 927–931. http://dx.doi.org/10.1038/nature08858.

Okano, M., Xie, S., & Li, E. (1998). Cloning and characterization of a family of novel mammalian DNA (cytosine-5) methyltransferases. *Nature Genetics*, *19*(3), 219–220. http://dx.doi.org/10.1038/890.

Roulois, D., Loo Yau, H., Singhania, R., Wang, Y., Danesh, A., Shen, S. Y., et al. (2015). DNA-demethylating agents target colorectal cancer cells by inducing viral mimicry by endogenous transcripts. *Cell*, *162*(5), 961–973. http://dx.doi.org/10.1016/j.cell.2015.07.056.

Saito, Y., Friedman, J. M., Chihara, Y., Egger, G., Chuang, J. C., & Liang, G. (2009). Epigenetic therapy upregulates the tumor suppressor microrna-126 and its host gene egfl7 in human cancer cells. *Biochemical and Biophysical Research Communications*, *379*(3), 726–731. http://dx.doi.org/10.1016/j.bbrc.2008.12.098.

Saito, Y., Liang, G., Egger, G., Friedman, J. M., Chuang, J. C., Coetzee, G. A., et al. (2006). Specific activation of microrna-127 with downregulation of the proto-oncogene bcl6 by chromatin-modifying drugs in human cancer cells. *Cancer Cell*, *9*(6), 435–443.

Saito, Y., Nakaoka, T., Sakai, K., Muramatsu, T., Toshimitsu, K., Kimura, M., et al. (2016). Inhibition of DNA methylation suppresses intestinal tumor organoids by inducing an anti-viral response. *Scientific Reports*, *6*, 25311. http://dx.doi.org/10.1038/srep25311.

Saito, Y., Suzuki, H., Tsugawa, H., Nakagawa, I., Matsuzaki, J., Kanai, Y., et al. (2009). Chromatin remodeling at alu repeats by epigenetic treatment activates silenced microrna-512-5p with downregulation of mcl-1 in human gastric cancer cells. *Oncogene*, *28*(30), 2738–2744. http://dx.doi.org/10.1038/onc.2009.140.

Sato, T., Vries, R. G., Snippert, H. J., van de Wetering, M., Barker, N., Stange, D. E., et al. (2009). Single lgr5 stem cells build crypt-villus structures in vitro without a mesenchymal niche. *Nature*, *459*(7244), 262–265. http://dx.doi.org/10.1038/nature07935.

Stengel, S., Fiebig, U., Kurth, R., & Denner, J. (2010). Regulation of human endogenous retrovirus-k expression in melanomas by cpg methylation. *Genes, Chromosomes & Cancer, 49*(5), 401–411. http://dx.doi.org/10.1002/gcc.20751.

Strissel, P. L., Ruebner, M., Thiel, F., Wachter, D., Ekici, A. B., Wolf, F., et al. (2012). Reactivation of codogenic endogenous retroviral (erv) envelope genes in human endometrial carcinoma and prestages: Emergence of new molecular targets. *Oncotarget, 3*(10), 1204–1219. http://dx.doi.org/10.18632/oncotarget.679.

Suzuki, K., Suzuki, I., Leodolter, A., Alonso, S., Horiuchi, S., Yamashita, K., et al. (2006). Global DNA demethylation in gastrointestinal cancer is age dependent and precedes genomic damage. *Cancer Cell, 9*(3), 199–207. http://dx.doi.org/10.1016/j.ccr.2006.02.016.

Tsai, H. C., Li, H., Van Neste, L., Cai, Y., Robert, C., Rassool, F. V., et al. (2012). Transient low doses of DNA-demethylating agents exert durable antitumor effects on hematological and epithelial tumor cells. *Cancer Cell, 21*(3), 430–446. http://dx.doi.org/10.1016/j.ccr.2011.12.029.

van de Wetering, M., Francies, H. E., Francis, J. M., Bounova, G., Iorio, F., Pronk, A., et al. (2015). Prospective derivation of a living organoid biobank of colorectal cancer patients. *Cell, 161*(4), 933–945. http://dx.doi.org/10.1016/j.cell.2015.03.053.

Wrangle, J., Wang, W., Koch, A., Easwaran, H., Mohammad, H. P., Vendetti, F., et al. (2013). Alterations of immune response of non-small cell lung cancer with azacytidine. *Oncotarget, 4*(11), 2067–2079. http://dx.doi.org/10.18632/oncotarget.1542.

Yamashita, M., Toyota, M., Suzuki, H., Nojima, M., Yamamoto, E., Kamimae, S., et al. (2010). DNA methylation of interferon regulatory factors in gastric cancer and non-cancerous gastric mucosae. *Cancer Science, 101*(7), 1708–1716. http://dx.doi.org/10.1111/j.1349-7006.2010.01581.x.

Yao, J. Y., Zhang, L., Zhang, X., He, Z. Y., Ma, Y., Hui, L. J., et al. (2010). H3k27 trimethylation is an early epigenetic event of p16ink4a silencing for regaining tumorigenesis in fusion reprogrammed hepatoma cells. *The Journal of Biological Chemistry, 285*(24), 18828–18837. http://dx.doi.org/10.1074/jbc.M109.077974.

Yoo, C. B., Chuang, J. C., Byun, H. M., Egger, G., Yang, A. S., Dubeau, L., et al. (2008). Long-term epigenetic therapy with oral zebularine has minimal side effects and prevents intestinal tumors in mice. *Cancer Prevention Research (Philadelphia, Pa), 1*(4), 233–240. http://dx.doi.org/10.1158/1940-6207.CAPR-07-0008.

Yoo, C. B., & Jones, P. A. (2006). Epigenetic therapy of cancer: Past, present and future. *Nature Reviews. Drug Discovery, 5*(1), 37–50. http://dx.doi.org/10.1038/nrd1930.

Zahnow, C. A., Topper, M., Stone, M., Murray-Stewart, T., Li, H., Baylin, S. B., et al. (2016). Inhibitors of DNA methylation, histone deacetylation, and histone demethylation: A perfect combination for cancer therapy. *Advances in Cancer Research, 130*, 55–111. http://dx.doi.org/10.1016/bs.acr.2016.01.007.

Zeng, M., Hu, Z., Shi, X., Li, X., Zhan, X., Li, X. D., et al. (2014). Mavs, cgas, and endogenous retroviruses in t-independent b cell responses. *Science, 346*(6216), 1486–1492. http://dx.doi.org/10.1126/science.346.6216.1486.

CHAPTER NINE

Chromatin Remodeling and Plant Immunity

W. Chen*, Q. Zhu*, Y. Liu*, Q. Zhang*,†,1
*State Key Laboratory for Conservation and Utilization of Subtropical Agro-Bioresources, College of Life Sciences, South China Agricultural University, Guangzhou, China
†Key Laboratory of Plant Functional Genomics and Biotechnology of Guangdong Provincial Higher Education Institutions, College of Life Sciences, South China Agricultural University, Guangzhou, China
[1]Corresponding author: e-mail address: zqy@scau.edu.cn

Contents

Abstract

Chromatin remodeling, an important facet of the regulation of gene expression in eukaryotes, is performed by two major types of multisubunit complexes, covalent histone- or DNA-modifying complexes, and ATP-dependent chromosome remodeling complexes. Snf2 family DNA-dependent ATPases constitute the catalytic subunits of ATP-dependent chromosome remodeling complexes, which accounts for energy supply during chromatin remodeling. Increasing evidence indicates a critical role of chromatin remodeling in the establishment of long-lasting, even transgenerational immune memory in plants, which is supported by the findings that DNA methylation, histone deacetylation, and histone methylation can prime the promoters of immune-related genes required for disease defense. So what are the links between Snf2-mediated ATP-dependent chromosome remodeling and plant immunity, and what mechanisms might support its involvement in disease resistance?

Advances in Protein Chemistry and Structural Biology, Volume 106
ISSN 1876-1623
http://dx.doi.org/10.1016/bs.apcsb.2016.08.006

1. INTRODUCTION

Epigenetics is defined as the study of any potentially stable and, ideally, heritable change in gene expression or cellular phenotype that occurs without changes in DNA sequence (Goldberg, Allis, & Bernstein, 2007). Epigenetic regulation assumes, at least at the chromatin level, four major forms, DNA methylation, covalent histone modifications, ATP-dependent chromatin remodeling, and noncoding RNA-mediated chromatin remodeling (Goldberg et al., 2007) (Fig. 1). All these epigenetic mechanisms work together to provide an organism with prompt, stringent regulations of gene expression in response to environmental signals.

In the absence of a circulatory system and mobile immune cells, plants rely on a multilayered innate immune system to achieve highly specific, adaptive immunity and establish immune memory. The first line of defense is the immunity triggered by pathogen-/microbe-associated molecular patterns (PAMPs/MAMPs), within which cell wall components, short peptides, and lipopolysaccharides derived from the pathogen are recognized by pattern recognition receptors (Boller & Felix, 2009). To circumvent PAMP/MAMP-triggered immunity (PTI/MTI), some pathogens may

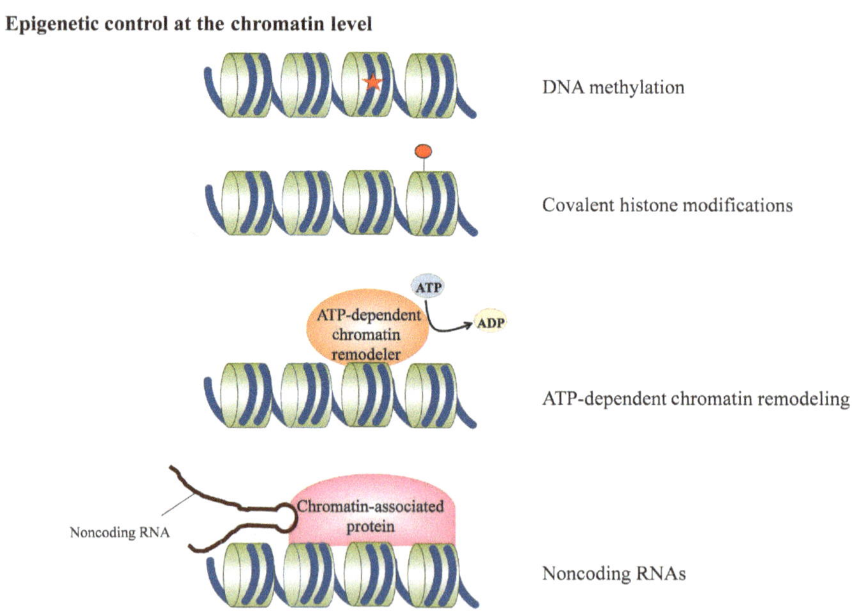

Fig. 1 Four main types of epigenetic regulations at the chromatin level.

directly deliver effector molecules into host cells, leading to effector-triggered immunity (ETI) activated by intracellular immune receptors known as resistance (R) proteins in plants (Jones & Dangl, 2006). Additionally, the local infection also induces broad-spectrum resistance against future infection in uninfected, distal plant tissues, which is called systemic acquired resistance (SAR) (Durrant & Dong, 2004). Emerging evidences have shown that the stimulation of PTI, ETI, and SAR requires epigenetic regulation of gene expression. For example, histone acetylation, histone methylation, and histone ubiquitination have been implicated in PTI, ETI, and SAR of *Arabidopsis* and rice (Alvarez-Venegas, Abdallat, Guo, Alfano, & Avramova, 2007; Berr et al., 2010; De-La-Pena, Rangel-Cano, & Alvarez-Venegas, 2012; Dhawan et al., 2009; Ding, Bellizzi Mdel, Ning, Meyers, & Wang, 2012; Kim, Lai, Fan, & Chen, 2008; Li et al., 2013; Palma et al., 2010; Singh et al., 2013; Singh, Roy, Singh, & Nandi, 2014; Singh, Yekondi, et al., 2014; Wang et al., 2010; Zhou, Zhang, Duan, Miki, & Wu, 2005; Zou, Yang, Shi, Dong, & Hua, 2014). Furthermore, H3 and H4 acetylation, and H3K4 methylation were found to contribute to the long-lasting immune memory seen in SAR (Jaskiewicz, Conrath, & Peterhänsel, 2011), and pathogen-induced DNA rearrangements could help establish transgenerational immune memory in plants (Boyko et al., 2007; Lucht et al., 2002; Molinier, Ries, Zipfel, & Hohn, 2006; Ries et al., 2000; Walbot, 1996).

Although numerous associations exist between plant immunity and DNA methylation, covalent histone modifications or noncoding RNAs, the critical role of ATP-dependent chromatin-remodeling complexes in plant immunity remains largely unknown, but has been recognized by several links between Snf2 proteins and immune response in plants. As the effects of DNA methylation, covalent histone modifications and noncoding RNAs on plant disease defense have been intensively reviewed in the past a few years, this review is primarily focused on the Snf2-mediated ATP-dependent chromatin-remodeling events that regulate defense cascades.

2. Snf2 PROTEINS AND ATP-DEPENDENT CHROMATIN REMODELING

Eukaryotic genomic DNA is packaged by histones and associated proteins and RNA into chromatin, of which a basic unit is a nucleosome consisting of DNA wrapped around a histone octamer (147 bp in 1.7 turns). This dynamic structure is stringently regulated according to the needs of the cell, thus leading to the altered DNA accessibility to regulatory factors during replication and transcription. The ATP-dependent chromatin-remodeling system provides a means of generating such changes in chromatin structure. ATP-dependent chromatin remodeling is principally mediated by the ATP-dependent chromatin-remodeling complexes. The individual subunits of the first ATP-dependent chromatin-remodeling complex, the yeast SWI/SNF, were originally identified through two screens for yeast mutants with faulty mating-type switching (SWI) and sucrose nonfermenting (SNF), respectively (Winston & Carlson, 1992). The genes identified in each screen were designated as either SWI or SNF genes, and eventually demonstrated to be members of a multisubunit complex, termed the SWI/SNF complex (Roberts & Orkin, 2004). Further, microarray analyses of yeast *swi/snf* mutants indicated that SWI/SNFs act to both activate and repress transcription of 5% of yeast genes (Sudarsanam, Iyer, Brown, & Winston, 2000). The central catalytic cores of chromatin-remodeling complexes are members of the SNF2 superfamily of DNA-dependent ATPases, which endow these complexes with the ability to disrupt the DNA–histone interaction through ATP hydrolysis. By using the energy stored in ATP, Snf2 ATPases help chromatin remodelers carry out a broad range of chromatin transformations, including changes in chromatin assembly, nucleosomal DNA conformation, and histone composition (Fig. 2; Clapier & Cairns, 2009). Of note, the binding of the chromatin-remodeling complexes to chromatin is ATP independent.

3. CLASSIFICATION, SUBUNIT COMPOSITION, AND FUNCTIONS OF PLANT Snf2 PROTEINS

Snf2 proteins are highly conserved among eukaryotes, and thus have been identified within the eukaryotic genomes (Flaus, Martin, Barton, & Owen-Hughes, 2006). In the plant kingdom, it has been documented that 41 and 39 Snf2 proteins are encoded in the genomes of *Arabidopsis* and rice (*Oryza sativa* L.), respectively (Knizewski, Ginalski, & Jerzmanowski, 2008; Li et al., 2011). The Snf2 family is composed of the proteins with a helicase-like region similar to the yeast (*Saccharomyces cerevisiae*) Snf2p and can be classified into 24 distinct subfamilies forming six larger groups (Flaus et al., 2006). Following this algorithm, the Snf2 proteins from *Arabidopsis* and rice have been assigned to these groups of subfamilies (Fig. 3; Knizewski et al., 2008), and the common primary architectures of these Snf2 proteins are shown in Fig. 3, on a subgroup basis (Fig. 3).

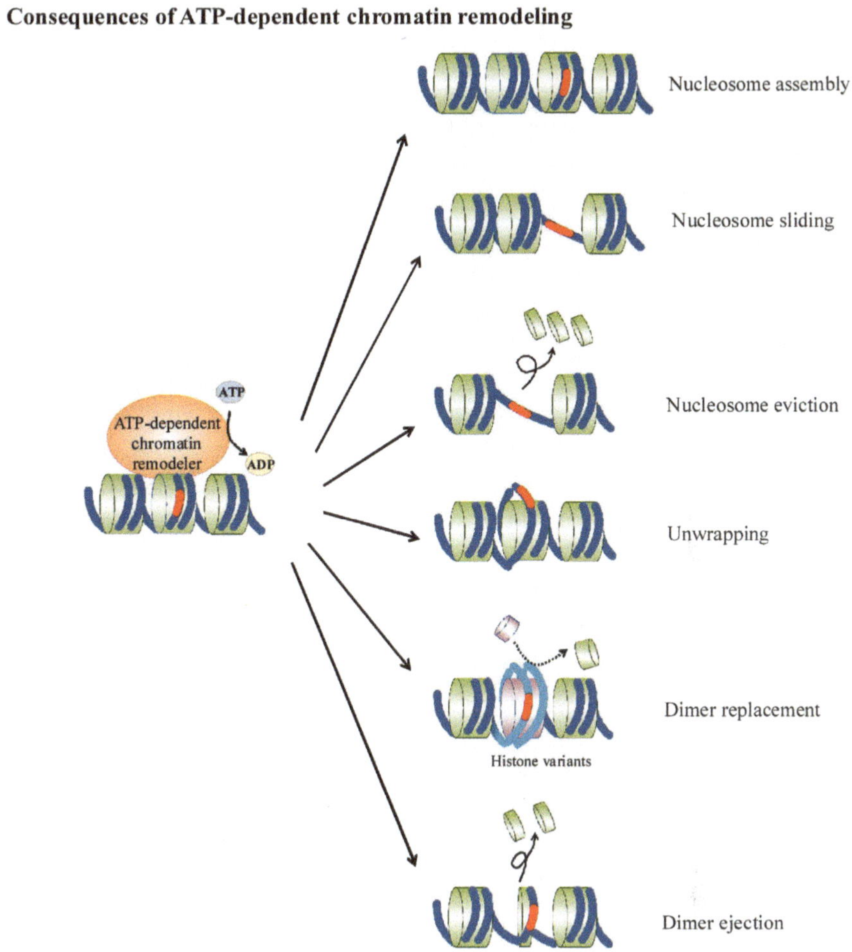

Fig. 2 Consequences of chromatin remodeling. Upon ATP addition, chromatin remodelers bound on the chromatin change the nucleosome conformation, which results in the exposure of the binding site (*red*) for transcriptional regulators (octamer sliding, octama eviction, or localized unwrapping), or the altered composition of nucleosomes (histone replacement or histone ejection). The binding of the chromatin-remodeling complex to chromatin is ATP independent. *Adapted from Clapier, C. R., & Cairns, B. R. (2009). The biology of chromatin remodeling complexes.* Annual Review of Biochemistry, *78, 273–304.*

The helicase-like region of the Snf2 proteins comprises two characteristic domains, SNF2_N and HELICc (Flaus et al., 2006; Jerzmanowski, 2007). The SNF2_N (SNF2 family N-terminal) domain, a variant of the typical DEXD/H domain with a conserved C-terminal extension of approximately 100 amino acids, contains an ATP-binding pocket which

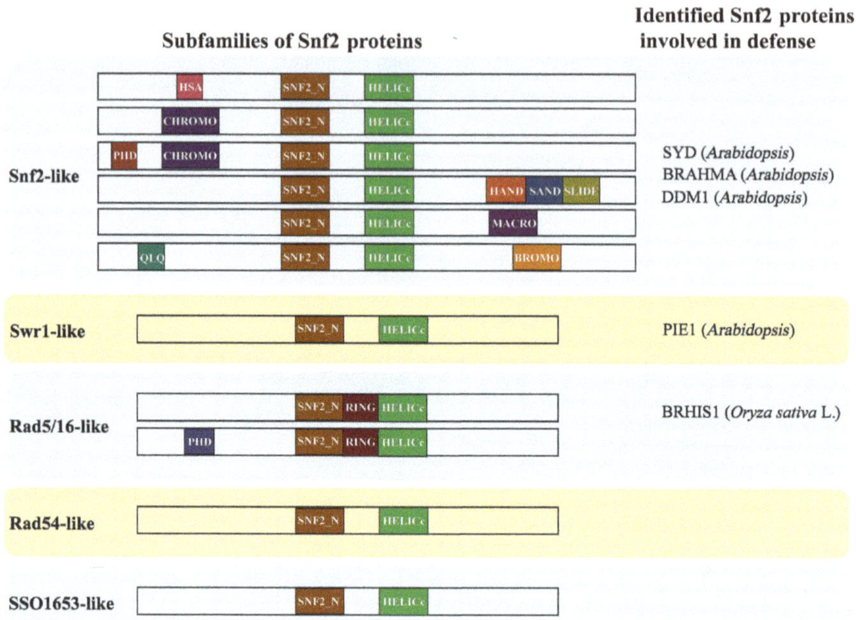

Fig. 3 Common schematic architecture of the Snf2 family proteins in Arabidopsis and rice (Knizewski et al., 2008; Li et al., 2011) and the list of Snf2 proteins that have been identified with relation to disease defense in plants. Domains are denoted by *colored boxes*. *BROMO*, bromodomain; *CHROMO*, chromatin organization modifier domain; *HAND*, named because its 4-helical structure resembles an open hand; *HSA*, helicase/SANT-associated domain; *MACRO*, ADP-ribose binding module; PHD, plant homeodomain finger; *QLQ*, containing the conserved Gln-Leu-Gln motif; *RING*, RING finger E3 ubiquitin ligase domain; *SAND*, c-Myb DNA-binding like; *SLIDE*, SANT-like but with several insertions; *SNF2_N* and *HELICc*, duplicated recA-like helicase domains.

mediates ATP hydrolysis (Jerzmanowski, 2007). The C-terminally located Helic C domain is connected to SNF2_N by a flexible spacer of variable length (Jerzmanowski, 2007). Although the helicase-like region is sufficient to support the classification of the Snf2 proteins, most of the Snf2 proteins possess additional sequences that form accessory domains differing among some subfamilies (Flaus et al., 2006; Knizewski et al., 2008). Based on the limited information on the *Arabidopsis* Snf2 family and even much less on rice Snf2s, these two model plant genomes share most of the accessory motifs found in the animal Snf2 family, but some of the Snf2 proteins are specific to plants and fungi. For example, the RING finger is a characteristic domain of the Ris1 members in the Rad5/16-like subgroup, a subfamily only found in plant and fungus genomes (Flaus et al., 2006; Knizewski et al., 2008). The rice Snf2 family even has some domain yet unidentified in *Arabidopsis*, such as the RRM domain (Li et al., 2011), on the basis of the in silico predictions from CDD (Marchler-Bauer et al., 2009) at NCBI.

While the available information about plant Snf2 proteins is still limited, several Snf2 proteins and their related chromatin-remodeling complexes have been identified and characterized in *Arabidopsis*, rice, and maize (Table 1). PKL (PICKLE) and PKR2 (PICKLE RELATED2), two Mi-2 members harboring the CHROMO and SANT domains involved in histone methylation, act redundantly to maintain root cell identity by regulating the H3K27me3 (trimethylation of histone H3 at lysine 27) levels of the promoter regions of their target genes in *Arabidopsis* (Aichinger et al., 2009; Table 1). PKL is also associated with the H3-K9 and H3-K27 methylation levels at the promoters of *ABI3* and *ABI5* (Perruc et al., 2007), two osmotolerance-related genes, and the H3K27me3 levels at the promoters of the marker genes functioning in light/dark, brassinosteroid and gibberellin signaling (Zhang et al., 2014). Recently, BRHIS1 (BIT-responsive histone-interacting SNF2 ATPase 1), a novel Ris1 member containing a RING finger domain involved in the ubiquitination pathway, has been found to regulate rice innate immunity by affecting the histone monoubiquitination state of the promoter regions of the BRHIS1 target genes (Li, Jiang, Ji, Liu, & Zhang, 2015). Taken together, there seems to be a good correlation between the accessory motifs and functions of the plant Snf2 proteins. The presence of accessory motifs also strengthens the evolutionary grouping of plant Snf2 proteins into subfamilies (Knizewski et al., 2008). Considering that Snf2 proteins interact with other proteins to form larger complexes, the function of Snf2 motors may also be adapted by accessory motifs within these complexes for different purposes (Flaus et al., 2006).

4. Snf2 PROTEINS INVOLVED IN PLANT IMMUNITY

Thus far, most of the Snf2 proteins encoded in the plant genomes are intensively investigated for their roles in development, and only five plant Snf2 proteins and their related chromatin-remodeling complexes have been discovered to function in pathogen resistance (Fig. 3 and Table 1).

4.1 Brahma and Splayed

Brahma (BRM) and Splayed (SYD) are the two most well-studied plant Snf2 proteins, both of which belong to the Snf2 subfamily. Early investigation

Table 1 Snf2 Proteins and Their Related Chromatin-Remodeling Complexes Have Been Identified and Characterized in *Arabidopsis*, Rice, and Maize

Plant Snf2 Proteins Identified	Additional Domains	Known Functions in Development and Immunity
BRAHMA (*Arabidopsis*)	QLQ, HSA, BROMO	Growth arrest (Hurtado, Farrona, & Reyes, 2006); ABA hypersensitivity during postgermination growth arrest (Han et al., 2012); Vegetative development and flowering time regulation (Li, Chen, et al., 2015); Root stem cell niche maintenance (Yang et al., 2015); Inflorescence architecture (Zhao et al., 2015); Formation and/or maintenance of boundary cells during embryogenesis (Kwon et al., 2006); Pathogen resistance (Bezhani et al., 2007)
CHR11 (*Arabidopsis*)	COIL, HAND, SANT, SLIDE	Haploid nuclear proliferation during megagametogenesis and cell expansion during the sporophytic phase (Huanca-Mamani, Garcia-Aguilar, León-Martínez, Grossniklaus, & Vielle-Calzada, 2005); Maintenance of the vegetative phase (Li et al., 2012)
AtCHR12/23 (*Arabidopsis*)	QLQ, HSA	Embryogenesis and stem cell maintenance (Sang et al., 2012); Salt tolerance (Folta et al., 2014); Seed germination (Leeggangers, Folta, Muras, Nap, & Mlynarova, 2015)
CHR17 (*Arabidopsis*)	COIL, HAND, SANT, SLIDE	Maintenance of the vegetative phase (Li et al., 2012)
DDM1 (*Arabidopsis*)	COIL	Leaf senescence (Cho et al., 2016); Pathogen resistance (Stokes, Kunkel, & Richards, 2002; Yi & Richards, 2007)
DRD1 (*Arabidopsis*)	N/A	Leaf senescence (Cho et al., 2016)

PICKLE (*Arabidopsis*)	PHD, CHROMO, HAND, SANT, SLIDE	Gibberellin-modulated developmental switch from embryonic to vegetative development (Ogas, Kaufmann, Henderson, & Somerville, 1999); Auxin-mediated lateral root formation (Fukaki, Taniguchi, & Tasaka, 2006); ABA-mediated seed germination (Perruc, Kinoshita, & Lopez-Molina, 2007) Determination of cell identity (Aichinger et al., 2009); Integrates light/darkness, BR, and GA signals during Skotomorphogenic Growth (Zhang, Jing, Jiang, & Lin, 2014)
PIE1 (*Arabidopsis*)	HSA, MYB	Flowering time regulation and petal development (Noh & Amasino, 2003); Pathogen resistance (March-Díaz et al., 2008)
PKR2 (*Arabidopsis*)	CHROMO, HAND, SANT, SLIDE	Determination of cell identity (Aichinger et al., 2009)
SYD (*Arabidopsis*)	QLQ, HSA	Meristem maintenance during reproductive development and proper carpel and ovule development (Wagner & Meyerowitz, 2002); Formation and/or maintenance of boundary cells during embryogenesis (Kwon et al., 2006); Pathogen resistance (Walley et al., 2008)
ALT1 (Rice)	RING	Alkaline tolerance (Guo et al., 2014)
BRHIS1 (Rice)	RING	Pathogen resistance (Li, Jiang, et al., 2015)
ENL1 (Rice)	PFD	Syncytial endosperm development (Hara et al., 2015)
RMR1 (Maize)	N/A	Transgenerational paramutation (Hale, Stonaker, Gross, & Hollick, 2007)

indicated that BRM and SYD regulate vegetative growth and reproductive development (Kwon & Wagner, 2007). In addition, subsequent studies have revealed that SYD is recruited to the promoters of selected genes downstream of the jasmonate and ethylene (ET) signaling pathways, and also required for resistance against the necrotrophic pathogen *Botrytis cinerea* in *Arabidopsis* (Walley et al., 2008). What is more, basal activation of pathogenesis-related (*PR*) genes was observed in noninfected *brm101* mutant plants (Bezhani et al., 2007), indicating the role of BRM in maintaining basal repression of the SA (salicylic acid)-mediated disease defense in *Arabidopsis*. Despite these significant findings, how BRM and SYD regulate the expression of SA-responsive genes at the chromatin level have yet to be determined in response to pathogen infection.

4.2 Decrease in DNA Methylation 1

DNA Methylation 1 (DDM1), the member of the Lsh subfamily in *Arabidopsis*, plays a critical role in the maintenance of DNA and histone H3 methylation (Gendrel, Lippman, Yordan, Colot, & Martienssen, 2002; Jeddeloh, Bender, & Richards, 1998; Vongs, Kakutani, Martienssen, & Richards, 1993). The *Arabidopsis bal* mutant created in the *ddm1* background displayed *ddm1*-induced hypomethylation, leading to the tandem duplication of a 55-kb region containing six *R* genes (Yi & Richards, 2009). This DNA rearrangement event thus enabled the constitutive expression of selected disease resistance genes involved in the SA signaling pathway (Yi & Richards, 2007). However, the duplication of the *SNC1* (one of the six *R* genes) alone is necessary and sufficient for the phenotypic changes in *bal* (Yi & Richards, 2009), suggesting that these changes may be associated with the *ddm1*-induced genome instability (Alvarez, Nota, & Cambiagno, 2010; Spoel & Dong, 2012).

4.3 Photoperiod-Independent Early Flowering 1

The functional studies of PIE1, the only *Arabidopsis* representative in the Swr1 subfamily, have long been focused on its interaction with the histone variant H2A.Z, which is conserved in both yeast and human. This interaction has been demonstrated to function in flowering time regulation, petal development, and immune response (Table 1). In the PIE1-mediated disease defense model, the PIE1–H2A.Z interaction underpins the maintenance of the recessive state of SAR in the absence of pathogen challenge

(March-Díaz et al., 2008). Interestingly, H2A.Z, PIE1, and other components of the *Arabidopsis* SWR1c complex, the PIE1-containing ATP-dependent chromatin-remodeling unit, may have distinct roles in gene regulation related to disease resistance, suggesting a potential role of PIE1 in the regulation of the cross-talk between defense signaling pathways (Berriri, Gangappa, & Kumar, 2016). In spite of the fact that the SWR1c complex has been coupled to the plant immunity, the regulatory mechanism of this chromatin-remodeling event in response to pathogen infection remains largely unknown.

4.4 BIT-Responsive Histone-Interacting SNF2 ATPase 1

BRHIS1 was first identified in rice through a full-length cDNA suppression subtractive hybridization screen for genes responsive to BIT, a defense priming inducing compound (Li, Jiang, et al., 2015). This gene, encoding a Ris1 member, is downregulated by BIT (Li, Jiang, et al., 2015). Inhibition of the expression of *BRHIS1* exhibited increased resistance to blast pathogen in rice, suggesting the negative regulatory role of BRHIS1 in plant immunity. Interestingly, the RNA-seq and ChIP-seq analyses further showed that the SA marker genes, such as *OsNPR1*, *OsWRKY03*, and *OsWRKY71*, are not affected by *BRHIS1*, while the peroxidase genes and 30 other defense-related genes, including *OsPBZc* and *OsSIRK1*, are upregulated when *BRHIS1* expression is suppressed. Thus, this novel defense regulator represses plant immunity in an SA-independent manner (Li, Jiang, et al., 2015).

As a putative SWI/SNF2 class ATPase, BRHIS1 features chromatin remodeling. Co-IP analyses showed that BRHIS1 specifically interacts with certain monoubiquitinated forms of histone variants, H2A.Xa/H2A.Xb/H2A.3 and H2B.7. Moreover, the enrichment of the monoubiquitinated H2A.Xa/H2A.Xb/H2A.3 and H2B.7 at the *OsPBZc* and *OsSIRK1* promoter regions is correlated with the increased expression of *OsPBZc* and *OsSIRK1* induced by the BIT treatment, during which *BRHIS1* expression is suppressed. These data suggest that the BRHIS1 complex may produce an inaccessible chromatin state for certain defense genes by blunting further monoubiquitination of the local histone variants under normal growth conditions (Li, Jiang, et al., 2015). Upon the perception of signals from BIT or pathogen assaults, suppressed *BRHIS1* expression provokes the inaccessible chromatin to become accessible for defense gene expression through enhanced monoubiquitination of those targeted histone variants, eventually leading to disease defense priming in rice (Li, Jiang, et al., 2015).

This BRHIS1 complex-based model has revealed a negative control of disease defense priming by monoubiquitinated histone variants-mediated chromatin remodeling. Of note, the BRHIS1-interacting histone variants are in a monoubiquitinated state, which indicates plants may establish an expression-ready state at some poised promoters of defense genes to facilitate rapid activation of induced immune responses (Li, Jiang, et al., 2015). While the critical role of the BRHIS1 complex in plant immune response has been appreciated, it still remains unclear that how BRHIS1 is regulated in the BIT/pathogen signaling cascades and how BRHIS1 affects transcriptional rates. For example, BRHIS1 bears a RING finger domain known to mediate ubiquitination, which implicates that BRHIS1 may function in the process of histone ubiquitination. Additionally, given the reduced BRHIS1 binding concurrent with the increasing monoubiquitination of H2B.7 and H2A.3/H2A.Xa/H2A.Xb, there may be a possibility that the activity of BRHIS1 could be modulated by monoubiquitinated histone variants (Li, Jiang, et al., 2015).

5. CONCLUDING REMARKS

Despite growing evidence indicating that Snf2 proteins and their related ATP-dependent chromatin remodelers play irreplaceable roles in plant immunity, the underlying regulatory mechanisms are just beginning to be recognized. A recent report has revealed that the SWI/SNF chromatin-remodeling complex associates preferentially with H2Bub1 (monoubiquitylation of histoneH2Bon lysine 120)-rich chromatin for optimal transcription of a subset of genes that are selectively H2Bub1 dependent (Shema-Yaacoby et al., 2013). Consistent with this report, rice BRHIS1 is proposed to be recruited to the promoter regions of its target genes via the interaction between BRHIS1 and certain monoubiquitinated histone variants, including monoubiquitinated H2B.7 (Li, Jiang, et al., 2015). Along with the pioneering work in yeast and human (Martens & Winston, 2003), these reports have opened up a new direction for exploring the regulatory mechanisms underlying Snf2-mediated epigenetic control in plant immunity, based on the fact that promoter-targeting sustains chromatin-remodeling events performed by ATP-dependent chromatin remodelers (Martens & Winston, 2003; Peterson & Workman, 2000). In addition, the study on BRHIS1 indicates a novel model for ATP-dependent

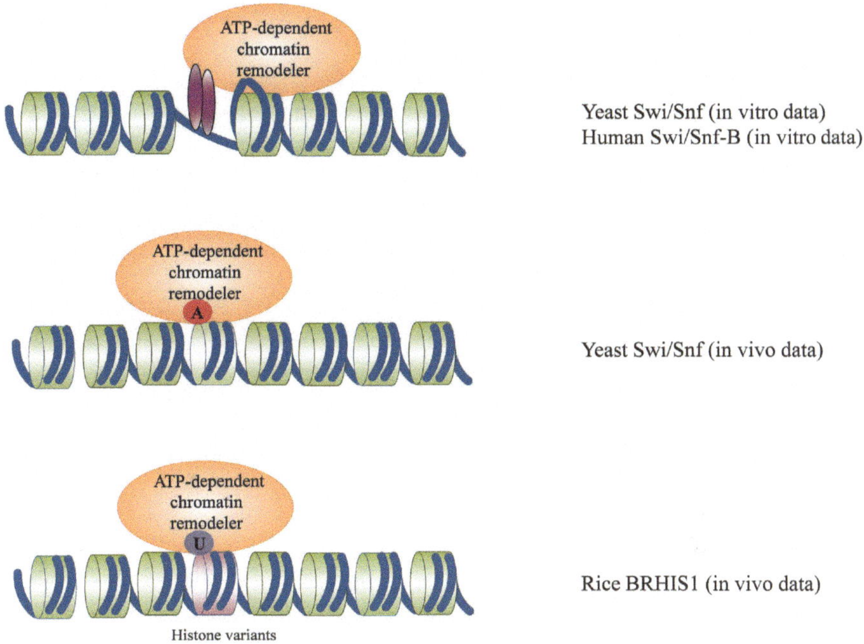

Fig. 4 Models for the recruitment of SNF2-containing chromatin-remodeling complexes to target promoters. These complexes may be recruited to *ss*-regulatory DNA regions by sequence-specific transcriptional activators/repressors (*upper panel*), or by specific histone modifications (*middle panel*)/histone variants (*bottom panel*) present at these sites. *Upper panel: Adapted from Peterson, C. L., & Workman, J. L. (2000). Promoter targeting and chromatin remodeling by the SWI/SNF complex.* Current Opinion in Genetics & Development, 10 *(2), 187–192.*

chromatin remodeler recruitment to target promoters, wherein certain histone variants provide the promoter selectivity (Fig. 4).

Brief inspection of the two working models currently available in Snf2-mediated epigenetic control of plant immunity, the PIE 1 model and the BRHIS1 model, reveals distinct epigenetic machinery employed by these two model plants for disease defense (Fig. 5). H2A.Z deposition by PIE1 suppresses, while monoubiquitination of histone variants, H2B.7 and H2A.3/H2A.Xa/H2A.Xb, by relieved BRHIS1 binding, provokes defense gene expression (Fig. 5). Furthermore, the PIE1 target genes are SA-dependent, while the BRHIS1 target genes are selectively irresponsive to SA (Fig. 5). Therefore, Snf2 proteins might act differently between monocotyledons and dicotyledons, even they share similar molecular architectures. In view of plant immunity, SNF2-mediated chromatin remodeling still remains mysterious.

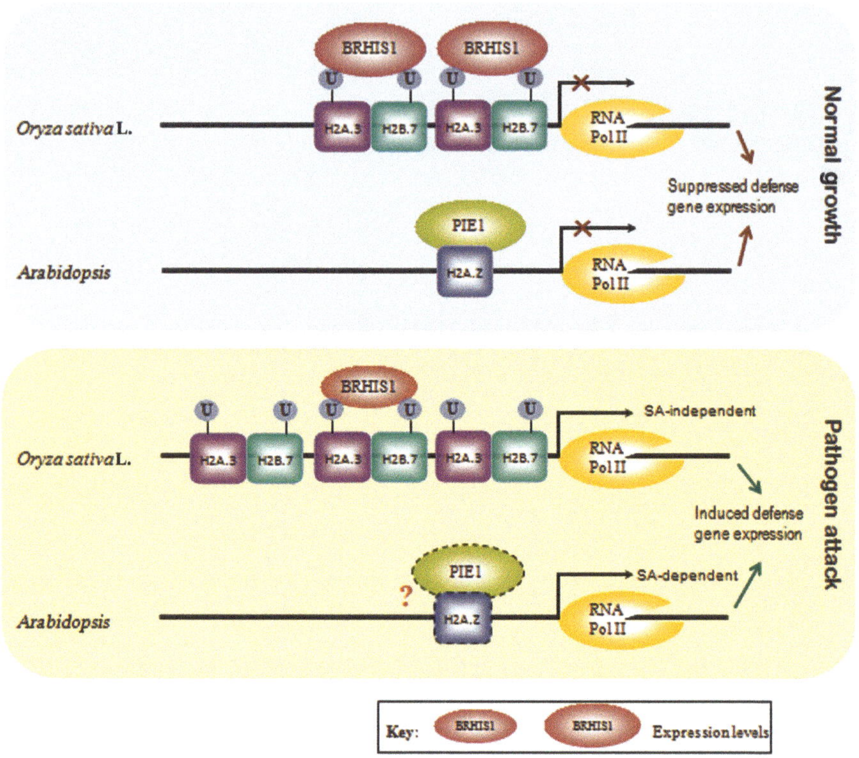

Fig. 5 Models for the involvement of histone variants, histone modifications, and ATP-dependent chromatin-remodeling complexes in plant immunity. In rice, BRHIS1 interacts with the monoubiquitinated histone variants H2B.7 and H2A.3/H2A.Xa/H2A.Xb (H2A.3 is used as an example shown here). Under normal growth conditions, the high-levelly expressed BRHIS1 complex binds to the histone variants and obstructs further monoubiquitinations, and thus constantly restricts the promoter activity to a basal level. Upon pathogen attack, the H2A.Xa and H2B.7 expression is upregulated, while BRHIS1 expression is decreased, thus leading to more deposition of H2A.3/H2A.Xa/H2A.Xb and H2B.7, and to the relief of the BRHIS1 complex binding on the chromatin. This process increases the monoubiquitination status of the histone variants, and thus induces stronger expression of disease defense genes. In *Arabidopsis*, PIE1 interacts with the histone variant H2A.Z to maintain a repressive state of the SAR response under normal growth conditions. However, it remains unknown yet that is how this interaction get involved in the induction of defense response upon pathogen attack.

REFERENCES

Aichinger, E., Villar, C. B., Farrona, S., Reyes, J. C., Hennig, L., & Köhler, C. (2009). CHD3 proteins and polycomb group proteins antagonistically determine cell identity in Arabidopsis. *PLoS Genetics*, *5*(8), e1000605.

Alvarez, M. E., Nota, F., & Cambiagno, D. A. (2010). Epigenetic control of plant immunity. *Molecular Plant Pathology*, *11*(4), 563–576.

Alvarez-Venegas, R., Abdallat, A. A., Guo, M., Alfano, J. R., & Avramova, Z. (2007). Epigenetic control of a transcription factor at the cross section of two antagonistic pathways. *Epigenetics*, *2*(2), 106–113.

Berr, A., Mccallum, E. J., Alioua, A., Heintz, D., Heitz, T., & Shen, W. H. (2010). *Arabidopsis* histone methyltransferase SET DOMAIN GROUP8 mediates induction of the jasmonate/ethylene pathway genes in plant defense response to necrotrophic fungi. *Plant Physiology*, *154*(3), 1403–1414.

Berriri, S., Gangappa, S. N., & Kumar, S. V. (2016). SWR1 chromatin-remodelling complex subunits and H2A.Z have non-overlapping functions in immunity and gene regulation in Arabidopsis. *Molecular Plant*, *9*, 1051–1065. http://dx.doi.org/10.1016/j.molp.2016.04.003.

Bezhani, S., Winter, C., Hershman, S., Wagner, J. D., Kennedy, J. F., Kwon, C. S., et al. (2007). Unique, shared, and redundant roles for the Arabidopsis SWI/SNF chromatin remodeling ATPases BRAHMA and SPLAYED. *Plant Cell*, *19*(2), 403–416.

Boller, T., & Felix, G. (2009). A renaissance of elicitors: Perception of microbe-associated molecular patterns and danger signals by pattern-recognition receptors. *Annual Review of Plant Biology*, *60*, 379–406.

Boyko, A., Kathiria, P., Zemp, F. J., Yao, Y., Pogribny, I., & Kovalchuk, I. (2007). Transgenerational changes in the genome stability and methylation in pathogen-infected plants: (Virus-induced plant genome instability). *Nucleic Acids Research*, *35*(5), 1714–1725.

Cho, E. J., Choi, S. H., Kim, J. H., Kim, J. E., Lee, M. H., Chung, B. Y., et al. (2016). A mutation in plant-specific SWI2/SNF2-like chromatin-remodeling proteins, DRD1 and DDM1, delays leaf senescence in Arabidopsis thaliana. *PloS One*, *11*(1), e0146826.

Clapier, C. R., & Cairns, B. R. (2009). The biology of chromatin remodeling complexes. *Annual Review of Biochemistry*, *78*, 273–304.

De-La-Pena, C., Rangel-Cano, A., & Alvarez-Venegas, R. (2012). Regulation of disease-responsive genes mediated by epigenetic factors: Interaction of *Arabidopsis-Pseudomonas*. *Molecular Plant Pathology*, *13*(4), 388–398.

Dhawan, R., Luo, H., Foerster, A. M., Abuqamar, S., Du, H. N., Briggs, S. D., et al. (2009). Histone Monoubiquitination1 interacts with a subunit of the mediator complex and regulates defense against necrotrophic fungal pathogens in *Arabidopsis*. *Plant Cell*, *21*(3), 1000–1019.

Ding, B., Bellizzi Mdel, R., Ning, Y., Meyers, B. C., & Wang, G. L. (2012). HDT701, a histone H4 deacetylase, negatively regulates plant innate immunity by modulating histone H4 acetylation of defense-related genes in rice. *Plant Cell*, *24*(9), 3783–3794.

Durrant, W. E., & Dong, X. (2004). Systemic acquired resistance. *Annual Review of Phytopathology*, *42*, 185–209.

Flaus, A., Martin, D. M., Barton, G. J., & Owen-Hughes, T. (2006). Identification of multiple distinct Snf2 subfamilies with conserved structural motifs. *Nucleic Acids Research*, *34*(10), 2887–2905.

Folta, A., Severing, E. I., Krauskopf, J., van de Geest, H., Verver, J., Nap, J. P., et al. (2014). Over-expression of Arabidopsis AtCHR23 chromatin remodeling ATPase results in increased variability of growth and gene expression. *BMC Plant Biology*, *14*, 76.

Fukaki, H., Taniguchi, N., & Tasaka, M. (2006). PICKLE is required for SOLITARY-ROOT/IAA14-mediated repression of ARF7 and ARF19 activity during Arabidopsis lateral root initiation. *The Plant Journal*, *48*(3), 380–389.

Gendrel, A. V., Lippman, Z., Yordan, C., Colot, V., & Martienssen, R. A. (2002). Dependence of heterochromatic histone H3 methylation patterns on the Arabidopsis gene DDM1. *Science*, *297*(5588), 1871–1873.

Goldberg, A. D., Allis, C. D., & Bernstein, E. (2007). Epigenetics: A landscape takes shape. *Cell*, *128*(4), 635–638.

Guo, M., Wang, R., Wang, J., Hua, K., Wang, Y., Liu, X., et al. (2014). ALT1, a Snf2 family chromatin remodeling ATPase, negatively regulates alkaline tolerance through enhanced defense against oxidative stress in rice. *PloS One*, *9*(12), e112515.

Hale, C. J., Stonaker, J. L., Gross, S. M., & Hollick, J. B. (2007). A novel Snf2 protein maintains trans-generational regulatory states established by paramutation in maize. *PLoS Biology*, *5*(10), e275.

Han, S. K., Sang, Y., Rodrigues, A., BIOL425 F2010, Wu, M. F., Rodriguez, P. L., et al. (2012). The SWI2/SNF2 chromatin remodeling ATPase BRAHMA represses abscisic acid responses in the absence of the stress stimulus in Arabidopsis. *Plant Cell*, *24*(12), 4892–4906.

Hara, T., Katoh, H., Ogawa, D., Kagaya, Y., Sato, Y., Kitano, H., et al. (2015). Rice SNF2 family helicase ENL1 is essential for syncytial endosperm development. *The Plant Journal*, *81*(1), 1–12.

Huanca-Mamani, W., Garcia-Aguilar, M., León-Martínez, G., Grossniklaus, U., & Vielle-Calzada, J. P. (2005). CHR11, a chromatin-remodeling factor essential for nuclear proliferation during female gametogenesis in Arabidopsis thaliana. *Proceedings of the National Academy of Sciences of the United States of America*, *102*(47), 17231–17236.

Hurtado, L., Farrona, S., & Reyes, J. C. (2006). The putative SWI/SNF complex subunit BRAHMA activates flower homeotic genes in Arabidopsis thaliana. *Plant Molecular Biology*, *62*(1–2), 291–304.

Jaskiewicz, M., Conrath, U., & Peterhänsel, C. (2011). Chromatin modification acts as a memory for systemic acquired resistance in the plant stress response. *EMBO Reports*, *12*(1), 50–55.

Jeddeloh, J. A., Bender, J., & Richards, E. J. (1998). The DNA methylation locus DDM1 is required for maintenance of gene silencing in Arabidopsis. *Genes & Development*, *12*(11), 1714–1725.

Jerzmanowski, A. (2007). SWI/SNF chromatin remodeling and linker histones in plants. *Biochimica et Biophysica Acta*, *1769*(5-6), 330–345.

Jones, J. D., & Dangl, J. L. (2006). The plant immune system. *Nature*, *16*(444), 323–329.

Kim, K. C., Lai, Z., Fan, B., & Chen, Z. (2008). Arabidopsis WRKY38 and WRKY62 transcription factors interact with histone deacetylase 19 in basal defense. *Plant Cell*, *20*(9), 2357–2371.

Knizewski, L., Ginalski, K., & Jerzmanowski, A. (2008). Snf2 proteins in plants: Gene silencing and beyond. *Trends in Plant Science*, *13*(10), 557–565.

Kwon, C. S., Hibara, K., Pfluger, J., Bezhani, S., Metha, H., Aida, M., et al. (2006). A role for chromatin remodeling in regulation of CUC gene expression in the Arabidopsis cotyledon boundary. *Development*, *133*(16), 3223–3230.

Kwon, C. S., & Wagner, D. (2007). Unwinding chromatin for development and growth: A few genes at a time. *Trends in Genetics*, *23*(8), 403–412.

Leeggangers, H. A., Folta, A., Muras, A., Nap, J. P., & Mlynarova, L. (2015). Reduced seed germination in Arabidopsis over-expressing SWI/SNF2 ATPase genes. *Physiologia Plantarum*, *153*(2), 318–326.

Li, C., Chen, C., Gao, L., Yang, S., Nguyen, V., Shi, X., et al. (2015). The Arabidopsis SWI2/SNF2 chromatin remodeler BRAHMA regulates polycomb function during vegetative development and directly activates the flowering repressor gene SVP. *PLoS Genetics*, *11*(1), e1004944.

Li, T., Chen, X., Zhong, X., Zhao, Y., Liu, X., Zhou, S., et al. (2013). Jumonji C domain protein JMJ705-mediated removal of histone H3 lysine 27 trimethylation is involved in defense-related gene activation in rice. *Plant Cell*, *25*(11), 4725–4736.

Li, X., Jiang, Y., Ji, Z., Liu, Y., & Zhang, Q. (2015). BRHIS1 suppresses rice innate immunity through binding to monoubiquitinated H2A and H2B variants. *EMBO Reports*, *16*(9), 1192–1202.

Li, X., Wang, C., Nie, P., Lu, X., Wang, M., Liu, W., et al. (2011). Characterization and expression analysis of the SNF2 family genes in response to phytohormones and abiotic stresses in rice. *Biologia Plantarum*, *55*(4), 625–633.

Li, G., Zhang, J., Li, J., Yang, Z., Huang, H., & Xu, L. (2012). Imitation Switch chromatin remodeling factors and their interacting RINGLET proteins act together in controlling the plant vegetative phase in Arabidopsis. *The Plant Journal*, *72*(2), 261–270.

Lucht, J. M., Mauch-Mani, B., Steiner, H. Y., Metraux, J. P., Ryals, J., & Hohn, B. (2002). Pathogen stress increases somatic recombination frequency in Arabidopsis. *Nature Genetics*, *30*(3), 311–314.

March-Díaz, R., Garcia-Dominguez, M., Lozano-Juste, J., Leon, J., Florencio, F. J., & Reyes, J. C. (2008). Histone H2A.Z and homologues of components of the SWR1 complex are required to control immunity in Arabidopsis. *The Plant Journal*, *53*(3), 475–487.

Marchler-Bauer, A., Marchler-Bauer, A., Anderson, J. B., Chitsaz, F., Derbyshire, M. K., DeWeese-Scott, C., et al. (2009). CDD: Specific functional annotation with the conserved domain database. *Nucleic acids research*, *37*(Database issue), D205–D210.

Martens, J. A., & Winston, F. (2003). Recent advances in understanding chromatin remodeling by Swi/Snf complexes. *Current Opinion in Genetics & Development*, *13*(2), 136–142.

Molinier, J., Ries, G., Zipfel, C., & Hohn, B. (2006). Transgeneration memory of stress in plants. *Nature*, *442*(7106), 1046–1049.

Noh, Y. S., & Amasino, R. M. (2003). PIE1, an ISWI family gene, is required for FLC activation and floral repression in Arabidopsis. *Plant Cell*, *15*(7), 1671–1682.

Ogas, J., Kaufmann, S., Henderson, J., & Somerville, C. (1999). PICKLE is a CHD3 chromatin-remodeling factor that regulates the transition from embryonic to vegetative development in Arabidopsis. *Proceedings of the National Academy of Sciences of the United States of America*, *96*(24), 13839–13844.

Palma, K., Thorgrimsen, S., Malinovsky, F. G., Fiil, B. K., Nielsen, H. B., Brodersen, P., et al. (2010). Autoimmunity in *Arabidopsis* acd11 is mediated by epigenetic regulation of an immune receptor. *PLoS Pathogens*, *6*(10), e1001137.

Perruc, E., Kinoshita, N., & Lopez-Molina, L. (2007). The role of chromatin-remodeling factor PKL in balancing osmotic stress responses during Arabidopsis seed germination. *The Plant Journal*, *52*(5), 927–936.

Peterson, C. L., & Workman, J. L. (2000). Promoter targeting and chromatin remodeling by the SWI/SNF complex. *Current Opinion in Genetics & Development*, *10*(2), 187–192.

Ries, G., Heller, W., Puchta, H., Sandermann, H., Seidlitz, H. K., & Hohn, B. (2000). Elevated UV-B radiation reduces genome stability in plants. *Nature*, *406*(6791), 98–101.

Roberts, C. W., & Orkin, S. H. (2004). The SWI/SNF complex—Chromatin and cancer. *Nature Reviews. Cancer*, *4*(2), 133–142.

Sang, Y., Silva-Ortega, C. O., Wu, S., Yamaguchi, N., Wu, M. F., Pfluger, J., et al. (2012). Mutations in two non-canonical Arabidopsis SWI2/SNF2 chromatin remodeling ATPases cause embryogenesis and stem cell maintenance defects. *The Plant Journal*, *72*(6), 1000–1014.

Shema-Yaacoby, E., Nikolov, M., Haj-Yahya, M., Siman, P., Allemand, E., Yamaguchi, Y., et al. (2013). Systematic identification of proteins binding to chromatin-embedded ubiquitylated H2B reveals recruitment of SWI/SNF to regulate transcription. *Cell Reports*, *4*(3), 601–608.

Singh, V., Roy, S., Giri, M. K., Chaturvedi, R., Chowdhury, Z., Shah, J., et al. (2013). *Arabidopsis thaliana* FLOWERING LOCUS D is required for systemic acquired resistance. *Molecular Plant-Microbe Interactions*, *26*(9), 1079–1088.

Singh, V., Roy, S., Singh, D., & Nandi, A. K. (2014). *Arabidopsis* flowering locus D influences systemic-acquired-resistance- induced expression and histone modifications of WRKY genes. *Journal of Biosciences*, *39*(1), 119–126.

Singh, P., Yekondi, S., Chen, P. W., Tsai, C. H., Yu, C. W., Wu, K., et al. (2014). Environmental history modulates Arabidopsis pattern-triggered immunity in a HISTONE ACETYLTRANSFERASE1-dependent manner. *Plant Cell, 26*(6), 2676–2688.

Spoel, S. H., & Dong, X. (2012). How do plants achieve immunity? Defence without specialized immune cells. *Nature Reviews. Immunology, 12*(2), 89–100.

Stokes, T. L., Kunkel, B. N., & Richards, E. J. (2002). Epigenetic variation in Arabidopsis disease resistance. *Genes & Development, 16*(2), 171–182.

Sudarsanam, P., Iyer, V. R., Brown, P. O., & Winston, F. (2000). Whole-genome expression analysis of *snf/swi* mutants of *Saccharomyces cerevisiae. Proceedings of the National Academy of Sciences of the United States of America, 97*(7), 3364–3369.

Vongs, A., Kakutani, T., Martienssen, R. A., & Richards, E. J. (1993). Arabidopsis thaliana DNA methylation mutants. *Science, 260*(5116), 1926–1928.

Wagner, D., & Meyerowitz, E. M. (2002). SPLAYED, a novel SWI/SNF ATPase homolog, controls reproductive development in Arabidopsis. *Current Biology, 12*(2), 85–94.

Walbot, V. (1996). Sources and consequences of phenotypic and genotypic plasticity in flowering plants. *Trends in Plant Science, 1*(1), 27–32.

Walley, J. W., Rowe, H. C., Xiao, Y., Chehab, E. W., Kliebenstein, D. J., Wagner, D., et al. (2008). The chromatin remodeler SPLAYED regulates specific stress signaling pathways. *PLoS Pathogens, 4*(12), e1000237.

Wang, C., Gao, F., Wu, J., Dai, J., Wei, C., & Li, Y. (2010). Arabidopsis putative deacetylase AtSRT2 regulates basal defense by suppressing PAD4, EDS5 and SID2 expression. *Plant & Cell Physiology, 51*(8), 1291–1299.

Winston, F., & Carlson, M. (1992). Yeast SNF/SWI transcriptional activators and the SPT/SIN chromatin connection. *Trends in Genetics, 8*(11), 387–391.

Yang, S., Li, C., Zhao, L., Gao, S., Lu, J., Zhao, M., et al. (2015). The Arabidopsis SWI2/SNF2 chromatin remodeling ATPase BRAHMA targets directly to PINs and is required for root stem cell niche maintenance. *Plant Cell, 27*(6), 1670–1680.

Yi, H., & Richards, E. J. (2007). A cluster of disease resistance genes in Arabidopsis is coordinately regulated by transcriptional activation and RNA silencing. *Plant Cell, 19*(9), 2929–2939.

Yi, H., & Richards, E. J. (2009). Gene duplication and hypermutation of the pathogen resistance gene *SNC1* in the Arabidopsis *bal* variant. *Genetics, 183*(4), 1227–1234.

Zhang, D., Jing, Y., Jiang, Z., & Lin, R. (2014). The chromatin-remodeling factor PICKLE integrates brassinosteroid and gibberellin signaling during skotomorphogenic growth in Arabidopsis. *Plant Cell, 26*(6), 2472–2485.

Zhao, M., Yang, S., Chen, C. Y., Li, C., Shan, W., Lu, W., et al. (2015). Arabidopsis BREVIPEDICELLUS interacts with the SWI2/SNF2 chromatin remodeling ATPase BRAHMA to regulate KNAT2 and KNAT6 expression in control of inflorescence architecture. *PLoS Genetics, 11*(3), e1005125.

Zhou, C., Zhang, L., Duan, J., Miki, B., & Wu, K. (2005). HISTONE DEACETYLASE19 is involved in jasmonic acid and ethylene signaling of pathogen response in Arabidopsis. *Plant Cell, 17*(4), 1196–1204.

Zou, B., Yang, D. L., Shi, Z., Dong, H., & Hua, J. (2014). Monoubiquitination of histone 2B at the disease resistance gene locus regulates its expression and impacts immune responses in *Arabidopsis. Plant Physiology, 165*(1), 309–318.

AUTHOR INDEX

Note: Page numbers followed by "*f*" indicate figures, and "*t*" indicate tables.

B

C

D

E

G

H

I

J

K

L

M

Q

R

S

T

U

V

W

X

Y

Z

SUBJECT INDEX

Note: Page numbers followed by "*f*" indicate figures, and "*t*" indicate tables.

D

E

F

G

H

T

U

W

X

FSC
www.fsc.org
MIX
Papier | Fördert
gute Waldnutzung
FSC® C083411

Kolibri 360 GmbH
Ferdinand-Jühlke-Straße 7
99095 Erfurt, Deutschland
produktsicherheit@kolibri360.de